METHODS IN MOLECULAR BIOLOGY™

Series Editor
John M. Walker
School of Life Sciences
University of Hertfordshire
Hatfield, Hertfordshire, AL10 9AB, UK

For other titles published in this series, go to
www.springer.com/series/7651

Bioinformatics Methods in Clinical Research

Edited by

Rune Matthiesen

Institute of Molecular Pathology and Immunology of the University of Porto (IPATIMUP), University of Porto, Porto, Portugal

Humana Press

Editor
Rune Matthiesen
Universidade do Porto
Inst. Patologia e Imunologia
Molecular
(IPATIMUP)
Rua Dr. Roberto Frias s/n
4200-465 Porto
Portugal
rmatthiesen@ipatimup.pt

ISSN 1064-3745
ISBN 978-1-61779-670-8 e-ISBN 978-1-60327-194-3
DOI 10.1007/978-1-60327-194-3

Printed on acid-free paper

springer.com

Preface

This book discusses the latest developments in clinical omics research and describes in detail a subset of the algorithms used in publicly available software tools. The book should be considered as an omics-bioinformatics resource. However, it is not just a pure bioinformatics resource filled with complex equations; it describes to some extent the biological background and also discusses experimental methods. The advantages and drawbacks of the various experimental methods in relation to data analysis will be reviewed as well. In other words, the intention is to establish a bridge between theory and practice. Practical examples showing methods, results, and conclusions from data mining strategies will be given in some cases. It is not possible to cover all areas of omics techniques and bioinformatics algorithms in one book. However, an important subset is described and discussed from both the experimental and the bioinformatics views. The book starts out by discussing various successful examples in which omics techniques have been used in a clinically related study. An important buzz word in omics is *biomarkers*. The word "biomarker" has different meanings depending on the context in which it is used. Here, it is used in a clinical context and should be interpreted as "a substance whose specific level indicates a particular cellular or clinical state." In theory, one could easily imagine cases where one biomarker is found at different levels, at different intervals that indicate various states. An even more complex example would be a set of biomarkers and their corresponding set of concentration levels, which could be used for classifying a specific cellular or clinical state. In complex cases, more elaborate models based on machine learning and statistics are essential for identifying interrelationships between biomarkers.

The introduction chapter is therefore followed by an introductory overview of machine learning, which can be and has been extensively applied to many omics data analysis problems. The subsequent chapter discusses statistics, algorithms, and experimental consideration in genomics, transcriptomics, proteomics, and metabolomics. One of the challenges for bioinformatics in the future is to incorporate and integrate information from all omics subareas to obtain a unified view of the biological samples. This is exactly the aim of systems biology. Systems biology is a broad field involving data storage, controlled vocabulary, data mining, interaction studies, data correlation, and modeling of biochemical pathways. The data input comes from various "omics" fields such as genomics, transcriptomics, proteomics, interactomics, and metabolomics. Metabolomics can be further divided into subcategories such as peptidomics, glycomics, and lipidomics. The term "systems biology" has raised some discussion since more conservative scientists prefer a strict usage where prediction and mathematical modeling should, at a minimum, be part of a systems biology study.

The last chapters mainly concentrate on automatic ways to retrieve information for a biological study. Chapter 15 describes automated ways to correlate experimental findings with annotated features in publicly available databases. It describes how automated methods can help in experimental design and in setting the final results from omics studies into a larger context. Chapter 16 focuses on text mining to retrieve more extended information about the system under study.

It is true that many omics techniques are currently not cost-effective enough to be clinical applicable, but that is very likely going to change in the near future, which means that integrated bioinformatics solutions will be highly valuable.

Rune Matthiesen

Contents

Contributors

ANTÓNIO AMORIM • *Instituto de Patologia e Imunologia Molecular da Universidad do Porto - IPATIMUP, Porto, Portugal*

ANA M. ARANSAY • *Functional Genomics Unit, Parque Technológico de Bizkaia, Derio, Bizkaia, Spain*

USUE ARIZ • *Metabolomics, Parque Technológico de Bizkaia, Derio, Bizkaia, Spain*

RUBÉN ARMAÑANZAS • *"Intelligent Systems Group," Donostia - San Sebastián, Basque Country, Spain*

HANS CHRISTIAN BECK • *Teknologisk Institut, Kolding, Denmark*

ENDIKA BENGOETXEA • *"Intelligent Systems Group," Donostia - San Sebastián, Basque Country, Spain*

BORJA CALVO • *"Intelligent Systems Group," Donostia - San Sebastián, Basque Country, Spain*

ANA SOFIA CARVALHO • *Instituto de Patologia e Imunologia Molecular da Universidad do Porto – IPATIMUP, Porto, Portugal*

MARIA L. MARTÍNEZ CHANTAR • *Metabolomics, Parque Technológico de Bizkaia, Derio, Bizkaia, Spain*

FERNANDO J. CORRALES • *Proteomics, Genomics and Bioinformatics, Center for Applied Medical Research, University of Navarra, Spain*

SUSMITA DATTA • *Department of Bioinformatics and Biostatistics, School of Public Health and Information Sciences, University of Louisville, Louisville, KY, USA*

ITZIAR FRADES • *Bioinformatics, Parque Technológico de Bizkaia, Derio, Bizkaia, Spain*

EWA GUBB • *Bioinformatics, Parque Technológico de Bizkaia, Derio, Bizkaia, Spain*

ELIZABETH GURUCEAGA • *CEIT, Centro de Estudios e Investigaciones Técnicas de Gipuzkoa, San Sebastian, Spain*

MICHAEL HACKENBERG • *Bioinformatics, Parque Technológico de Bizkaia, Derio, Bizkaia, Spain*

IÑAKI INZA • *"Intelligent Systems Group," Donostia - San Sebastián, Basque Country, Spain*

MARTIN KRALLINGER • *Centro Nacional de Investigaciones Oncológicas, Madrid, Spain*

PEDRO LARRAÑAGA • *"Intelligent Systems Group," Donostia - San Sebastián, Basque Country, Spain*

GORKA LASSO • *Bioinformatics, Parque Technológico de Bizkaia, Derio, Bizkaia, Spain*

FLORIAN LEITNER • *Centro Nacional de Investigaciones Oncológicas, Madrid, Spain*

JOSÉ A. LOZANO • *"Intelligent Systems Group," Donostia - San Sebastián, Basque Country, Spain*

SHELLY C. LU • *Professor in the Division of Gastrointestinal and Liver Diseases, Keck School of Medicine, University Southern California, Los Angeles*

JOSE MARIA MATO • *Metabolomics, Parque Technológico de Bizkaia, Derio, Bizkaia, Spain*

RUNE MATTHIESEN • *Instituto de Patologia e Imunologia Molecular da Universidad do Porto – IPATIMUP, Porto, Portugal*

VASYL PIHUR • *Department of Bioinformatics and Biostatistics, School of Public Health and Information Sciences, University of Louisville, Louisville, KY, USA*

MANUELA M. REGUEIRO • *Department of Biological Sciences, Florida International University, Miami, FL, ETATS-UNIS*

ANGEL RUBIO • *CEIT, Centro de Estudios e Investigaciones Técnicas de Gipuzkoa, San Sebastian, Spain*

INGO RUCZINSKI • *Biostatistics, Johns Hopkins Bloomberg School of Public Health, Baltimore, MD, USA*

ROBERT B. SCHARPF • *Biostatistics, Johns Hopkins Bloomberg School of Public Health, Baltimore, MD, USA*

VICTOR SEGURA • *Proteomics, Genomics and Bioinformatics, Center for Applied Medical Research, University of Navarra, Spain*

ALFONSO VALENCIAN • *Centro Nacional de Investigaciones Oncol'ogicas, Madrid, Spain*

MARÍA DM. VIVANCO • *Cell Biology, Parque Technológico de Bizkaia, Derio, Bizkaia, Spain*

DAVID S. WISHART • *Departments of Computing Science and Biological Sciences, University of Alberta, Edmonton, Alberta, Canada; National Institute for Nanotechnology, Edmonton, Alberta, Canada*

Chapter 1

Introduction to Omics

Ewa Gubb and Rune Matthiesen

Abstract

Exploiting the potential of omics for clinical diagnosis, prognosis, and therapeutic purposes has currently been receiving a lot of attention. In recent years, most of the effort has been put into demonstrating the possible clinical applications of the various omics fields. The cost-effectiveness analysis has been, so far, rather neglected. The cost of omics-derived applications is still very high, but future technological improvements are likely to overcome this problem.

In this chapter, we will give a general background of the main omics fields and try to provide some examples of the most successful applications of omics that might be used in clinical diagnosis and in a therapeutic context.

Key words: Clinical research, bioinformatics, omics, machine learning, diagnosis, therapeutic.

1. Omics and Omes

"Omics" refers to various branches of science dealing with "omes," the latter being complete collections of objects or the whole systems under study, such as genome, proteome, metabolome, etc. Both "omics" and "ome" have only recently graduated from their humble origins as simple suffixes of dubious etymology to fully fledged, generally accepted nouns. The oldest and best known of the omes family of names is, of course, "genome." The term was introduced in 1920 by the German botanist Hans Winkler, from *gen* ("gene") + *(chromos)om* ("chromosome"). The word "chromosome" is even older, derived from the German *chromosom*, coined in 1888 by Wilhelm von Waldeyer-Hartz (1836–1921), from the Greek *khroma* ("color") + *soma* ("body"), as chromosomes are easily stained with basic dyes.

R. Matthiesen (ed.), *Bioinformatics Methods in Clinical Research*, Methods in Molecular Biology 593,
DOI 10.1007/978-1-60327-194-3_1, © Humana Press, a part of Springer Science+Business Media, LLC 2010

Genome is now defined as the total genetic content contained in a haploid set of chromosomes in eukaryotes, in a single chromosome in bacteria, in the DNA or RNA of viruses, or, more simply, as an organism's genetic material [*The American Heritage Dictionary of the English Language*, 4th Ed., from Dictionary.com's website (1)]. The suffix "-ome" is now in widespread use and is thought to have originated as a *backformation* from "genome" (1). As "genome" is understood to encompass the complete genetic content of an organism, so by analogy, the suffix came to suggest a complete set, a sum total of objects and interactions (of a certain type) in the analyzed system. In case of dynamic omes, such as metabolome, proteome, or transcriptome, this might mean the contents of such a set only under given conditions: a certain developmental stage or a temporary state caused by a naturally occurring or experimentally introduced perturbation.

In theory, only the genome can be considered to be a static, unchanging set. One can argue, of course, that the occurrence of somatic mutations and recombination (resulting in substitutions, deletions, duplications, etc.) makes the system far from absolutely unchangeable, even though the changes might be local and the complete set of genes would still stay the same.

The suffix "-omics" came into use relatively recently, formed again by analogy to the ending of the word "genomics." "Genomics" itself was introduced by Victor McKusick and Frank Ruddle in 1987, as the name for the new journal they had started at the time (3). It has since become the household name for the study of an organism's entire genome, by now traditionally understood to include determining the entire DNA sequences of organisms and their detailed genetic mapping. Functional genomics, with its studies of gene expression patterns under varying conditions, might be considered to be its dynamic offspring, as it could come into being only in the wake of the success of the various genome sequencing projects.

2. Description of Some Omics

Many omics terms are currently in existence, some very new, some already dropping out of use, and some, hopefully, never to be used at all. If you look at the full list of omics (4), you will find some real oddities. Words like "arenayomics" [the study of "arenay" (RNA)] must surely be a joke. However, there are others, apparently created in all seriousness, but that often seem superfluous. "Cytomics" and "cellomics" (already firmly entrenched) might be comfortably accommodated within "cellular proteomics"; "biome" and "biolome" could be profitably dropped in favor of the well-established "system biology."

It is difficult to see why anyone would want to use an awkward term like "hormonomics" when we have "endocrinology" and "endocrine system." If we must have "textomics" (must we?), why do we need "bibliomics," etc.? However, a language is a living and constantly changing entity and will either absorb or discard such new constructs. Some might acquire subtly different meanings or survive as they are; others will disappear. It will be interesting to see what happens. In the meantime, there is not much point in becoming a linguistic Don Quixote trying to fight omics windmills, so we'll let them be.

Here we are only going to concern ourselves with the main biological and biomedical omics. The omics described below are not necessarily distinct branches of science; they overlap or contain each other in some cases. We are not attempting the description of all possible omics: We'll only discuss the terms most relevant to biomedical research.

2.1. Genomics, Functional Genomics, and Transcriptomics

As mentioned above, genomics is the study of an organism's entire genome. It is the first of the omics branches to have been defined, initially mainly involved in analyzing the data coming from DNA sequencing projects. The first genome (of bacteriophage MS2) was sequenced in 1976 (5), the first full bacterial genome (*Haemophilus influenzae*) in 1995 (6, 7). Genomics really came into its own with the completion of the Human Genome Project, the international, collaborative research program with the aim to supply precise mapping and the complete sequence of human genome. The project was initiated in 1990 with funding from the National Institutes of Health in the United States and the Wellcome Trust in the United Kingdom, with research groups in many countries participating, forming the International Human Genome Mapping Consortium (HGPMC). A private company, Celera Genomics, joined the race in 1998. This created much controversy over Celera's plans to sell human genome data and use of HGPMC-generated resources. However, some claimed that it seemed to spur the public sequencing groups into even more concerted effort and so accelerated achieving the goal. The first drafts of the human genome were published in the journals *Nature* (HGPMC) and *Science* (The Celera Genomics Sequencing Team) in February 2001 (8–10). The full sequence was completed and published in April 2003 (11, 12), 50 years after the discovery of the DNA structure (13). Improved drafts were announced in 2005; around 92% of the sequence is available at the moment. Since then, several genomes of model organisms have also been sequenced, and many other full genomic sequences have been added to the sequence collections [for the current statistics, see the NCBI website (14)]. Completed sequences are now available for the worm *Caenorhabditis elegans*, the fruit fly *Drosophila melanogaster*, and the mouse *Mus musculus*; many others are at the draft stage. Ten new genome sequences of various *Drosophila* species have been published recently [*see* (15) for discussion], bringing the total of fruit fly genomes sequenced to 13.

The function of the genes and their regulation and expression patterns are of the outmost importance for understanding both normal and aberrant processes in living organisms: That's the field of action for *functional genomics*. The term "transcriptomics" is used for the detailed studies of transcription, that is, the expression levels of mRNAs in a given organism, tissue, etc. (under specific set of conditions), and can be considered a part of, or an extension of, functional genomics. One of the key methods used in functional genomics is microarray technology, capable of examining the expression of thousands of genes in a single experiment. Microarray experiments can be employed for "visualizing the genes likely to be used in a particular tissue at a particular time under a particular set of conditions" (16), resulting in gene expression profiles. A microarray is usually constructed on a small glass, silicon, or nylon slide (chip) and contains many DNA (cDNA, oligonucleotides, etc.) samples arranged in a regular pattern. The basic assumption is that transcript abundance can be inferred from the amount of labeled (e.g., with a fluorescent dye) RNA (or other substance, depending on array type) hybridized to such complementary probes. Analysis of the results might find genes with similar expression profiles (functionally related genes). A comparison of results for the same genes under different conditions and/or at different developmental stages might supply information on transcription regulation (17–19). To be able to visualize and interpret the results, clustering analysis is usually performed to partition data into some meaningful groups with common characteristics. Various algorithms can be employed to achieve that goal, for example, SOTA (Self-Organizing Tree Algorithm), SOM (Self-Organizing Map), Hierarchical Clustering, K-Means, and SVM (Supported Vector Machine). Machine learning is often used to construct such clusters, using supervised and unsupervised clustering techniques (20, 21).

SNP microarrays contain SNP (single-nucleotide polymorphism) variations of one or more genes. When a sample to be analyzed is applied to the chip, the spots showing hybridization correspond to the particular gene variants present in the sample (22, 23). In genotyping applications [e.g., (24, 25)], microarrays are used to identify the single-nucleotide polymorphisms that might be related to genetic predisposition to disease or some other type of genetic variation. SNP microarrays can be used to profile somatic mutations (e.g., in cancer, during infection) such as loss of heterozygosity (26–28). In biomedical research, the most important application of such arrays is comparing specific regions of the genome between cohorts, for example, matched cohorts with and without a disease (29, 30).

SNP microarray technique is used by the International HapMap Project, whose aim is to develop a haplotype map of the human genome (31, 32). Large numbers of diseases are related to the effects of many different DNA variants in combination with environmental factors. The project catalogs genetic similarities and differences in human beings. Using the information in the HapMap, it might be possible to find genes that affect disease and analyze individual responses to medication and environmental factors. The HapMap ENCODE resequencing and genotyping project aims to produce a comprehensive set of genotypes across large genomic regions (33).

The latest explosion of new, promising research in the regulation of gene expression has been triggered by the discovery of RNA interference (34). Both small interfering RNA (siRNA) and microRNA (miRNA) are now being studied as sequence-specific posttranscriptional regulators of gene expression (35), regulating the translation and degradation of mRNAs. This opens new horizons in biomedical research: Some specific siRNAs have been introduced into animal and human cells, achieving successful expression silencing of chosen genes (36). miRNAs have also been reported to play a role in human tumorogenesis (37–40). Silencing RNAs are likely to become important tools in the treatment of viral infections, cancer, and other diseases (35, 41, 42).

The comparative genomic hybridization (CGH, using DNA–DNA hybridization) method can be used for the analysis of copy number changes of genes in abnormal and normal cell populations. Such CGH-derived data on chromosomal aberrations in cancer are accessible in the NCI and NCBI SKY/M-FISH & CGH database and the Cancer Chromosomes database (43).

Comparative genomics is the field in which the genome sequences of different species are compared: This supplies valuable information on genome evolution and allows animal models of human diseases to be built by finding homologous genes in nonhuman organisms (44, 45).

2.2. Proteomics

Proteomics is a branch of science dealing with the large-scale study of proteins, their structures, and their functions, namely, the study of a proteome. The word "proteome" is a portmanteau of "protein" and "genome." One possible definition of proteome is "the set of proteins produced during an organism's life" (46). A narrower and possibly more useful definition would take into account just the proteins present under strictly defined experimental or environmental conditions or at a certain developmental stage.

Proteomics has to face a very complex task of analyzing a dynamic system of a constantly changing protein set of an organism, differing not only between various developmental stages, but also in different tissues, cell types, and intracellular compartments. Environmental stresses also produce changes in the proteome. These changes might be initiated and express themselves on many levels, in a variety of ways. Protein concentration and content can be influenced by transcriptional and/or translational regulation, posttranslational modifications, activity regulation (activation, inhibition, protein–protein interactions, proteolysis, etc.), and intracellular as well as extracellular transport. The presence, absence, or changes in activity of certain proteins can be associated with some pathological conditions and may be useful as disease biomarkers, improving medical diagnosis, possibly opening new ways to the prevention and treatment of various diseases (46–50).

Analysis of proteomes [e.g., on a cellular, subcellular, or organ level (51–54)] is now rapidly becoming more successful, thanks to considerable improvements in techniques such as MS (mass spectrometry), MS/MS (tandem MS), and protein microarrays, used in combination with some traditional or newly improved separation methods [such as 2D electrophoresis with immobilized pH gradients (IPG-Dalt) (55), capillary electrophoresis, capillary electrophoresis–isoelectric focusing, difference gel electrophoresis, liquid chromatography, etc. (56)].

Mass spectrometry is used to find the molecular masses of proteins and their constituent peptides, leading to protein identification using database-dependent or -independent methods. Accurate protein sequences, with posttranslational modifications taken into account, are obtained using tandem mass spectrometry (57, 58). In most experiments, the proteins are digested into peptides before being analyzed in a mass spectrometer. The peptides are first separated by chromatography and then injected into a mass spectrometer for ionization and subsequent separation in one or more mass analyzers: This gives us elution profiles (retention times) in addition to *m/z* (mass-over-charge) ratios for each of the peptides. The intensity, at a specific mass and retention time characteristic for a specific peptide, can be used for quantitative analysis (59).

At this stage, the proteins might be identified using a peptide mass fingerprinting method (PMF, MS): This is often used for relatively simple protein mixtures. Theoretical peptide sequences are computationally constructed on the basis of observed peptide masses and protein candidates (containing such sequences) found by searching a protein database.

For more complex samples, chosen peptides can also be subjected to collision-induced fragmentation (MS/MS) and resulting data used to find the exact amino acid sequence. There are quite a few programs dedicated to MS data processing and analysis, both commercial and publicly available [such as Mascot, VEMS, X!Tandem, Phenyx, etc.; also see the ExPASy Proteomics tools website for a more comprehensive list (60)]. Different FASTA format databases can be used for sequence searches; the most extensive can be found at the EBI, NCBI, and Swiss-Prot websites, with the IPI database constituting the top-level resource of nonredundant protein sequence databases.

Most importantly for proteomic profiling, it is possible to determine the absolute or relative abundance of individual proteins; such quantitative data can be obtained in MS by peptide intensity profiling or by stable isotope labeling (61). To be properly interpreted, MS data have to undergo sophisticated processing on several levels. It is difficult to overestimate the importance of applying appropriate statistical and computational methods in this multistage process [for a review, *see* (62)]. Good, reliable bioinformatics software tools are critical for accurately interpreting the enormous amounts of data delivered by mass spectrometry methods, both for protein identification and for quantitative studies. Fortunately, quite a few such tools are already publicly available and are being constantly improved (63–66). A review of various methods for quantification in MS, listing software packages employing those methods, has recently been published (67, 68).

Most proteins in living organisms are subject to dynamic posttranslational modifications (PTMs). Such modifications play an important role in the regulation of protein activity, transcription, and translation levels, etc. They might also be important in the pathogenesis of some diseases, such as autoimmune diseases (69), heart disease (70), Alzheimer's disease, multiple sclerosis, malaria, and cancer (71–74). PTMs have been the subject of many studies in the past and will no doubt continue being examined in the future. An analysis of posttranslational modifications of histones is one practical example, where modifications of human histones (such as acetylations, methylations, and ubiquitinations) were found using LC-MSMS technology and the VEMS software package (61). Posttranslational modifications will change the mass of proteins and peptides analyzed by MS and can be, in theory, easily identified. The challenge here is the fact that considering posttranslational modifications dramatically increases the demands on computational power required. There are quite a few software tools successfully coping with the problem, using a variety of approaches, including PTMfinder (75, 76), MODi (77), VEMS (61, 63), and ProSight PTM (78).

At a basic level, protein microarrays, now widely used in proteomics studies, are conceptually similar to DNA/RNA microarrays. In practice, they are somewhat different in their nature [(*see* (79) for a succinct overview and prospects discussion].

Quantitative arrays (80), with their arrangement of antibodies spotted on a chip to capture specific proteins in a complex sample, are more reminiscent of classical genomic arrays than functional arrays. In this technique, proteins specifically bound to antibody spots (previously fixed to a chip in a regular pattern) are usually visualized by binding the second antibody carrying a fluorescent label. A variation of this method, reverse-phase arrays (81), uses an arrangement of small aliquots of protein sample (extract, lysate) spotted onto chips, to which labeled antibodies can be hybridized. Crude extracts can be directly used with this method, and only one "layer" of antibody per chip needs to be applied. By quantifying the amount of label bound, one can infer the level of expression of specific proteins under given conditions in the examined (subsection of) proteome.

Functional protein arrays, consisting of purified proteins affixed to a chip, are used for a variety of protein activity and structural studies (82). Protein interactions with other proteins, nucleic acids, lipids, and carbohydrates, as well as with small (e.g., enzyme substrates, or possibly drug-like) molecules can be studied using such arrays. This kind of microarray has potentially many applications, not just in pure science (assuming there is such a thing), but also in drug design and in finely tuned, personalized disease diagnosis and treatment. Some examples of interesting applications of protein arrays are the types used in the studies of autoimmune diseases. Protein arrays might be constructed using collection of autoantigens (known or potential) and probed with serum containing labeled autoantibodies (83, 84). At the moment, the only treatment for most of the serious autoimmune diseases (such as Crohn's, multiple sclerosis, Parkinson's, lupus, etc.) is nonspecific and usually involves immune system suppression, introducing an increased risk of infections and malignancies. If specific epitopes for these disorders could be found, then some better-targeted therapies could be devised, without interfering with the rest of the immune system [for a review of methods used in autoimmune diseases investigation, *see* (85)].

The whole field of protein microarrays is now advancing at a tremendous speed, with new systems constantly being developed in search of more stable, reproducible, and easily individualized arrays. One example is the "on-surface translation" array type. Nucleic acid programmable protein arrays (NAPPA) have cDNA printed on slides and are then transcribed and translated in situ using rabbit reticulocytes (86). Another array, DAPA (DNA array to protein array), follows a similar idea but makes prints of the protein arrays on separate chips, keeping the DNA templates for reuse (87). Large-scale screening and drug research projects might consider using a variation on the theme: suspension array technology (SAT), also referred to as multiplexed bead array technique. These are optically encoded (e.g., with fluorescent dye combinations), micron-sized polymer particles; suspension microarrays can contain hundreds of thousands to millions of individual probes, created to enable highly multiplexed analysis of complex samples (88, 89).

In spite of (or maybe because of) the highly dynamic state of the technology, there are many problems still to be tackled: from designing and producing an exhaustive range of stable, pure protein microarrays with their attendant databases and analysis handling software to the solving of the organizational, social, and ethical conundrums these exciting techniques bring. In clinical practice, only a few of the advanced techniques are being used at the moment; there is a question of cost and information dissemination, but above all, that of proper, large-scale clinical validation. For such techniques to be used in "field conditions," a certain downscaling might also be required, and more specialized clinical kits will have to be produced.

Collating and maintaining the databases of proteins annotated with information on their origins, functions, interactions, modifications, etc. is one of the most important objects of proteomics. Several international collaborations have been set up to create and maintain such databases, as well as to support the development of necessary tools and coordinate and promote fundamental research, such as the Human Proteome Organization (HUPO) and the EuPA [the federation of European national proteomics societies (90)].

A recent study comparing the extent of coverage obtained by MS-based proteomics to that obtained with microarray expression analysis (51) suggests that proteomics and transcriptomics methods are similar in their ability to supply a comprehensive measure of gene expression.

2.3. Metabolomics

Metabolomics research deals with the identification, quantification, and characterization of the small molecule metabolites in the metabolome (which can be defined as the set of all small molecule metabolites found in a specific cell, organ, or organism) (91).

Small changes in the proteome are often visible as much more dramatic differences in the metabolomic fingerprint. From a clinical point of view, it should be possible to identify disease-related metabolic changes in animal models and human patients and also analyze and predict the efficacy and side effects of drugs by observing the changes in the metabolome (92).

Some metabolic disorders, such as alkaptonuria (mainly deficiency of homogentisic acid oxidase), pentosuria (deficiency of L-xylulose reductase), cystinuria (inadequate reabsorption of cystine during the filtering process in the kidneys), and albinism have been known since early 1900s. Research into those diseases was pioneered at that time by Sir Archibald Garrod, who also introduced the idea that some diseases were "inborn errors of metabolism." His book of the same title was published in 1909 and revised in 1923.

Nowadays we know many more (around 150) genetically inherited metabolic disorders (93). Some better-known examples of these are cystic fibrosis, hypothyroidism, sickle cell anemia, phenylketonuria, and Tay-Sachs disease; their metabolic and genetic signatures are now quite well characterized (94–97). Perhaps not surprisingly, even those "simple" Mendelian diseases are still being intensely studied. Some turned out not to be so simple after all: For example, there is considerable expression variability in some cases, possibly caused by environmental factors, multiple alleles, or modifier genes (98).

Metabolomics is in many ways very different from genomics and, to an extent, from proteomics. Apart from the studied molecules being smaller, there are also fewer. A lower number (compared to, say, protein content of an organism) of metabolites, estimated to be in the range of about 2,500–3,000, also makes metabolic profiles a bit less challenging to analyze than a bewildering variety of proteins produced at any time, whose numbers may very well go into the hundreds of thousands. In clinical applications, it is an important factor that both experimental samples for metabolomics studies and samples for diagnostic purposes are easy to obtain and, in most cases, are done so in a noninvasive manner (mainly from body fluids).

The metabolomics fingerprint can also reflect the lifestyle, diet, and effects of other environmental factors much more readily than profiles obtained by other omics methods. Some metabolomic disease markers have been around for many years, if not centuries (e.g., glucose levels in diabetes, cholesterol for risk of heart disease), and the measurements of their levels for the diagnosis of certain diseases are firmly established and routinely performed. Such a quantitative analysis approach, nowadays often broadened to several metabolites in a pathway or performed for whole classes of compounds, is one of the two mainstream methodologies in modern metabolomics, the second being true metabolic fingerprinting (99, 100). In this last approach, the patterns of metabolites in control and perturbed systems (by disease, toxins, or other factors) are compared. This is done using methods such as nuclear magnetic resonance (NMR), MS, and various types of chromatography. The analysis is performed employing statistical tools such as hierarchical cluster analysis or principal component analysis and uses various screening databases (101). An interesting review of metabolic disorders and their specific biomarkers has been recently published in the *AAPS Journal* (93). Another general paper, by Dettmer et al. (102), discusses mass spectrometry–based metabolomics, with its specific approaches for sample preparation, separation, and analysis, with an emphasis on metabolic fingerprinting. With the wealth of data accumulating in the field of genomics and proteomics, an integrative approach carries great promise; such a methodology was used, for example, in studies of fatty liver disease (103, 104).

The advanced methods available today, such as MS and nuclear magnetic resonance (NMR), are supported by many database and tool resources such as KEGG [(pathways, protein network, the gene and the chemical information (105)], BioCyc [cellular networks and genome information (106)], Reactome [reactions, pathways (107)], OMMBID (Metabolic and Molecular Bases of Inherited Disease), and OMIM [human genes and genetic disorders (108)]. Recently, a new metabolomics database, HMDB (Human Metabolome Database), was published as a "first draft" version (91). According to the authors, it is a "multi-purpose bioinformatics – cheminformatics – medical informatics database with a strong focus on quantitative, analytic or molecular-scale information about metabolites, their associated enzymes or transporters and their disease-related properties."

3. Biomarkers

To follow HUPO's glossary definition (109), biomarkers can be described as substances (or, we might perhaps say, more general characteristics) "used to indicate or measure a biological process (for instance, levels of a specific protein in blood or spinal fluid, genetic mutations, or brain abnormalities observed in a PET scan or other imaging test)." It also adds that "detecting biomarkers specific to a disease can aid in the identification, diagnosis, and treatment of affected individuals and people who may be at risk but do not yet exhibit symptoms." The official NIH definition of a biomarker is "a characteristic that is objectively measured and evaluated as an indicator of normal biologic processes, pathogenic processes, or pharmacologic responses to a therapeutic intervention." Although some genome-level biomarkers have been found for some of the Mendelian (simple, single-gene) diseases (110), this is much more difficult for the majority of multifactorial disorders, although considerable progress is being made in this field (111). The proteomics approach seems to be very promising at the moment (112–114); there are also some exciting new developments in miRNA profiling (40, 115) and metabolomic biomarkers research (116).

In many cases, one should expect biomarker combinations or patterns, rather than single biomarkers, to be of real practical value in medical classification, diagnosis, and treatment. This is almost certainly true for disorders with a high degree of heterogeneity and complex disease progress, such as multiple sclerosis (117–120) or multiple myeloma (121). Using biomarkers not only for diagnosing but also for monitoring and predicting the possible progress of serious diseases and categorizing them into subclasses is particularly important (122, 123). This is especially true for the disorders for which the treatment itself carries a high risk and its deployment might be problematic, not desirable, or even unnecessary under some circumstances.

One of the more difficult and potentially enormously important fields in biomarker research is studying predisease biomarkers. These are usually understood to be some early warning changes in normal metabolic profile well before any noticeable physical symptoms. Trying to establish "typical" metabolic profiles in generally healthy individuals would be an essential step in achieving this; the task is made extremely difficult by individual variation, environmental and developmental differences, etc.

It is interesting to see what the actual clinical recommendations and practice in the field of biomarkers are at the moment; as an example, let's look at biomarkers in the breast cancer field. In October 2007, the American Society of Clinical Oncology (ASCO) published the "2007 Update of Recommendations for the Use of Tumor Markers in Breast Cancer" (124). Thirteen categories were considered, six of which were new compared to the previous recommendations issued in 2000. The following categories showed evidence of clinical utility and were recommended for use in practice: CA 15-3 (cancer antigen), CA 27.29 (cancer antigen), carcinoembryonic antigen (CEA), estrogen receptor (ER), progesterone receptor (PgR), human epidermal growth factor receptor 2 (HER-2), urokinase plasminogen activator (uPA, new), plasminogen activator inhibitor 1(PAI-1, new), and certain multiparameter gene expression assays (Oncotype DX (Recurrence score, assay, new). Others, which "demonstrated insufficient evidence to support routine use in clinical practice," were DNA/ploidy by flow cytometry, P53, cathepsin D, cyclin E, proteomics, certain multiparameter assays, detection of bone marrow micrometastases, and circulating tumor cells. The report also states that some other multiparameter assays, such as the MammaPrint assay (125), the "Rotterdam Signature," and the Breast Cancer Gene Expression Ratio, are still under investigation.

Some of the most promising studies considered by ASCO were those of multiple protein biomarkers on tissue microarrays (126–128), and they "have identified subclasses of breast cancer with clinical implications." The report here underlines the need for better validation, as unfortunately "at present, none of the proteomic profiling techniques has been validated sufficiently to be used for patient care." MammaPrint and Oncotype DX (both already marketed and in use, if not necessarily recommended by ASCO) are examples of recent successes in biomarker research using gene expression profiling. MammaPrint is a 70-gene prognosis expression signature used to predict the risk of breast cancer recurrence (tumors stages 1 and 2), Oncotype DX is a 21-gene signature, also for predicting the likelihood of recurrence.

An interesting discussion on practical applications of these gene expression assays can be found in a recent *Clinical Laboratory News* article (129), where MammaPrint and Oncotype DX are compared from the viewpoint of a physician in the field. It seems that there are many factors influencing the decision of whether or not to use such a test, from doubts about its validation status to the kind of sample to be supplied (e.g., fresh or frozen tissue, etc.). Some more gene expression signatures potentially useful for the prognosis in early breast cancer are discussed by Sotiriou and Piccart (130); two listed there ("Amsterdam signature" and "Recurrence score") are undergoing more comprehensive clinical trials, although all the signatures listed there have been independently validated.

Some of the results emerging from the studies of DNA damage response pathways are among the recent proteomics success stories (131, 132). PARP (poly-ADP-ribose polymerase) has become one of the new targets for cancer treatment; preclinical studies indicated that PARP inhibitors selectively inhibit tumor cell proliferation. In November 2007, BiPar Sciences Inc. announced initiation of a Phase 2 study of its lead PARP inhibitor, BSI-201, in patients with triple-negative breast cancer. Another proteomics study (133, 134) found increased expression of cytoplasmic serine hydroxymethyltransferase (cSHMT), Tbx3 (T-box transcription factor 3), and utrophin in the plasma of ovarian and breast cancer patients, using samples taken at early stages of disease. Measuring expression levels of these proteins could be used in a program of multiparameter monitoring of ovarian and breast cancer (importantly, in their early stages), with the possible use of cSHMT as a prognostic marker.

Considering that more than 200 articles on breast cancer biomarkers alone have been published between 1996 and the end of 2007, the number of new tests recommended or in practical use is not that large. However, there are many reasons to be optimistic: Some of the papers were on methods only (hopefully to be improved and more widely employed). Many others, although they presented excellent biomarker candidates, were not sufficiently supported by appropriate clinical validation: the drawback that can and should be remedied.

The lack of proper large-scale clinical validation studies as a follow-up to the most promising research results, as already mentioned above, is probably the most critical factor (apart from the low specificity of some already-validated markers). For such studies, it is very important to obtain samples from large groups of matched controls and diseased individuals, taking into account the effects of medication and center-specific bias, etc. The reproducibility between different experiments and experimental centers should be improved. The various problems and possible solutions are now widely discussed (135–137), and hopefully common protocols and practices will be soon established and followed. Having said all that, there are, of course, quite a few well-known validated biomarkers used in clinical practice, some already of many years' standing; a few examples are given below.

Prostate-specific antigen (138–140) (PSA), first introduced about 15 years ago, is used as a marker in the diagnosis and management of prostate cancer; CA 15-3 (cancer antigen), CA 27.29 (cancer antigen), carcinoembryonic antigen (CEA), estrogen receptor (ER), progesterone receptor (PgR), and human epidermal growth factor receptor 2 (HER-2) were already mentioned in the "Recommendations for the Use of Tumor Markers in Breast Cancer" discussed above.

Human alpha-fetoprotein (AFP) is the main tumor marker (along with Human HCG) for diagnosing testicular cancer, hepatocellular carcinoma, and germ cell (nonseminoma) carcinoma (141–143). Chromogranin A is the marker found in increased amounts in patients with metastatic neuroendocrine tumors, although there are some contradictory reports on its specificity (144, 145).

There are definitely many encouraging stories around, not only in the field of biomarkers but also in practical therapy: Results of the PERCY Quattro trial have been recently published and discussed in "The Times They Are A-Changing," an editorial in *Cancer* (146). The trial compared medroxyprogesterone acetate (MPA), subcutaneous interferon-alpha (INF-a), subcutaneous interleukin-2 (IL-2), or a combination of the two cytokines for treatment of patients with metastatic renal cell carcinoma. Although far from absolutely conclusive, the trial showed some clear benefits of some of the treatments examined. The authors of the editorial underline the need for well-designed follow-up trials and pose many new questions to be answered, the asking of the questions being, indeed, the proof that the times are changing.

4. Integrative Approach

In practice, the methods from various fields, such as genomics, transcriptomics, proteomics, and metabolomics (as well as lipidomics, glycomics, etc.), are used in the investigation of human diseases. To determine the perturbations in the complex pathways involved in various disorders, one might search for some subsets of genes with similar expression profiles, forming clusters of functionally related genes, and also perform protein profiling (147–152). Some other omics combinations were successfully used in the past few years; for instance, functional genomics and metabolomics methods were combined in the study of neuroendocrine cancers (153). The research in multiple sclerosis biomarkers, using both proteomics and metabolomics, is described in a 2006 *Disease Markers* review (120).

The almost classical example of using genomics, transcriptomics, and proteomics methods in a single investigation was that of Mootha et al. (154), who analyzed Leigh syndrome. After performing genomic profiling, the group suggested 30 genes as candidates implicated in the disorder. The group constructed a mitochondrial gene expression signature, looking for genes with expression matching those suspected of being involved in the disease. A gene for LRPPRC (leucine-rich pentatricopeptide repeat-containing protein) matched the profile. Mitochondrial protein profiling was then performed, and in the follow-up analysis protein fragments matching the predicted protein for LRPPRC were found. The group then sequenced the LRPPRC gene and found a single base mutation present in all 22 patients they tested. This proved conclusively that LRPPRC gene mutation was responsible for Leigh syndrome disease.

In practice, in most such studies, some of the data would not come from a single individual or group study, but would be found in already existing publications or publicly accessible databases. Indeed, it seems that there are a lot of data just waiting to be rediscovered and reanalyzed and then followed by more omics research studies.

Several examples of different omics fields and some of the methods used in studying diseases in humans and animal models are summarized in **Table 1.1**.

Acknowledgments

This work has been partially supported by the Department of Industry, Tourism and Trade of the Government of the Autonomous Community of the Basque Country (Etortek Research Programs 2005/2006) and from the Innovation Technology Department of the Bizkaia County. Support for RM was provided from Ramon y Cajal (RYC-2006-001446).

Table 1.1
Examples of omics techniques used in clinical research

Disease/animal model/biochemical process studied	Omics fields and some of the methods used	Reference
Normal glucose metabolism, homeostasis, insulin sensitivity	Metabolic profiling, metabolomics, mass spectrometry (LC-MS/MS), radioimmunoassay, hexokinase assay	(155)
Leigh syndrome, mitochondrial complex I deficiency	Proteomics, PAGE, mass spectrometry (LC-MS/MS), genomics, homozygosity mapping, Affymetrix GeneChip mapping	(156)
Kidney cancer	Proteomics, metabolic profiling, PAGE, MS, immunoblotting	(157)
Alzheimer's disease, Parkinson disease, and multiple sclerosis	Metabolomics, plasma mass spectrometry	(158)
Various cancers	Genomics, transcriptomics, RNA interference (RNAi)	(159)
Plant storage proteins, allergens	Proteomics, affinity columns, PAGE	(160)
Diabetes, obesity, coronary heart disease	Functional genomics, metabonomics, NMR spectroscopy, mass spectrometry	(161, 162)
Type II diabetes and dyslipidemia	Metabonomics, biofluid NMR spectroscopy	(163)
Muscular dystrophy in mice	Metabolomics, NMR	(164)
Amyotrophic lateral sclerosis in a mouse model	Genomics, proteomics, immunochemistry, genetic engineering, gene silencing	(165)
Crohn's disease and ulcerative colitis	Genomics, expression microarrays, quantitative RT-PCR	(166)
Rheumatoid arthritis, hypertension, Crohn's, coronary artery disease, bipolar disorder, diabetes	Genomics, genome-wide association, genotyping, GeneChip arrays	(167)
Phenylketonuria	Genomics, population genetics, metabolomics	(94)
Gene expression in human liver	Genomics, expression profiling, genotyping	(168)
Crohn's disease and ulcerative colitis	Proteomics, genomics, protein microarrays	(169)
Parkinson disease	Metabolomics, high-performance liquid chromatography, electrochemical coulometric array detection	(170)
Coronary disease	Lipidomics, liquid chromatography-mass spectrometry	(171)
Ovarian cancer	Glycomics, mass spectrometry (MALDI-FTMS)	(172)

References

1. Valkova N, Kultz D. (2006) *Biochim Biophys Acta* 1764:1007–1020. http://www.etymonline.com/index.php
2. Takatalo MS, Kouvonen P, Corthals G, Nyman TA, Ronnholm RH. (2006) *Proteomics* 6:3502–3508. http://en.wikipedia.org/wiki/-omics
3. Kuska B. (1998) Beer, Bethesda, and biology: how "genomics" came into being. *J Natl Cancer Inst* 90:93.
4. Zhang JF, He SM, Cai JJ, Cao XJ, Sun RX, Fu Y, Zeng R, Gao W. (2005) *Genom Proteom Bioinformat* 3:231–237.
5. Fiers W, Contreras R, Duerinck F, et al. (1976) Complete nucleotide sequence of bacteriophage MS2 RNA: primary and secondary structure of the replicase gene. *Nature* 260:500–507.
6. Nowak R. (1995) Bacterial genome sequence bagged. *Science* 269:468–470.
7. Fleischmann RD, Adams MD, White O, Clayton RA, Kirkness EF, Kerlavage AR, Bult CJ, Tomb JF, Dougherty BA, Merrick JM, et al. (1995) Whole-genome random sequencing and assembly of Haemophilus influenzae Rd. *Science* 269:496–512.
8. Sachidanandam R, et al. (2001) A map of human genome sequence variation containing 1.42 million single nucleotide polymorphisms. *Nature* 409:928–933.
9. McPherson JD, Marra M, Hillier L, et al. (2001) A physical map of the human genome. *Nature* 409:934–941.
10. Venter JC, Adams MD, Myers EW, et al. (2001) The sequence of the human genome. *Science* 291:1304–1351.
11. Collins FS, Morgan M, Patrinos A. (2003) The Human Genome Project: lessons from large-scale biology. *Science* 300: 286–290.
12. Arnold J, Hilton N. (2003) Genome sequencing: Revelations from a bread mould. *Nature* 422:821–822.
13. Watson JD, Crick FH. (1953) Molecular structure of nucleic acids; a structure for deoxyribose nucleic acid. *Nature* 171:737–738.
14. Chong PK, Gan CS, Pham TK, Wright PC. (2006) *J Proteome Res* 5:1232–1240. http://www.ncbi.nlm.nih.gov/genomes/static/gpstat.html
15. Ashburner M. (2007) Drosophila Genomes by the Baker's Dozen. *Genetics* 177: 1263–1268.
16. Gibson G. (2003) Microarray analysis: genome-scale hypothesis scanning. *PLoS Biol* 1:E15.
17. Nguyen DH, D'Haeseleer P. (2006) Deciphering principles of transcription regulation in eukaryotic genomes. *Mol Syst Biol* 2:2006.0012.
18. Landry CR, Oh J, Hartl DL, et al. (2006) Genome-wide scan reveals that genetic variation for transcriptional plasticity in yeast is biased towards multi-copy and dispensable genes. *Gene* 366:343–351.
19. Stern S, Dror T, Stolovicki E, et al. (2007) Genome-wide transcriptional plasticity underlies cellular adaptation to novel challenge. *Mol Syst Biol* 3:106.
20. Leban G, Bratko I, Petrovic U, et al. (2005) VizRank: finding informative data projections in functional genomics by machine learning. *Bioinformatics* 21:413–414.
21. Wilkinson DJ. (2007) Bayesian methods in bioinformatics and computational systems biology. *Brief Bioinform* 8:109–116.
22. Syvanen AC. (1994) Detection of point mutations in human genes by the solid-phase minisequencing method. *Clin Chim Acta* 226:225–236.
23. Guo Z, Guilfoyle RA, Thiel AJ, et al. (1994) Direct fluorescence analysis of genetic polymorphisms by hybridization with oligonucleotide arrays on glass supports. *Nucleic Acids Res* 22:5456–5465.
24. Pastinen T, Raitio M, Lindroos K, et al. (2000) A system for specific, high-throughput genotyping by allele-specific primer extension on microarrays. *Genome Res* 10:1031–1042.
25. Hirschhorn JN, Sklar P, Lindblad-Toh K, et al. (2000) SBE-TAGS: an array-based method for efficient single-nucleotide polymorphism genotyping. *Proc Natl Acad Sci USA* 97:12164–12169.
26. Forche A, May G, Magee PT. (2005) Demonstration of loss of heterozygosity by single-nucleotide polymorphism microarray analysis and alterations in strain morphology in Candida albicans strains during infection. *Eukaryot Cell* 4:156–165.
27. Irving JA, Bloodworth L, Bown NP, et al. (2005) Loss of heterozygosity in childhood acute lymphoblastic leukemia detected by genome-wide microarray single nucleotide polymorphism analysis. *Cancer Res* 65: 3053–3058.
28. Jacobs S, Thompson ER, Nannya Y, et al. (2007) Genome-wide, high-resolution detection of copy number, loss of heterozygosity, and genotypes from formalin-fixed, paraffin-embedded tumor tissue using microarrays. *Cancer Res* 67:2544–2551.

29. Oostenbrug LE, Nolte IM, Oosterom E, et al. (2006) CARD15 in inflammatory bowel disease and Crohn's disease phenotypes: an association study and pooled analysis. *Dig Liver Dis* 38:834–845.
30. Duerr RH, Taylor KD, Brant SR, et al. (2006) A genome-wide association study identifies IL23R as an inflammatory bowel disease gene. *Science* 314:1461–1463.
31. Frazer KA, Ballinger DG, Cox DR, et al. (2007) A second generation human haplotype map of over 3.1 million SNPs. *Nature* 449:851–861.
32. Everberg H, Clough J, Henderson P, Jergil B, Tjerneld F, Ramirez IB. (2006) *J Chromatogr A* 1118:244–252. http://www.hapmap.org/
33. Birney E, Stamatoyannopoulos JA, Dutta A, et al. (2007) Identification and analysis of functional elements in 1% of the human genome by the ENCODE pilot project. *Nature* 447:799–816.
34. Fire A, Xu S, Montgomery MK, et al. (1998) Potent and specific genetic interference by double-stranded RNA in Caenorhabditis elegans. *Nature* 391:806–811.
35. Rossi JJ. (2004) Medicine: a cholesterol connection in RNAi. *Nature* 432:155–156.
36. Soutschek J, Akinc A, Bramlage B, et al. (2004) Therapeutic silencing of an endogenous gene by systemic administration of modified siRNAs. *Nature* 432: 173–178.
37. Hutchinson E. (2006) Expression profiling: Small but influential. *Nat Rev Cancer* 6:345.
38. Yanaihara N, Caplen N, Bowman E, et al. (2006) Unique microRNA molecular profiles in lung cancer diagnosis and prognosis. *Cancer Cell* 9:189–198.
39. Meltzer PS. (2005) Cancer genomics: small RNAs with big impacts. *Nature* 435: 745–746.
40. Blenkiron C, Goldstein LD, Thorne NP, et al. (2007) MicroRNA expression profiling of human breast cancer identifies new markers of tumor subtype. *Genome Biol* 8:R214.
41. Pruijn GJ. (2006) The RNA interference pathway: a new target for autoimmunity. *Arthritis Res Ther* 8:110.
42. Miller VM, Gouvion CM, Davidson BL, et al. (2004) Targeting Alzheimer's disease genes with RNA interference: an efficient strategy for silencing mutant alleles. *Nucleic Acids Res* 32:661–668.
43. Knutsen T, Gobu V, Knaus R, et al. (2005) The interactive online SKY/M-FISH & CGH database and the Entrez cancer chromosomes search database: linkage of chromosomal aberrations with the genome sequence. *Genes Chromosomes Cancer* 44:52–64.
44. Hardison RC. (2003) Comparative genomics. *PLoS Biol* 1:E58.
45. Bergman CM, Pfeiffer BD, Rincon-Limas DE, et al. (2002). Assessing the impact of comparative genomic sequence data on the functional annotation of the Drosophila genome. *Genome Biol* 3:RESEARCH0086.
46. Fermin D, Allen BB, Blackwell TW, Menon R, Adamski M, Xu Y, Ulintz P, Omenn GS, States DJ. (2006) *Genome Biol* 7:R35.
47. Sabbioni G, Sepai O, Norppa H, et al. (2007) Comparison of biomarkers in workers exposed to 2,4,6-trinitrotoluene. *Biomarkers* 12:21–37.
48. Lakhan SE. (2006) Schizophrenia proteomics: biomarkers on the path to laboratory medicine ? *Diagn Pathol* 1:11.
49. Hunter DJ, Kraft P, Jacobs KB, et al. (2007) A genome-wide association study identifies alleles in FGFR2 associated with risk of sporadic postmenopausal breast cancer. *Nat Genet* 39:870–874.
50. Easton DF, Pooley KA, Dunning AM, et al. (2007) Genome-wide association study identifies novel breast cancer susceptibility loci. *Nature* 447:1087–1093.
51. Cox J, Mann M. (2007) Is proteomics the new genomics ? *Cell* 130:395–398.
52. Yates JR, 3rd, Gilchrist A, Howell KE, et al. (2005) Proteomics of organelles and large cellular structures. *Nat Rev Mol Cell Biol* 6: 702–714.
53. Zheng J, Gao X, Beretta L, He F. (2006) The Human Liver Proteome Project (HLPP) workshop during the 4th HUPO World Congress. *Proteomics* 6:1716–1718.
54. Hamacher M, Stephan C, Bluggel M, et al. (2006) The HUPO Brain Proteome Project jamboree: centralised summary of the pilot studies. *Proteomics* 6:1719–1721.
55. Gorg A, Obermaier C, Boguth G, et al. (2000) The current state of two-dimensional electrophoresis with immobilized pH gradients. *Electrophoresis* 21:1037–1053.
56. Pelzing M, Neususs C. (2005) Separation techniques hyphenated to electrospray-tandem mass spectrometry in proteomics: capillary electrophoresis versus nanoliquid chromatography. *Electrophoresis* 26:2717–2728.
57. Seet BT, Dikic I, Zhou MM, et al. (2006) Reading protein modifications with interaction domains. *Nat Rev Mol Cell Biol* 7: 473–483.
58. Aebersold R, Mann M. (2003) Mass spectrometry-based proteomics. *Nature* 422:198–207.

59. Hattan SJ, Parker KC. (2006) Methodology utilizing MS signal intensity and LC retention time for quantitative analysis and precursor ion selection in proteomic LC-MALDI analyses. *Anal Chem* 78:7986–7996.
60. Wan Y, Yang A, Chen T. (2006) *Anal Chem* 78:432–437. http://us.expasy.org/tools/
61. Beck HC, Nielsen EC, Matthiesen R, et al. (2006) Quantitative proteomic analysis of post-translational modifications of human histones. *Mol Cell Proteomics* 5:1314–1325.
62. Listgarten J, Emili A. (2005) Statistical and computational methods for comparative proteomic profiling using liquid chromatography-tandem mass spectrometry. *Mol Cell Proteomics* 4:419–434.
63. Matthiesen R, Trelle MB, Hojrup P, et al. (2005) VEMS 3.0: algorithms and computational tools for tandem mass spectrometry based identification of post-translational modifications in proteins. *J Proteome Res* 4:2338–2347.
64. Tokheim AM, Martin BL. (2006) *Proteins* 64:28–33. http://msquant.sourceforge.net/
65. MacCoss MJ, Wu CC, Liu H, et al. (2003) A correlation algorithm for the automated quantitative analysis of shotgun proteomics data. *Anal Chem* 75:6912–6921.
66. Venable JD, Dong MQ, Wohlschlegel J, et al. (2004) Automated approach for quantitative analysis of complex peptide mixtures from tandem mass spectra. *Nat Methods* 1: 39–45.
67. Matthiesen R. (2007) Methods, algorithms and tools in computational proteomics: a practical point of view. *Proteomics* 7: 2815–2832.
68. Mueller LN, Brusniak MY, Mani DR, et al. (2008) An Assessment of Software Solutions for the Analysis of Mass Spectrometry Based Quantitative Proteomics Data. *J Proteome Res* 7:51–61.
69. Doyle HA, Mamula MJ. (2005) Posttranslational modifications of self-antigens. *Ann N Y Acad Sci* 1050:1–9.
70. Yuan C, Ravi R, Murphy AM. (2005) Discovery of disease-induced post-translational modifications in cardiac contractile proteins. *Curr Opin Mol Ther* 7:234–239.
71. Biroccio A, Del Boccio P, Panella M, et al. (2006) Differential post-translational modifications of transthyretin in Alzheimer's disease: a study of the cerebral spinal fluid. *Proteomics* 6:2305–2313.
72. Kim JK, Mastronardi FG, Wood DD, et al. (2003) Multiple sclerosis: an important role for post-translational modifications of myelin basic protein in pathogenesis. *Mol Cell Proteomics* 2:453–462.
73. Anderton SM. (2004) Post-translational modifications of self antigens: implications for autoimmunity. *Curr Opin Immunol* 16:753–758.
74. Eastman RT, Buckner FS, Yokoyama K, et al. (2006) Thematic review series: lipid post-translational modifications. Fighting parasitic disease by blocking protein farnesylation. *J Lipid Res* 47:233–240.
75. Lamerz J, Selle H, Scapozza L, et al. (2005) Correlation-associated peptide networks of human cerebrospinal fluid. *Proteomics* 5:2789–2798.
76. Tanner S, Payne SH, Dasari S, et al. (2008) Accurate Annotation of Peptide Modifications through Unrestrictive Database Search. *J Proteome Res* 7:170–181.
77. Kim S, Na S, Sim JW, et al. (2006) MODi: a powerful and convenient web server for identifying multiple post-translational peptide modifications from tandem mass spectra. *Nucleic Acids Res* 34: W258–W263.
78. Zamdborg L, LeDuc RD, Glowacz KJ, et al. (2007) ProSight PTM 2.0: improved protein identification and characterization for top down mass spectrometry. *Nucleic Acids Res* 35:W701–W706.
79. Griffiths J. (2007) The way of array. *Anal Chem* 79:8833.
80. Lv LL, Liu BC. (2007) High-throughput antibody microarrays for quantitative proteomic analysis. *Expert Rev Proteomics* 4:505–513.
81. Espina V, Wulfkuhle JD, Calvert VS, et al. (2007) Reverse phase protein microarrays for monitoring biological responses. *Methods Mol Biol* 383:321–336.
82. LaBaer J, Ramachandran N. (2005) Protein microarrays as tools for functional proteomics. *Curr Opin Chem Biol* 9: 14–19.
83. Joos TO, Schrenk M, Hopfl P, et al. (2000) A microarray enzyme-linked immunosorbent assay for autoimmune diagnostics. *Electrophoresis* 21:2641–2650.
84. Robinson WH, DiGennaro C, Hueber W, et al. (2002) Autoantigen microarrays for multiplex characterization of autoantibody responses. *Nat Med* 8:295–301.
85. Balboni I, Chan SM, Kattah M, et al. (2006) Multiplexed protein array platforms for analysis of autoimmune diseases. *Annu Rev Immunol* 24:391–418.
86. Ramachandran N, Hainsworth E, Bhullar B, et al. (2004) Self-assembling protein microarrays. *Science* 305:86–90.
87. Taussig MJ, Stoevesandt O, Borrebaeck CA, et al. (2007) ProteomeBinders: planning a

European resource of affinity reagents for analysis of the human proteome. *Nat Methods* 4:13–17.
88. Nolan JP, Sklar LA. (2002) Suspension array technology: evolution of the flat-array paradigm. *Trends Biotechnol* 20:9–12.
89. Wang L, Cole KD, Peterson A, et al. (2007) Monoclonal antibody selection for interleukin-4 quantification using suspension arrays and forward-phase protein microarrays. *J Proteome Res* 6:4720–4727.
90. McLaughlin T, Siepen JA, Selley J, Lynch JA, Lau KW, Yin H, Gaskell SJ, Hubbard SJ. (2006) *Nucleic Acids Res* 34:D649–D654. http://www.eupa.org/
91. Wishart DS, Tzur D, Knox C, et al. (2007) HMDB: the Human Metabolome Database. *Nucleic Acids Res* 35:D521–D526.
92. Salek RM, Maguire ML, Bentley E, et al. (2007) A metabolomic comparison of urinary changes in type 2 diabetes in mouse, rat, and human. *Physiol Genomics* 29:99–108.
93. Vangala S, Tonelli A. (2007) Biomarkers, metabonomics, and drug development: can inborn errors of metabolism help in understanding drug toxicity? *AAPS J* 9: E284–E297.
94. Scriver CR. (2007) The PAH gene, phenylketonuria, and a paradigm shift. *Hum Mutat* 28:831–845.
95. Peters T, Thaete C, Wolf S, Popp A, et al. (2003) A mouse model for cystinuria type I. *Hum Mol Genet* 12:2109–2120.
96. Weiss KM. (1996) Variation in the human genome, Introduction. *Ciba Found Symp* 197:1–5.
97. Scriver CR, Byck S, Prevost L, et al. (1996) The phenylalanine hydroxylase locus: a marker for the history of phenylketonuria and human genetic diversity. PAH Mutation Analysis Consortium. *Ciba Found Symp* 197:73–90; discussion 90–66.
98. Botstein D, Risch N. (2003) Discovering genotypes underlying human phenotypes: past successes for mendelian disease, future approaches for complex disease. *Nat Genet* 33(Suppl):228–237.
99. Dettmer K, Hammock BD. (2004) Metabolomics–a new exciting field within the "omics" sciences. *Environ Health Perspect* 112:A396–A397.
100. Hollywood K, Brison DR, Goodacre R. (2006) Metabolomics: current technologies and future trends. *Proteomics* 6:4716–4723.
101. Baumgartner C, Baumgartner D. (2006) Biomarker discovery, disease classification, and similarity query processing on high-throughput MS/MS data of inborn errors of metabolism. *J Biomol Screen* 11:90–99.
102. Dettmer K, Aronov PA, Hammock BD. (2007) Mass spectrometry-based metabolomics. *Mass Spectrom Rev* 26: 51–78.
103. Griffin JL, Scott J, Nicholson JK. (2007) The influence of pharmacogenetics on fatty liver disease in the wistar and kyoto rats: a combined transcriptomic and metabonomic study. *J Proteome Res* 6:54–61.
104. Griffin JL, Bonney SA, Mann C, et al. (2004) An integrated reverse functional genomic and metabolic approach to understanding orotic acid-induced fatty liver. *Physiol Genomics* 17:140–149.
105. Kanehisa M, Goto S, Kawashima S, et al. (2004) The KEGG resource for deciphering the genome. *Nucleic Acids Res* 32: D277–D280.
106. Krummenacker M, Paley S, Mueller L, et al. (2005) Querying and computing with BioCyc databases. *Bioinformatics* 21: 3454–3455.
107. Joshi-Tope G, Gillespie M, Vastrik I, et al. (2005) Reactome: a knowledgebase of biological pathways. *Nucleic Acids Res* 33:D428–D432.
108. McKusick VA. (2007) Mendelian Inheritance in Man and its online version, OMIM. *Am J Hum Genet* 80:588–604.
109. Steely HT, Dillow GW, Bian L, Grundstad J, Braun TA, Casavant TL, McCartney MD, Clark AF. (2006) *Mol Vis* 12:372–383. http://www.hupo.org/overview/glossary/
110. Cambien F, Tiret L. (2007) Genetics of cardiovascular diseases: from single mutations to the whole genome. *Circulation* 116: 1714–1724.
111. Kingsmore SF, Lindquist IE, Mudge J, et al. (2007) Genome-Wide Association Studies: Progress in Identifying Genetic Biomarkers in Common, Complex Diseases. *Biomarker Insights* 2:283–292.
112. Srinivas PR, Verma M, Zhao Y, et al. (2002) Proteomics for cancer biomarker discovery. *Clin Chem* 48:1160–1169.
113. Meyer HE, Stuhler K. (2007) High-performance Proteomics as a Tool in Biomarker Discovery. *Proteomics* 7(Suppl 1):18–26.
114. Vosseller K. (2007) Proteomics of Alzheimer's disease: Unveiling protein dysregulation in complex neuronal systems. *Proteomics Clin Appl* 1:1351–1361.
115. Iorio MV, Visone R, Di Leva G, et al. (2007) MicroRNA signatures in human ovarian cancer. *Cancer Res* 67:8699–8707.
116. Goodenowe DB, Cook LL, Liu J, et al. (2007) Peripheral ethanolamine plasmalogen deficiency: a logical causative factor in

Alzheimer's disease and dementia. *J Lipid Res* 48:2485–2498.
117. Martin R, Bielekova B, Hohlfeld R, et al. (2006) Biomarkers in multiple sclerosis. *Dis Markers* 22:183–185.
118. Weinshenker BG, Wingerchuk DM, Pittock SJ, et al. (2006) NMO-IgG: a specific biomarker for neuromyelitis optica. *Dis Markers* 22:197–206.
119. Berger T, Reindl M. (2006) Biomarkers in multiple sclerosis: role of antibodies. *Dis Markers* 22:207–212.
120. O'Connor KC, Roy SM, Becker CH, et al. (2006) Comprehensive phenotyping in multiple sclerosis: discovery based proteomics and the current understanding of putative biomarkers. *Dis Markers* 22:213–225.
121. Bhattacharyya S, Epstein J, Suva LJ. (2006) Biomarkers that discriminate multiple myeloma patients with or without skeletal involvement detected using SELDI-TOF mass spectrometry and statistical and machine learning tools. *Dis Markers* 22: 245–255.
122. Hoshida Y, Brunet JP, Tamayo P, et al. (2007) Subclass mapping: identifying common subtypes in independent disease data sets. *PLoS ONE* 2:e1195.
123. Liu JJ, Cutler G, Li W, et al. (2005) Multiclass cancer classification and biomarker discovery using GA-based algorithms. *Bioinformatics* 21:2691–2697.
124. Harris L, Fritsche H, Mennel R, et al. (2007) American Society of Clinical Oncology 2007 update of recommendations for the use of tumor markers in breast cancer. *J Clin Oncol* 25:5287–5312.
125. van 't Veer LJ, Dai H, van de Vijver MJ, et al. (2002) Gene expression profiling predicts clinical outcome of breast cancer. *Nature* 415:530–536.
126. El-Rehim DMA, Ball G, Pinder SE, et al. (2005) High-throughput protein expression analysis using tissue microarray technology of a large well-characterised series identifies biologically distinct classes of breast cancer confirming recent cDNA expression analyses. *Int J Cancer* 116:340–350.
127. Makretsov NA, Huntsman DG, Nielsen TO, et al. (2004) Hierarchical clustering analysis of tissue microarray immunostaining data identifies prognostically significant groups of breast carcinoma. *Clin Cancer Res* 10: 6143–6151.
128. Nielsen TO, Hsu FD, Jensen K, et al. (2004) Immunohistochemical and clinical characterization of the basal-like subtype of invasive breast carcinoma. *Clin Cancer Res* 10: 5367–5374.
129. Levenson D. (2007) Gene Expression Profile Tests for Breast Cancer Recurrence. *Clin Lab News* 33:4–5.
130. Sotiriou C, Piccart MJ. (2007) Taking gene-expression profiling to the clinic: when will molecular signatures become relevant to patient care? *Nat Rev Cancer* 7:545–553.
131. McCabe N, Turner NC, Lord CJ, et al. (2006) Deficiency in the repair of DNA damage by homologous recombination and sensitivity to poly(ADP-ribose) polymerase inhibition. *Cancer Res* 66:8109–8115.
132. O'Connor M. (2006) Proteomics Success Story. Novel Biomarkers for DNA Damage Response Pathways: Insights and Applications for Cancer Therapy. *Proteomics* 6: 69–71.
133. Souchelnytskyi S, Lomnytska M, Dubrovska A, et al. (2006) Proteomics Success Story. Towards Early Detection of Breast and Ovarian Cancer: Plasma Proteomics as a Tool to Find Novel Markers. *Proteomics* 6: 65–68.
134. Lomnytska M, Dubrovska A, Hellman U, et al. (2006) Increased expression of cSHMT, Tbx3 and utrophin in plasma of ovarian and breast cancer patients. *Int J Cancer* 118:412–421.
135. Brenner DE, Normolle DP. (2007) Biomarkers for cancer risk, early detection, and prognosis: the validation conundrum. *Cancer Epidemiol Biomarkers Prev* 16:1918–1920.
136. Coombes KR, Morris JS, Hu J, et al. (2005) Serum proteomics profiling–a young technology begins to mature. *Nat Biotechnol* 23:291–292.
137. Wang SJ, Cohen N, Katz DA, et al. (2006) Retrospective validation of genomic biomarkers– what are the questions, challenges and strategies for developing useful relationships to clinical outcomes– workshop summary. *Pharmacogenomics J* 6:82–88.
138. Wang MC, Valenzuela LA, Murphy GP, et al. (1979) Purification of a human prostate specific antigen. *Invest Urol* 17:159–163.
139. Papsidero LD, Wang MC, Valenzuela LA, et al. (1980) A prostate antigen in sera of prostatic cancer patients. *Cancer Res* 40: 2428–2432.
140. Diamandis EP. (2000) Prostate-specific antigen: a cancer fighter and a valuable messenger? *Clin Chem* 46:896–900.
141. Wang MC, Valenzuela LA, Murphy GP, et al. (2002) Purification of a human prostate specific antigen. 1979. *J Urol* 167:960–964; discussion 64–65.
142. Liu FC, Chang DM, Lai JH, et al. (2007) Autoimmune hepatitis with raised alpha-fetoprotein level as the presenting symp-

toms of systemic lupus erythematosus: a case report. *Rheumatol Int* 27:489–491.
143. Supriatna Y, Kishimoto T, Furuya M, et al. (2007) Expression of liver-enriched nuclear factors and their isoforms in alpha-fetoprotein-producing gastric carcinoma cells. *Exp Mol Pathol* 82:316–321.
144. Campana D, Nori F, Piscitelli L, et al. (2007) Chromogranin A: is it a useful marker of neuroendocrine tumors? *J Clin Oncol* 25: 1967–1973.
145. Zatelli MC, Torta M, Leon A, et al. (2007) Chromogranin A as a marker of neuroendocrine neoplasia: an Italian Multicenter Study. *Endocr Relat Cancer* 14:473–482.
146. Bradley DA, Redman BG. (2007) The times they are a-changin' (Bob Dylan, 1964). *Cancer* 110:2366–2369.
147. Ma Q, Abel K, Sripichai O, et al. (2007) Beta-globin gene cluster polymorphisms are strongly associated with severity of HbE/beta(0)-thalassemia. *Clin Genet* 72: 497–505.
148. Erlich PM, Lunetta KL, Cupples LA, et al. (2006) Polymorphisms in the PON gene cluster are associated with Alzheimer disease. *Hum Mol Genet* 15:77–85.
149. Selwood SP, Parvathy S, Cordell B, et al. (2007) Gene expression profile of the PDAPP mouse model for Alzheimer's disease with and without Apolipoprotein E. *Neurobiol Aging* 30:574–90.
150. Prentice H, Webster KA. (2004) Genomic and proteomic profiles of heart disease. *Trends Cardiovasc Med* 14:282–288.
151. Sanchez-Carbayo M, Socci ND, Richstone L, et al. (2007) Genomic and proteomic profiles reveal the association of gelsolin to TP53 status and bladder cancer progression. *Am J Pathol* 171:1650–1658.
152. McRedmond JP, Park SD, Reilly DF, et al. (2004) Integration of proteomics and genomics in platelets: a profile of platelet proteins and platelet-specific genes. *Mol Cell Proteomics* 3:133–144.
153. Ippolito JE, Xu J, Jain S, et al. (2005) An integrated functional genomics and metabolomics approach for defining poor prognosis in human neuroendocrine cancers. *Proc Natl Acad Sci USA* 102:9901–9906.
154. Mootha VK, Lepage P, Miller K, et al. (2003) Identification of a gene causing human cytochrome c oxidase deficiency by integrative genomics. *Proc Natl Acad Sci USA* 100:605–610.
155. Shaham O, Wei R, Wang TJ, et al. (2008) Metabolic profiling of the human response to a glucose challenge reveals distinct axes of insulin sensitivity. *Mol Syst Biol* 4:214.
156. Pagliarini DJ, Calvo SE, Chang B, et al. (2008) A mitochondrial protein compendium elucidates complex I disease biology. *Cell* 134:112–123.
157. Perroud B, Lee J, Valkova N, et al. (2006) Pathway analysis of kidney cancer using proteomics and metabolic profiling. *Mol Cancer* 5:64.
158. Alimonti A, Ristori G, Giubilei F, et al. (2007) Serum chemical elements and oxidative status in Alzheimer's disease, Parkinson disease and multiple sclerosis. *Neurotoxicology* 28:450–456.
159. Pai SI, Lin YY, Macaes B, et al. (2006) Prospects of RNA interference therapy for cancer. *Gene Ther* 13:464–477.
160. Yano H, Kuroda S. (2008) Introduction of the disulfide proteome: application of a technique for the analysis of plant storage proteins as well as allergens. *J Proteome Res* 7:3071–3079.
161. Griffin JL, Vidal-Puig A. (2008) Current challenges in metabolomics for diabetes research: a vital functional genomic tool or just a ploy for gaining funding? *Physiol Genomics* 34:1–5.
162. Brindle JT, Antti H, Holmes E, et al. (2002) Rapid and noninvasive diagnosis of the presence and severity of coronary heart disease using 1H-NMR-based metabonomics. *Nat Med* 8:1439–1444.
163. Ringeissen S, Connor SC, Brown HR, et al. (2003) Potential urinary and plasma biomarkers of peroxisome proliferation in the rat: identification of N-methylnicotinamide and N-methyl-4-pyridone-3-carboxamide by 1H nuclear magnetic resonance and high performance liquid chromatography. *Biomarkers* 8:240–271.
164. Griffin JL. (2006) Understanding mouse models of disease through metabolomics. *Curr Opin Chem Biol* 10:309–315.
165. Saito Y, Yokota T, Mitani T, et al. (2005) Transgenic small interfering RNA halts amyotrophic lateral sclerosis in a mouse model. *J Biol Chem* 280:42826–42830.
166. Wu F, Dassopoulos T, Cope L, et al. (2007) Genome-wide gene expression differences in Crohn's disease and ulcerative colitis from endoscopic pinch biopsies: insights into distinctive pathogenesis. *Inflamm Bowel Dis* 13:807–821.
167. The Welcome Trust Case Control Consortium. (2007) Genome-wide association study of 14,000 cases of seven common diseases and 3,000 shared controls. *Nature* 447: 661–678.
168. Schadt EE, Molony C, Chudin E, et al. (2008) Mapping the genetic architecture of

gene expression in human liver. *PLoS Biol* 6:e107.

169. Kader HA, Tchernev VT, Satyaraj E, et al. (2005) Protein microarray analysis of disease activity in pediatric inflammatory bowel disease demonstrates elevated serum PLGF, IL-7, TGF-beta1, and IL-12p40 levels in Crohn's disease and ulcerative colitis patients in remission versus active disease. *Am J Gastroenterol* 100: 414–423.
170. Bogdanov M, Matson WR, Wang L, et al. (2008) Metabolomic profiling to develop blood biomarkers for Parkinson's disease. *Brain* 131:389–396.
171. Bergheanu SC, Reijmers T, Zwinderman AH, et al. (2008) Lipidomic approach to evaluate rosuvastatin and atorvastatin at various dosages: investigating differential effects among statins. *Curr Med Res Opin* 24: 2477–2487.
172. Leiserowitz GS, Lebrilla C, Miyamoto S, et al. (2008) Glycomics analysis of serum: a potential new biomarker for ovarian cancer? *Int J Gynecol Cancer* 18:470–475.

Chapter 2

Machine Learning: An Indispensable Tool in Bioinformatics

Iñaki Inza, Borja Calvo, Rubén Armañanzas, Endika Bengoetxea, Pedro Larrañaga, and José A. Lozano

Abstract

The increase in the number and complexity of biological databases has raised the need for modern and powerful data analysis tools and techniques. In order to fulfill these requirements, the machine learning discipline has become an everyday tool in bio-laboratories. The use of machine learning techniques has been extended to a wide spectrum of bioinformatics applications. It is broadly used to investigate the underlying mechanisms and interactions between biological molecules in many diseases, and it is an essential tool in any biomarker discovery process.

In this chapter, we provide a basic taxonomy of machine learning algorithms, and the characteristics of main data preprocessing, supervised classification, and clustering techniques are shown. Feature selection, classifier evaluation, and two supervised classification topics that have a deep impact on current bioinformatics are presented. We make the interested reader aware of a set of popular web resources, open source software tools, and benchmarking data repositories that are frequently used by the machine learning community.

Key words: Machine learning, data mining, bioinformatics, data preprocessing, supervised classification, clustering, classifier evaluation, feature selection, gene expression data analysis, mass spectrometry data analysis.

1. Introduction

The development of high-throughput data acquisition technologies in biological sciences in the last 5 to 10 years, together with advances in digital storage, computing, and information and communication technologies in the 1990s, has begun to transform biology from a data-poor into a data-rich science. While previous lab technologies that monitored different molecules could

R. Matthiesen (ed.), *Bioinformatics Methods in Clinical Research*, Methods in Molecular Biology 593,
DOI 10.1007/978-1-60327-194-3_2,

quantify a limited number of measurements, current devices are able to screen an amount of molecules nonenvisaged by biologists 20 years ago. This phenomenon is gradually transforming biology from classic hypothesis-driven approaches, in which a single answer to a single question is provided, to a data-driven research, in which many answers are given at a time and we have to seek the hypothesis that best explains them.

As a reaction to the exponential growth in the amount of biological data to handle, the incipient discipline of *bioinformatics* stores, retrieves, analyzes and assists in understanding biological information. The development of methods for the analysis of this massive (and constantly increasing) amount of information is one of the key challenges in bioinformatics. This analysis step – also known as *computational biology* – faces the challenge of extracting biological knowledge from all the in-house and publicly available data. Furthermore, the knowledge should be formulated in a transparent and coherent way if it is to be understood and studied by bio-experts.

In order to fulfill the requirements of the analysis of the biodata available, bioinformatics has found an excellent and mature ally in the *data mining* field. Thanks to the advances in computational power and storage of the previous decade, the data mining field achieved a notable degree of maturity in the late 1990s, and its usefulness has largely been proven in different application areas such as banking, weather forecasting, and marketing. Data mining has also demonstrated its usefulness in different medical applications, resulting in the well-known *evidence-based medicine* and *medical informatics* fields. At present, the time has come for its application in biology. The participation of data mining specialists or statisticians is broadly accepted in multidisciplinary groups working in the field of bioinformatics. Although the term *data mining* can be interpreted as having a number of different meanings within a wide range of contexts, when related to bioinformatics, it refers to the set of techniques and working trends aimed at discovering useful relationships and patterns in biological data that were previously undetected. **Figure 2.1** illustrates all the different steps that are included in a classical data mining approach that is fully valid too for the analysis of biodata, which combines techniques from the domains of *statistics, computer science,* and *artificial intelligence.*

Due to the nature and characteristics of the diverse techniques that are applied for biological data acquisition, and depending on the specificity of the domain, the biodata might require a number of preparative steps prior to its analysis. These steps are illustrated in the first three steps in **Fig. 2.1**. They are usually related to the selection and cleaning, preprocessing, and transformation of the original data. Once data have been prepared for analysis, the *machine learning* field offers a range of modelization techniques

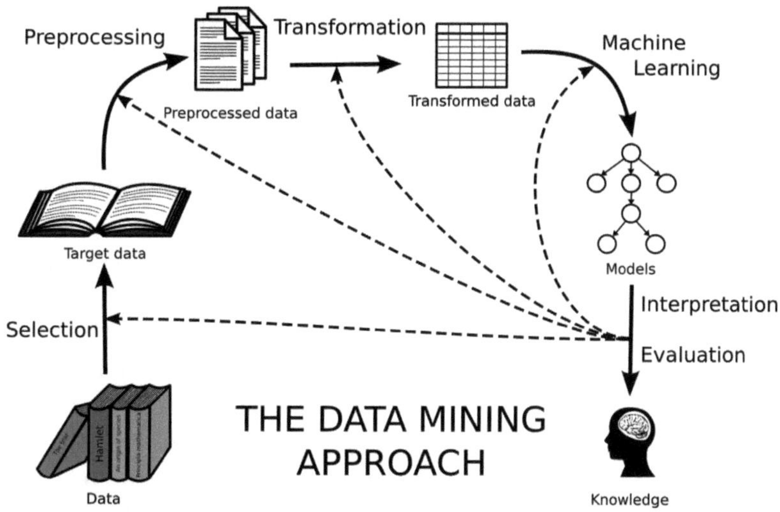

Fig. 2.1. The general chain of work of a common data mining task.

and algorithms for the automatic recognition of patterns in data, which have to be applied differently depending on the goals of the study and the nature of the available data.

Data mining techniques provide a robust means to evaluate the generalization power of extracted patterns on unseen data, although these must be further validated and interpreted by the domain expert. The implication of the bio-expert in the inspection and validation of extracted patterns is essential for a useful outcome of data mining since these patterns provide the possibility to formulate novel hypotheses to be further tested and new research trends to be opened. After all, the discovery of new *knowledge* is regarded as the ultimate desired result of the data mining chain of work shown in **Fig. 2.1**.

From now on, this chapter will focus on the machine learning discipline, which is the most representative task of many data mining applications. Machine learning methods are essentially computer programs that make use of sampled data or past experience information to provide solutions to a given problem. A wide spectrum of algorithms, commonly based on the artificial intelligence and statistics fields, have been proposed by the machine learning community in the last decades.

Prompramote et al. (1) point out a set of reasons to clear up the wide use of machine learning in several application domains, especially in bioinformatics:

- Experts are not always able to describe the factors they take into account when assessing a situation or when explaining the rules they apply in normal practice. Machine learning can serve as a valuable aid to extract the description of the hidden situation in terms of those factors and then propose the rules that better describe the expert's behavior.

- Due to the inherent complexity of biological organisms, experts are very often confronted with finding nondesired results. Unknown properties could be the cause of these results. The dynamic improvement of machine learning can cope with this problem and provide hints to further describe the properties or characteristics that are hidden to the expert.
- As new data and novel concept types are generated every day in molecular biology research, it is essential to apply techniques able to fit this fast-evolving nature. Machine learning can be adapted efficiently to these changing environments.
- Machine learning is able to deal with the abundance of missing and noisy data from many biological scenarios.
- Machine learning is able to deal with the huge volumes of data generated by novel high-throughput devices, in order to extract hidden relationships that exist and that are not noticeable to experts.
- In several biological scenarios, experts can only specify input–output data pairs, and they are not able to describe the general relationships between the different features that could serve to further describe how they interrelate. Machine learning is able to adjust its internal structure to the existing data, producing approximate models and results.

Machine learning methods are used to investigate the underlying mechanisms and the interactions between biological molecules in many diseases. They are also essential for the biomarker discovery process. The use of machine learning techniques has been broadly extended in the bioinformatics community, and successful applications in a wide spectrum of areas can be found. Mainly due to the availability of novel types of biology throughput data, the set of biology problems on which machine learning is applied is constantly growing. Two practical realities severely condition many bioinformatics applications (2): a limited number of samples (*curse of data set sparsity*) and several thousands of features characterizing each sample (*curse of dimensionality*). The development of machine learning techniques capable of dealing with these *curses* is currently a challenge for the bioinformatics community. **Figure 2.2**, which has been adapted and updated from the work of Larrañaga et al. (3), shows a general scheme of the current applications of machine learning techniques in bioinformatics.

According to the objectives of the study and the characteristics of the available data, machine learning algorithms can be roughly taxonomized in the following way:

- Supervised learning: Starting from a database of training data that consists of pairs of input cases and desired outputs, its goal is to construct a function (or model) to accurately

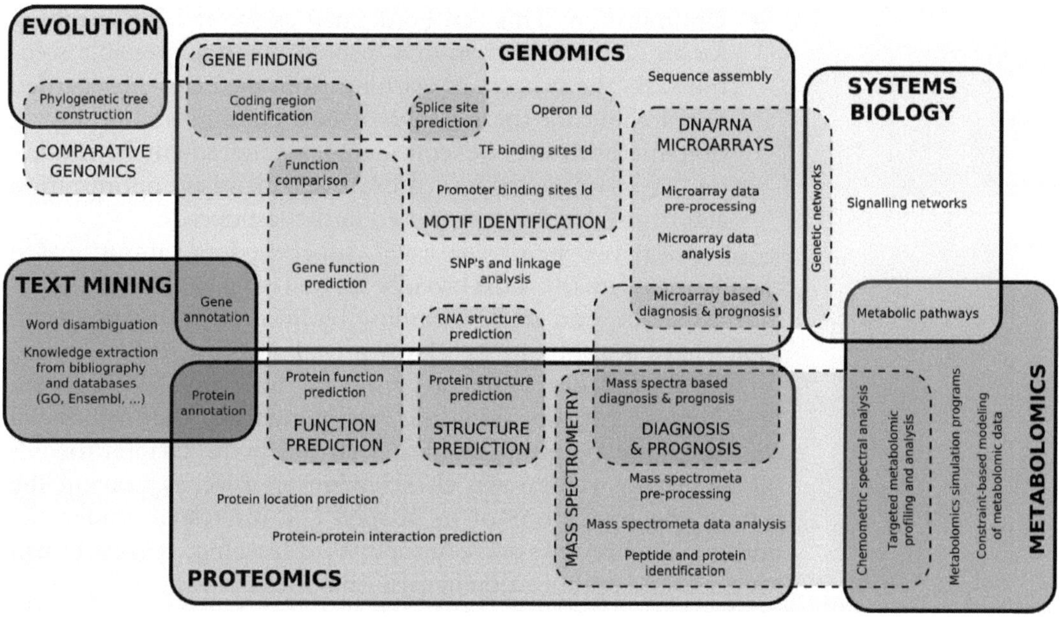

Fig. 2.2. General scheme of the current applications of machine learning techniques in bioinformatics.

predict the target output of future cases whose output value is unknown. When the target output is a continuous-value variable, the task is known as *regression*. Otherwise, when the output (or label) is defined as a finite set of discrete values, the task is known as *classification*.

- Unsupervised learning or clustering: Starting from a database of training data that consists of input cases, its goal is to partition the training samples into subsets (clusters) so that the data in each cluster show a high level of proximity. In contrast to supervised learning, the labels for the data are not used or are not available in clustering.
- Semisupervised learning: Starting from a database of training data that combines both labeled and unlabeled examples, the goal is to construct a model able to accurately predict the target output of future cases for which its output value is unknown. Typically, this database contains a small amount of labeled data together with a large amount of unlabeled data.
- Reinforcement learning: These algorithms are aimed at finding a policy that maps states of the world to actions. The actions are chosen among the options that an agent ought to take under those states, with the aim of maximizing some notion of long-term reward. Its main difference regarding the previous types of machine learning techniques is that input–output pairs are not present in a database, and its goal resides in online performance.

- Optimization: This can be defined as the task of searching for an optimal solution in a space of multiple possible solutions. As the process of learning from data can be regarded as searching for the model that best fits the data, optimization methods can be considered an ingredient in modeling. A broad collection of exact and heuristic optimization algorithms has been proposed in the last decade.

The first two items just listed, supervised and unsupervised classification, are the most broadly applied machine learning types in most application areas, including bioinformatics. Even if both topics have a solid and well-known tradition, the 1990s constituted a fruitful development period of different techniques on both topics, and they fulfill the requirements of the majority of classification experts and studies. That is why this chapter focuses on these two well-known classification approaches, leaving the rest of the topics out of its scope. The interested reader can find qualified reviews on semisupervised learning, reinforcement learning, and optimization in classical books of the machine learning literature (4, 5).

The rest of the chapter is organized as follows. The next section addresses the main techniques applied for data preparation and preprocessing. **Sections 3** and **4** provide an overview of supervised and unsupervised classification topics, respectively, highlighting the principal techniques of each approach. Finally, the interested reader is also directed to a set of web resources, open source software tools, and benchmarking data repositories that are frequently used by the machine learning community. Due to the authors' area of expertise, a special emphasis will be put on the application of the introduced techniques to the analysis of gene expression and mass spectrometry data throughout the chapter. The following references cover extensive reviews on the use of different machine learning techniques in gene expression (6, 7) and mass spectrometry (8, 9).

2. Engineering the Input; the First Analysis Step: Data Preprocessing

Machine learning involves far more than choosing a learning algorithm and running it over the data. Prior to any direct application of machine learning algorithms, it is essential to be conscious of the quality of the initial raw data available, and accordingly, we must discard the machine learning techniques that are not eligible or suitable. The lack of data quality will lead to poor quality in the mined results. As a result, the need to ensure a minimum quality of the data – which might require among other decisions, to discard a part of the original data – is critical, especially in the field of bioinformatics for several biological high-throughput devices such as DNA microarray or mass spectrometry–based studies, in which the preparation of the raw data could demand the majority of the data mining work.

The *data preprocessing* task is subdivided as a set of relevant steps that could improve the quality – success – when applying machine learning modelization techniques. These procedures are considered "engineering" the input data: They refine/depurate the data to make it more tractable for machine learning schemes. The human attention and time needed by these procedures are not negligible, and the data preprocessing step could be the most time-consuming task for certain data mining applications.

This section briefly describes the main properties and advantages of three well-known data preprocessing topics that are among the most usually applied. These are missing value imputation, data normalization, and discretization. Although several authors consider that the feature selection process belongs to the data preprocessing category, we will revise it as part of the basic supervised modelization scheme.

2.1. Missing Value Imputation

Multiple events can cause the loss of data for a particular problem: malfunctioning measurement equipment, deletion of the data due to inconsistencies with other recorded data, data not entered due to misunderstandings, etc. The first factor is especially critical in modern biological devices, and large amounts of missing data can occur in several biological domains.

Regardless of the reason for data loss, it is important to have a consistent criterion for dealing with the missing data. A simple choice could be the exclusion of the complete sample having any missing value, although this is not an advisable solution since it increases the risk of reaching invalid and nonsignificant conclusions. As an example, let us consider the case of the personal and economical effort required to obtain a DNA microarray sample. Another reason to apply an imputation method is that several classification algorithms cannot be applied on the event of missing values happening.

As the manual imputation of missing values is a tedious and commonly unfeasible approach, the machine learning community has proposed a number of alternatives to handle this situation. The most common approach is to use attribute mean/mode to fill in the missing value: This approach can be improved by imputing the mean/mode conditioned to the class label. More advanced approaches such as decision tree or Bayesian inference and imputation based on the expectation-maximization (EM) algorithm are also proposed in the related literature.

Due to the specificities of biodata, the bioinformatics community has proposed interesting imputation methods that are most suited according to the different data acquisition methods and nature of the data. For instance, the amount of missing data could be huge in DNA microarray data due to technical failure, low signal-to-noise ratio, and measurement error. That is why the gene expression researchers' community has focused its attention on the proposal of specific imputation methods for DNA microarray data (6).

2.2. Data Normalization

This type of data transformation consists of the process of removing statistical errors in repeated measured data. Data are scaled to fall within a small, specified range, thus allowing a fair comparison between different data samples. The normalization methods identify and remove the systematic effects and variations that usually occur due to the measurement procedure. In this way, a fair integration and comparison of different data samples are guaranteed. Common statistical normalization techniques include min-max normalization, *z*-score normalization, and normalization by decimal scaling (6). Both the DNA microarray and mass spectrometry bioinformatics communities have developed a broad spectrum of interesting normalization methods that are specially suited for the specificities of these domains.

2.3. Data Discretization

Some classification algorithms (e.g., general Bayesian networks) cannot handle attributes measured on a numerical scale. Therefore, if these techniques are to be applied, continuous data must be transformed. This demands the discretization of continuous-range attributes into a small number of distinct states. Although many authors argue that discretization brings about "loss of information" from the original data matrix, other researchers minimize this effect and encourage its use.

Nowadays, a broad range of discretization methods is available for data analysts. Since there are many ways to taxonomize discretization techniques, these can be categorized based on their use of the class label.

Unsupervised discretization methods quantize each attribute in the absence of any knowledge of the classes of the samples. The two most well-known unsupervised techniques are equal-width binning – based on dividing the range of the attribute into a predetermined number of equal-width intervals – and equal-frequency binning – based on dividing the range of the attribute into a predetermined number of intervals with an equal amount of instances.

On the other hand, *supervised discretization* methods take the class label into account for the discretization process. The most widely used algorithm of this category is "entropy-based discretization" (10), which has proven to obtain positive results in a broad range of problems. The goal of this algorithm is to find splits that minimize the class entropy over all possible boundaries, thus creating intervals with a majority of samples of a single class and a reduced number of samples of the rest of the classes.

Many DNA microarray researchers feel comfortable discretizing original continuous values in three intervals and interpreting them as "underexpression" (with respect to the reference sample), "baseline," and "overexpression."

3. Supervised Classification: The Class Prediction Approach

Supervised classification, also known as class prediction, is a key topic in the machine learning discipline. Its starting point is a training database formed by a set of N independent samples $D_N = \{(\mathbf{x}^1, c^1), \ldots, (\mathbf{x}^N, c^N)\}$ drawn from a joint, unknown probability distribution $p(\mathbf{x}, c)$. Each sample $(\mathbf{x}^i, c^i)$ is characterized by a group of d predictive variables or features $\{X_1, \ldots, X_d\}$ and a label or class variable of interest C, which "supervises" the whole ongoing process. We will limit our study to the case where the class variable is defined for a finite set of discrete values. Once the needed *preprocessing* steps are performed over the available data, a supervised classification algorithm uses the training database to induce a classifier whose aim is to predict the class value of future examples with an unknown class value.

Supervised classification is broadly used to solve very different bioinformatics problems such as protein secondary structure prediction, gene expression–based diagnosis, or splice site prediction. Current supervised classification techniques have been shown capable of obtaining satisfactory results.

Although the application of an algorithm to induce a classifier is the main step of the supervised classification discipline, two other aspects are vital in this overall process:

- The need to fairly estimate the predictive accuracy of the built model.
- The need for a dimensionality reduction process (e.g., feature selection), in order to improve the prediction accuracy or to handle a manageable number of attributes.

These two concepts are introduced in this section, together with an overview of the main supervised classification algorithms.

3.1. Main Classification Models

Motivated by the "no free lunch" assumption which ensures that there is not a single classification method that will be the best for all classification problems, a notable battery of supervised classification algorithms was proposed by the machine learning and statistics communities in the 1980s and 1990s. Among these, classification models of very diverse characteristics can be found, each defining a different decision surface to discriminate the classes of the problem. When the only objective is to optimize the predictive accuracy, the common methodology is to evaluate and compare the accuracy of a group of classifiers. However, other factors such as the classifier's transparency, simplicity, or interpretability could be crucial to selecting a final model. Since a description of all the available classification algorithms is beyond the scope of this chapter, we briefly present the main characteristics of four representative models with such different biases: classification trees, Bayesian classifiers, nearest neighbor, and support vector machines.

3.1.1. Classification Trees

Due to its simplicity, speed of classifying unlabeled samples, and intuitive graphical representation, classification trees is one of the most used and popular classification paradigms. The predictive model can be easily checked and understood by domain experts, and it is induced by a recursive top-down procedure. Each decision tree starts with a root node that gathers all training samples. The rest of the nodes are displayed in a sequence of internal nodes (or questions) that recursively divide the set of samples, until a terminal node (or leaf) that does the final prediction is accessed. **Figure 2.3** shows an example of a classification tree.

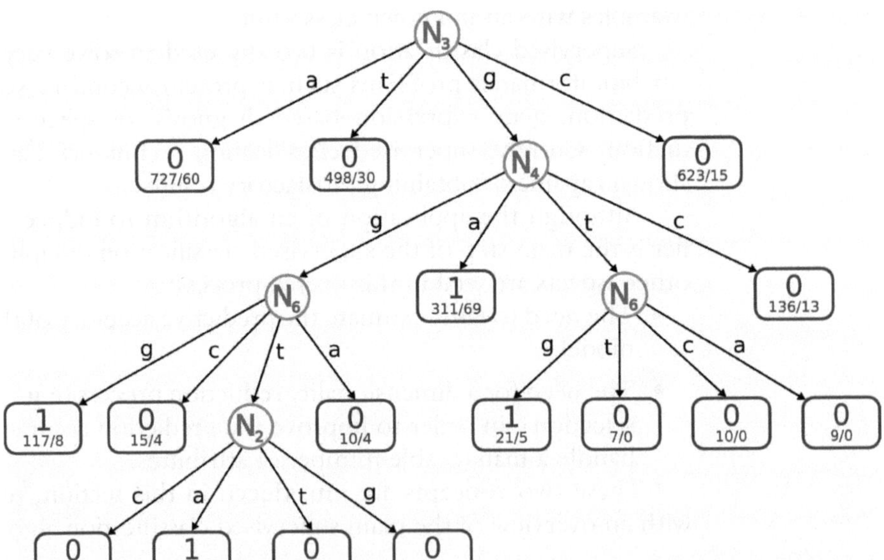

Fig. 2.3. Example of a decision tree constructed to identify donor splice sites. The model was generated from a data set where the class labels are true (1) and false (0) donor sites. The predictive variables represent the nucleotides around the 2-bp constant donor site (from N_1 to N_7). The circles represent the internal nodes and the rounded squares the terminal nodes. In each terminal node, the label of the majority class is indicated (1 or 0). Below this class label, each terminal node shows the number of donor sites in the training set that end up in the node (*left* figure), together with the number of samples that do not belong to the majority class (*right* figure).

Each internal node divides the instances based on the values of a specific informative variable that shows the highest correlation degree with the class label. The related literature proposes a broad range of metrics to measure this correlation degree, mainly based on information theory. Terminal nodes will ideally have samples of only one of the classes, although a mixture of classes is usually found. In order to avoid trees that are too specific and deep, after the tree passes through an initial growing phase, a pruning mechanism is applied in order to delete unrepresentative parts of the tree and to limit the effect of overfitting.

In spite of its popularity in many data analysis areas, in the case of bioinformatics problems – which usually have a limited number of samples per study – its use is not so extended. This could be explained due to its tendency to induce too simple and small trees when a small number of samples are provided.

Due to the instability of the basic formulation of this algorithm – small changes on the training set lead to very different trees – averaging processes are used to obtain more robust classifiers. Random forests average the prediction of a "forest" of decision trees built from resampled training sets of the original data set.

3.1.2. Bayesian Classifiers

This family of classifiers offers a broad range of possibilities to model $p(c|\ x_1, x_2, \ldots, x_d)$, which is the class distribution probability term conditioned to each possible value of the predictive variables. This term, in conjunction with the a priori probability of the class $p(c)$ and by means of Bayes' rule, is used to assign the most probable a posteriori class to a new unseen sample:

$$\gamma(x) = \arg\max_c p(c|x_1, x_2, \ldots, x_d) = \arg\max_c p(c)\, p(x_1, x_2, \ldots, x_d \mid c).$$

All the statistical parameters are computed from training data, commonly by their maximum-likelihood estimators.

Depending on the degree of complexity of the relationships between the variables of the problem to be modeled, an interesting battery of Bayesian classifiers can be found in the literature. *Naïve Bayes* is the most popular member of Bayesian classifiers. It assumes that all domain variables are independent when the class value is known. This assumption dramatically simplifies the exposed statistics, and only the univariate class-conditioned terms $p(x_i|c)$ are needed. Although this assumption is clearly violated in many situations (especially in many real problems with inherent complexity), the naïve Bayes classifier is able to obtain accurate enough results in many cases.

The *tree-augmented* classifier (11) goes one step further by learning a tree structure of dependences between domain variables. Besides the class label, each variable – except the tree root attribute – is conditioned by another predictor, and statistics of the form $p(x_i|c,x_j)$ have to be computed. This restriction in the

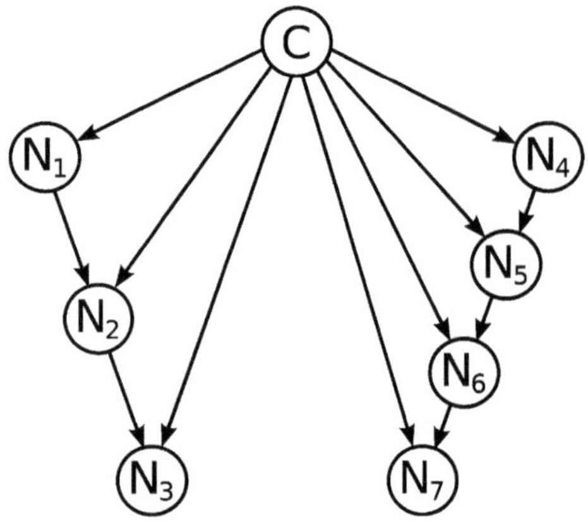

Fig. 2.4. Example of a Bayesian network constructed to identify donor splice sites. The model was generated from a data set where the class labels are true and false donor sites. The predictive variables represent the nucleotides around the 2-bp constant donor site (from N_1 to N_7). During the model-building process, a maximum number of two parents was used as a constraint.

number of parents is overcome by the *k-dependence Bayesian classifier*, which allows each predictive variable to be conditioned by up to *k* parent attributes.

The expressiveness of the graphical structure of a Bayesian classifier (*see* **Fig. 2.4** for an example), which is able to depict the conditional dependence relationships between the variables, is highly appreciated by domain experts, who are able to visually perceive the way in which the model operates. This property of Bayesian classifiers is increasing in popularity in the bioinformatics area. However, due to the nature of data from some bioinformatics tasks with a small number of samples and a large number of variables (e.g., gene expression domains), their application is severely restricted because the impossibility to compute reliable and robust statistics when complex relationships need to be learned from the scarce data. Because of its simplicity, and regardless of its lack of ability to represent too complex relationships among predictor variables of a problem, the naïve Bayes classifier is the most appropriate alternative in such scenarios.

3.1.3. The k-Nearest-Neighbor Paradigm

The basic formulation of the *k–nearest-neighbor* algorithm classifies an unlabeled sample by assigning it to the most frequent class among its *k* nearest samples. While a large battery of variations to follow this aim has been proposed, the majority-voting scheme among the *k* nearest samples for class prediction is the most commonly used. Other variants include the "distance-weighted nearest-neighbor" and the "nearest-hyperrectangle" methods. Implementations commonly use the Euclidean distance for numeric attributes and nominal-overlap for symbolic features.

More distance schemes are the Mahalanobis and the "modified value difference" metrics for numeric and symbolic features, respectively. See **Sections 4.7** and **6.4** in Witten and Frank (12) for a description of these alternatives and other variants. Also known as, "instance-based learning," or "lazy learning," this technique does not induce an explicit expression of the predictive model. Although able to obtain competitive predictive accuracies in many problems, it is discarded in many real situations where a descriptive knowledge discovery output is needed. This is due to the absence of an explicit model to be checked and observed by domain experts. The effect of the *k* parameter can be seen in **Fig. 2.5.**

3.1.4. Support Vector Machines

Support vector machines (*SVMs*) are one of the most popular classification techniques in use today. Its robust mathematical basis and

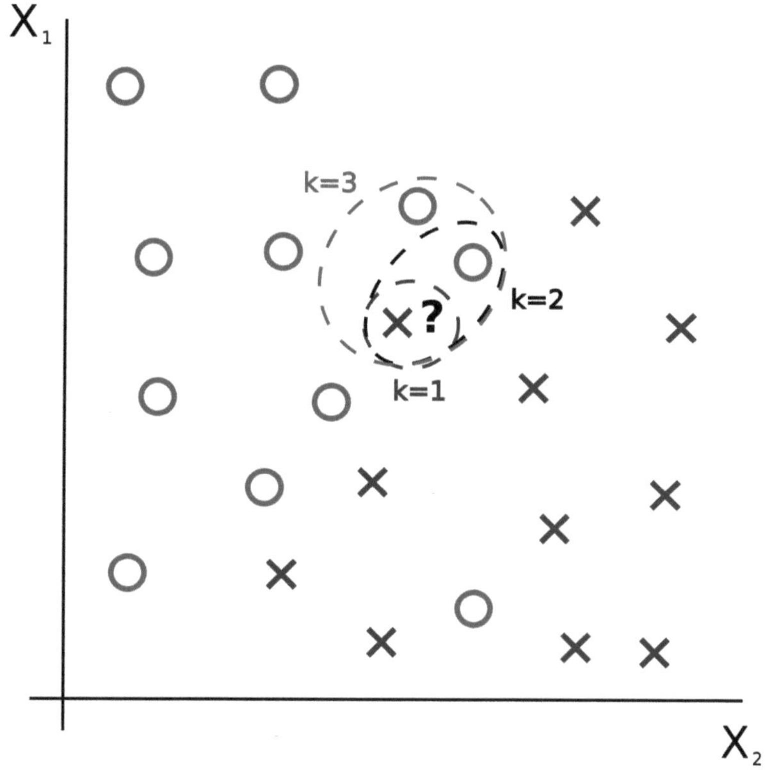

Fig. 2.5. Example of a *k*-nearest-neighbor classification. The problem consists of two variables, X_1 and X_2, and two classes, circle and cross. The circles and crosses represent the known examples, and the question mark a new instance that we need to classify. A 1-nearest neighbor classifies an unlabeled instance as the class of the known instance closest to the instance. In this case, a 1-nearest neighbor would classify the question mark as a cross. A 2-nearest neighbor looks at the two closest examples. In our example, we have a circle and a cross and thus have to choose a way to break the ties. A 3-nearest neighbor would classify the question mark as a circle (we have two circles and a cross). Setting the *k* at an odd value allows us to avoid ties in the class assignment.

the good accuracies that it demonstrates in many real tasks have placed it among practitioners' favorites. SVMs map input samples into a higher-dimensional space where a maximal separating hyperplane among the instances of different classes is constructed. The method works by constructing another two parallel hyperplanes on each side of this hyperplane. The SVM method tries to find the separating hyperplane that maximizes the area of separation between the two parallel hyperplanes. It is assumed that a larger separation between these parallel hyperplanes will imply a better predictive accuracy of the classifier. As the widest area of separation is, in fact, determined by a few samples that are close to both parallel hyperplanes, these samples are called *support vectors*. They are also the most difficult samples to be correctly classified. As in many situations, it is not possible to perfectly separate all the training points of different classes; the permitted distance between these misclassified points and the far side of the separation area is limited. Although SVM classifiers are popular due to the notable accuracy levels achieved in many bioinformatics problems, they are also criticized for the lack of expressiveness and comprehensibility of their mathematical concepts.

3.1.5. Ensemble Approaches

Although the most common approach is to use a single model for class prediction, the *combination of classifiers* with different biases is gaining popularity in the machine learning community. As each classifier defines its own decision surface to discriminate between problem classes, the combination could construct a more flexible and accurate decision surface. While the first approaches proposed in the literature were based on simple combinative models (majority vote, unanimity vote), more complex approaches are now demonstrating notable predictive accuracies. Among these we can cite the bagging, boosting, stacked generalization, random forest, or Bayesian combinative approaches. Due to the negative effect of small sample sizes on bioinformatics problems, model combination approaches are broadly used due to their ability to enhance the robustness of the final classifier (also known as the meta-classifier). On the other hand, the expressiveness and transparency of the induced final models are diminished.

3.2. Evaluation and Comparison of the Model Predictive Power

Since the assessment of the predictive accuracy of a classification model is a key issue in supervised classification, it is essential to measure the predictive power of our model over future unseen samples. This has been subject of deep research in the data analysis field during the last decades, resulting in an abundance of mature and robust concepts and techniques for model evaluation (13). Next, we review the most essential ones.

Given a two-class (positive and negative) problem, a *confusion matrix* such as the one presented in **Table 2.1** applies. This table gathers the basic statistics to assess the accuracy of a predictive

Table 2.1
Confusion matrix for a two-class problem

		Predicted class	
		+	–
Acutal class	+	a	b
	–	c	d

model, showing from qualitative and quantitative points of view a "photograph" of the hits and errors obtained by our model in an accuracy estimation procedure. Considering the counters *a, b, c,* and *d* is enough to compute the following key measures in model evaluation:

- Error rate, the portion of samples the model predicts incorrectly: $(b + c)/(a + b + c + d)$;
- True-positive rate or sensitivity, the portion of the positive samples the model predicts correctly: $a/(a + b)$;
- True-negative rate or specificity, the portion of the negative samples the model predicts correctly: $d/(c + d)$;
- False-negative rate or miss rate, the portion of the positive samples the classifier predicts falsely as negative: $b/(a + b)$;
- False-positive rate or false-alarm rate, the portion of the negative samples the classifier predicts falsely as positive: $c/(c + d)$.

These statistics are computed via an accuracy estimation technique. Since our working data set has a finite set of samples, evaluation involves splitting the available samples into several training and test sets. Since we know the class labels of the samples in the test sets, it is possible to evaluate the models induced by applying a particular classification algorithm by comparing the predictions that the model provides for the test cases. This computes the different accuracy scores. Obviously, the simplest way to estimate the predictive accuracy is to train the model over the whole data set and test it over the same instances. However, within the machine learning community, it is broadly accepted that this procedure, known as resubstitution error, leads to an optimistic bias. That is why machine learning researchers suggest a number of "honest" evaluation schemes, the most popular of which are the following:

- The *hold-out* method randomly divides the data set into a training set and a test set. The classification algorithm is induced in the training set and evaluated in the test set. This technique can be improved by applying different random train-test partitions. The latter is known as *repeated hold-out.*

- The *k-fold cross-validation* method involves partitioning the examples randomly into *k* folds or partitions. One partition is used as a test set and the remaining partitions form the training set. The process is repeated *k* times using each of the partitions as the test set. In *leave-one-out cross-validation,* a single observation is left out each time; i.e., it implies an *N-fold cross-validation* process, where *N* is the number of instances. *Stratified cross-validation* involves creating partitions so that the ratio of samples of each class in the folds is the same as in the whole data set. **Figure 2.6** shows a scheme of a fivefold cross-validation process.
- The *bootstrap* methodology has been adapted for accuracy estimation. This resampling technique involves sampling with replacement from the original data set to produce a group of *bootstrap data sets* of *N* instances each. Several variants of the bootstrap estimation can be used for accuracy assessment.

Receiver operating characteristic (ROC) curves are an interesting tool for representing the accuracy of a classifier. The ROC analysis evaluates the accuracy of an algorithm over a range of possible operating (or tuning) scenarios. A ROC curve is a plot of a model's true-positive rate against its false-positive rate: sensitivity versus 1-specifity. The ROC curve represents a plot of these two concepts for a number of values of a parameter (operating

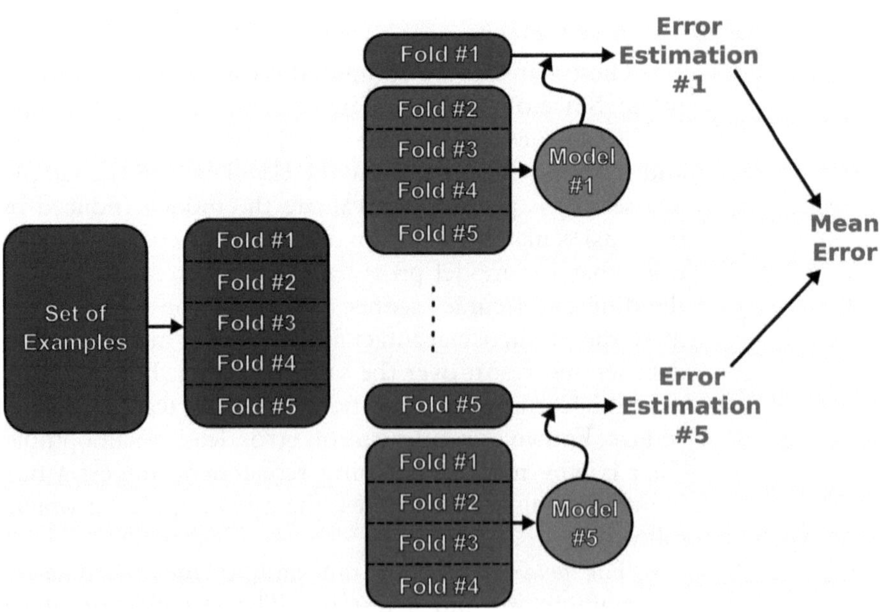

Fig. 2.6. Example of a 5-fold cross-validation process.

scenarios) of the classification algorithm. Examples of this free parameter are the different class misclassification costs or the variation in the class decision threshold of a probabilistic classifier. The area under the ROC curve can also be used for predictive accuracy estimation.

Due to the specificities of many bioinformatics problems that have an extremely low number of samples, the bioinformatics community has proposed novel predictive accuracy estimation methods with promising characteristics, such as bolstered error estimation (14).

Once the predictive accuracy of a group of classifiers in a specific domain has been estimated, an essential question is to perform a comparison between their accuracies. The statistics community has proposed (15) a varied and solid battery of parametric and nonparametric hypothesis tests to assess the degree of significance of the accuracy difference between compared algorithms. Several pioneering papers have alerted (13) the machine learning community about the need to apply statistical tests in order to complete a solid and reliable comparison between classification models. Going further than the classic comparison of classifiers in a single data set, novel conclusive references (16) establish the guidelines to perform a statistical comparison of classifiers in a group of data sets. The use of statistical tests has been extended in the bioinformatics community during recent years.

3.3. Feature Selection

It is well known by the machine learning community that the addition of variables to the classification model is not monotonic with respect to the predictive accuracy. Depending on the characteristics of the classification model, irrelevant and redundant features could worsen the prediction rate. As a natural answer to this problem, the feature selection (FS) problem can be defined as follows: Given a set of initial candidate features in a classification problem, select a subset of relevant features to build a robust model. Together with the improvement in computational and storage resources, a broad and varied range of interesting FS techniques has been proposed in the last 10–15 years, which has brought the FS topic to a high level of maturity and protagonism in many data analysis areas (17).

In contrast to other dimensionality reduction techniques such as those based on projection (e.g., principal component analysis) or compression (e.g., using information theory), FS techniques do not alter the original representation of the variables; they merely select a subset of them. Thus, they preserve the original semantics of the variables, hence offering the advantage of interpretability by a domain expert.

Besides the increase in accuracy, an FS procedure can bring several advantages to a supervised classification system such as decreasing the cost of data acquisition, improving the simplicity and understanding of the final classification model, and gaining deeper insight into the underlying processes that generated the data.

Although there are many ways to taxonomize FS techniques, these can be divided into three categories depending on how the FS search process interacts with the classification model. We thus have the filter, wrapper, and embedded approaches.

Filter techniques assess the relevance of features by looking only at the intrinsic characteristics of the data, and the interaction with the classification model is ignored. Most filter techniques are based on univariate feature relevance scores, which measure the correlation degree of each attribute with the class label. By means of a univariate metric, a ranking of features is established and low-scoring features are removed. Afterwards, this subset of high-ranked features is used to construct the final classification model. Although univariate metrics are computationally simple and fast, they ignore feature dependencies. Thus, a set of interesting multivariate filter techniques that take into consideration feature dependencies and redundancies has been proposed in the last years.

Wrapper techniques perform a search in the space of feature subsets by incorporating the classification algorithm within the process. The goodness of each subset is obtained by evaluating the predictive power of the classification algorithm when it is trained with the features included in the subset. As the cardinality of possible feature subsets is 2^n (where n is the number of initial attributes), a set of heuristic procedures has been proposed to conduct the search: sequential local techniques, genetic algorithms, ant-colony optimization approaches, etc. The main weaknesses of these techniques are that they have a higher risk of overfitting than filter techniques and they are very computationally intensive, especially if the classifier-building algorithm has a high computational cost.

Several classifier types (e.g., decision trees, decision rules) incorporate (embed) their own FS procedure in the model induction phase, and they do not make use of all initial variables to construct the final classifier. This FS modality is known as *embedded*. These techniques include the interaction with the classification model, and they have a lower computational cost than wrapper procedures.

As modern high-throughput biological devices are capable of monitoring a large number of features for each sample, the application of feature selection techniques in bioinformatics is an essential prerequisite for model building (18). As the magnitude of screened features is of several thousands in many problems, the direct application of any supervised modeling technique is unfeasible. This computational problem is worsened by the small sample sizes available for many bio-scenarios. While many feature selection techniques developed by the machine learning community are being used with success in bioinformatics research, the bio-community has also proposed during the last years an interesting set of techniques that fit the specificities of their data. The use of feature selection techniques is mandatory in any biomarker discovery process. The protagonism of feature selection is crucial in domains such as DNA microarray studies, sequence analysis, mass spectra, SNP analysis, or literature text mining (18).

4. Unsupervised Classification or Clustering: The Class Discovery Approach

Unsupervised classification – or clustering – is a key topic in the machine learning discipline. Its starting point is a training database formed by a set of N independent samples $D_N = (x^1, \ldots, x^N)$ drawn from a joint and unknown probability distribution $p(\mathbf{x}, c)$. Each sample is characterized by a group of d predictive variables or features $\{X_1, \ldots, X_d\}$ and C is a hidden variable that represents the cluster membership of each instance. In contrast to supervised classification, there is no label that denotes the class membership of an instance, and no information is available about the annotation of the database samples in the analysis. Clustering, which is also informally known as "class discovery," is applied when there is no class to be predicted, but rather when the instances are to be divided into natural groups. Once the appropriate *preprocessing* steps are performed over the available data, clustering techniques partition the set of samples into subsets according to the differences/similarities between them. The different objects are organized/taxonomized into groups such that the degree of association between two objects is maximal if they belong to the same group and minimal otherwise. Clustering reflects an attempt to discover the underlying mechanism from which instances originated.

A key concept in clustering is the type of distance measure that determines the similarity degree between samples. This will dramatically influence the shape and configuration of the induced clusters, and its election should be carefully studied. Usual distance functions are the Euclidean, Manhattan, Chebychev, or Mahalanobis.

The validation of a clustering structure, both from statistical and biological points of view, is a crucial task. Statistical validation can be performed by assessing the cluster coherence or by checking the robustness against the addition of noise. An intuitive criterion to be taken into account by any clustering algorithm is the minimization of dissimilarities of samples belonging to the same cluster (intracluster homogeneity), together with the maximization of the dissimilarities between the samples of different clusters (intercluster heterogeneity). Nevertheless, the problem of biological cluster validation is a highly demanded task by bioexperts that still remains an open challenge. Since a common characteristic of biological systems is the fact that they are not completely characterized, the election of the best cluster configuration is regarded as a difficult task for biologists. However, there are examples of recent methodologies (19) thought to validate clustering structures in different bioinformatics scenarios.

In many bio-scenarios, available samples are not annotated, which has led clustering to have been broadly used to solve different bioinformatics problems such as grouping homologous sequences into gene families, joining peaks that arise from the same peptide or protein in mass spectra experiments, or grouping similar gene expression profiles in DNA microarray experiments.

Clustering techniques play a central role in several bioinformatics problems, especially in the clustering of genes based on their expression profiles in a set of hybridizations. Based on the assumption that expressional similarity (i.e., co-expression) implies some kind of relationship, clustering techniques have opened a way for the study and annotation of sequences. As a natural extension to clustering, the recently revitalized *biclustering* topic has become a promising research area in bioinformatics (20). As it is known that not all the genes of a specific cluster have to be grouped into the same conditions, it seems natural to assume that several genes can only change their expression levels within a specified subset of conditions. This fact has motivated the development of specific biclustering algorithms for gene expression data.

In the following subsections, we briefly present the two principal families of clustering algorithms.

4.1. Partitional Clustering

Clustering algorithms that belong to this family assign each sample to a unique cluster, thus providing a *partition* of the set of points. In order to apply a partitional clustering algorithm, the user has to fix in advance the number of clusters in the partition. Although there are several heuristic methods for supporting the decision on the number of clusters (e.g., the Elbow method), this problem still remains open.

The *k-means algorithm* is the prototypical and best-known partitional clustering method. Its objective is to partition the set of samples into K clusters so that the within-group sum of squares is minimized. In its basic form, the algorithm is based on the alternation of two intuitive and fast steps. Before the iteration of these two steps starts, a random assignment of samples to K initial clusters is performed. In the first step, the samples are assigned to clusters, commonly to the cluster whose centroid is the closest by the Euclidean distance. In the second step, new cluster centroids are recalculated. The iteration of both steps is halted when no movement of an object to a different group will reduce the within-group sum of squares. The literature provides a high diversity of variations of the *K-means algorithm,* especially focused on improving the computing times. Its main drawback is that it does not return the same results in two different runs, since the final configuration of clusters depends on the initial random assignments of points to K initial clusters.

In *fuzzy* and *probabilistic* clustering, the samples are not forced to belong completely to one cluster. Via these approaches, each point has a degree of belonging to each of the clusters. Guided by the minimization of intracluster variance, the literature shows interesting fuzzy and probabilistic clustering methods, and the field is still open for further publication opportunities.

4.2. Hierarchical Clustering

This is the most broadly used clustering paradigm in bioinformatics. The output of a hierarchical clustering algorithm is a nested and hierarchical set of partitions/clusters represented by a tree diagram or *dendrogram,* with individual samples at one end (bottom) and a single cluster containing every element at the other (top). Agglomerative algorithms begin at the bottom of the tree, whereas divisive algorithms begin at the top. Agglomerative methods build the dendrogram from the individual samples by iteratively merging pairs of clusters. Divisive methods rarely are applied due to their inefficiency. Because of the transparency and high intuitive degree of the dendrogram, the expert can produce a partition into a desired number of disjoint groups by cutting the dendrogram at a given level. This capacity to decide the number of final clusters to be studied has popularized the use of hierarchical clustering among bio-experts.

A dissimilarity matrix with the distance between pairs of clusters is used to guide each step of the agglomerative merging process. A variety of distance measures between clusters is available in the literature. The most common measures are single-linkage (the distance between two groups is the distance between their closest members), complete-linkage (defined as the distance between the two farthest points), Ward's hierarchical clustering method (at each stage of the algorithm, the two groups that produce the smallest increase in the total within-group sum of squares are amalgamated), centroid distance (defined as the distance between the cluster means or centroids), median distance (distance between the medians of the clusters), and group average linkage (average of the dissimilarities between all pairs of individuals, one from each group).

5. Machine Learning Tools and Resources

Together with the improvement in computer storage and computation capacities, the machine learning community has developed a large number of interesting resources during the last decade. These common resources have crucially helped in the development of the field, and they have served as a useful basis to share experiences and results among different research groups.

Due to specific requirements of bioinformatics problems, the bioinformatics community has also contributed to this trend by developing a large number of applications and resources during the last five years. Three popular websites that collect a large amount of varied machine learning resources are Kdnuggets (21), Kmining (22), and the Google Group on Machine Learning (23). The interested practitioner can find in those references the latest data mining news, job offers, software, courses, etc.

We will limit this section to a set of useful and popular resources that have been proposed by the machine learning and data mining communities and that are being used by the bioinformatics community.

5.1. Open Source Software Tools

The MLC++ software (Machine Learning Library in C++) (24) was a pioneering initiative in the 1990s, providing free access to a battery of supervised classification models and performance evaluation techniques. This resulted in a dynamic initiative of the field, offering a base library to develop a large variety of machine learning techniques that appeared in different international publications during the last decade.

MLC++ served as an inspiration for more advanced and user-friendly initiatives during the last decade. Among these, we consider that WEKA (Waikato Environment for Knowledge Analysis) (16) and R-project (25) are nowadays the most influential and popular open source tools: Both offer a huge battery of techniques to cover a complete data mining process. While the algorithms covered by WEKA tend to have a heuristic bias, the R-project is more statistically oriented. As an essential component of the R-project, it is mandatory to reference the popular Bioconductor-project (26), which offers a powerful platform for the analysis and comprehension of genomic data.

Although there are recent initiatives to develop a more user-friendly interface for the powerful tools of the R-project, the intuitive and ease of use of the working environment offered by WEKA is highly appreciated by practitioners not familiarized with current data mining tools.

Other powerful and well-known machine learning free software tools developed by prestigious data mining research laboratories include RapidMiner (27) and Orange (28).

5.2. Benchmarking Data Sets

A common procedure among the developers of machine learning algorithms is to test and compare novel and original classifiers in established data sets. The UCI Machine Learning Repository (29) gathers a varied collection of classification data sets that have become a benchmark repository for machine learning practitioners. The UCI Knowledge Discovery in Databases Archive (30)is an online repository of large and complex data sets that proposes a set of varied, nontrivial data analysis challenges.

A novel and interesting initiative is the Swivel project (31), which is also known as the, "YouTube of data." Any registered user can upload his or her own data collection and correlate it with other data sets. The amount and variety of the collected data sets will surpass the expectations of any interested practitioner.

The interested researcher can find online repositories that collect preprocessed biological data sets ready to be loaded by machine learning software tools. The Kent Ridge Biomedical Data Set Repository (32) gathers a collection of benchmark gene expression and mass spectrometry databases to be mined by supervised classification techniques.

Acknowledgments

This work has been partially supported by the Etortek, Saiotek, and Research Groups 2007–2012 (IT-242-07) programs (Basque Government), the TIN2005-03824 and Consolider Ingenio 2010 – CSD2007-00018 projects (Spanish Ministry of Education and Science), and the COMBIOMED network in computational biomedicine (Carlos III Health Institute).

References

1. Prompramote S, Chen Y, Chen Y-PP. (2005) Machine learning in bioinformatics. In *Bioinformatics Technologies* (Chen Y-PP., ed.), Springer, Heidelberg, Germany, pp. 117–153.
2. Somorjai RL, Dolenko B, Baumgartner R. (2003) Class prediction and discovery using gene microarray and proteomics mass spectroscopy data: curses, caveats, cautions. *Bioinformatics* 19:1484–1491.
3. Larrañaga P, Calvo B, Santana R, Bielza C, Galdiano J, Inza I, Lozano JA, Armañanzas R, Santafé G, Pérez A, Robles V. (2006) Machine learning in bioinformatics. *Briefings in Bioinformatics* 7: 86–112.
4. Alpaydin E. (2004) *Introduction to Machine Learning*, MIT Press, Cambridge, MA.
5. Mitchell T. (1997) *Machine Learning*, McGraw Hill, New York.
6. Causton HC, Quackenbush J, Brazma A. (2003) *A Beginner's Guide. Microarray Gene Expression Data Analysis*, Blackwell Publishing, Oxford.

7. Parmigiani G, Garett ES, Irizarry RA, Zeger SL. (2003) *The Analysis of Gene Expression Data,* Springer-Verlag, New York.
8. Hilario M, Kalousis A, Pellegrini C, Muller M. (2006) Processing and classification of protein mass spectra. *Mass Spectrometry Rev* 25:409–449.
9. Shin H, Markey M. (2006) A machine learning perspective on the development of clinical decision support systems utilizing mass spectra of blood samples. *J Biomed Inform* 39:227–248.
10. Fayyad UM, Irani KB. (1993) Multi-interval discretization of continuous-valued attributes for classification learning. In *Proceedings of the 13th International Joint Conference on Artificial Intelligence,* pp. 1022–1029.
11. Friedman N, Geiger D, Goldszmidt M. (1997) Bayesian network classifiers. *Mach Learn* 29:131–163.
12. Witten IH, Frank E. (2005) *Data Mining. Practical Machine Learning Tools and Techniques (2nd ed.),* Morgan Kaufmann, San Francisco.
13. Dietterich TG. (1998) Approximate statistical test for comparing supervised classification learning algorithms. *Neural Comp* 10:1895–1923.
14. Sima C, Braga-Neto U, Dougherty E. (2005) Superior feature-set ranking for small samples using bolstered error estimation. *Bioinformatics* 21:1046–1054.
15. Kanji GK. (2006) *100 Statistical Tests,* SAGE Publications, Thousand Oaks, CA.
16. Demsar J. (2006) Statistical comparisons of classifiers over multiple data sets. *J Mach Learn Res* 7:1–30.
17. Liu H, Motoda H. (2007) *Computational Methods of Feature Selection,* Chapman and Hall–CRC Press, Boca Raton, FL.
18. Saeys Y, Inza I, Larrañaga P. (2007) A review of feature selection methods in bioinformatics. *Bioinformatics* 23:2507–2517.
19. Sheng Q, Moreau Y, De Smet F, Marchal K, De Moor B. (2005) Advances in cluster analysis of microarray data. In *Data Analysis and Visualization in Genomics and Proteomics* (Azuaje F, Dopazo J, Eds.), Wiley, New York, pp. 153–173.
20. Cheng Y, Church GM. (2000) Biclustering of expression data. In *Proceedings of the Eighth International Conference on Intelligent Systems for Molecular Biology,* pp. 93–103.
21. Kdnuggets: Data Mining, Web Mining and Knowledge Discovery (2008) http://www.kdnuggets.com
22. Kmining: Business Intelligence, Knowledge Discovery in Databases and Data Mining News (2008) http://www.kmining.com
23. Google Group – Machine Learning News (2008) http://groups.google.com/group/ML-news/
24. Kohavi R, Sommerfield D, Dougherty J. (1997) Data mining using MLC++, a machine learning library in C++. *Int J Artif Intell Tools* 6:537–566.
25. Dalgaard R. (2002) *Introductory Statistics with R,* Springer, New York.
26. Gentleman R, Carey VJ, Huber W, Irizarry RA, Dudoit S. (2005) *Bioinformatics and Computational Biology Solutions Using R and Bioconductor,* Springer, New York.
27. Mierswa I, Wurst M, Klinkerberg R, Scholz M, Euler T. (2006) YALE: Rapid prototyping for complex data mining tasks. In *Proceedings of the 12th ACM SIGKDD International Conference on Knowledge Discovery and Data Mining,* pp. 935–940.
28. Demsar J, Zupan B, Leban G. (2004) *Orange: From Experimental Machine Learning to Interactive Data Mining,* White Paper, Faculty of Computer and Information Science, University of Ljubljana, Slovenia.
29. Asunción A, Newman DJ. (2008) *UCI Machine Learning Repository,* University of California, Irvine, School of Information and Computer Sciences. http://archive.ics.uci.edu/ml/
30. Hettich S, Bay SD. (1999) *The UCI KDD Archive,* University of California, Irvine, School of Information and Computer Sciences. http://kdd.ics.uci.edu
31. Swivel project – Tasty Data Goodies (2008) http://www.swivel.com
32. Kent Ridge Biomedical Data Set Repository (2008) http://research.i2r.a-star.edu.sg/rp/

Chapter 3

SNP-PHAGE: High-Throughput SNP Discovery Pipeline

Ana M. Aransay, Rune Matthiesen, and Manuela M. Regueiro

Abstract

High-throughput genotyping technologies have become popular in studies that aim to reveal the genetics behind polygenic traits such as complex disease and the diverse response to some drug treatments. These technologies utilize bioinformatics tools to define strategies, analyze data, and estimate the final associations between certain genetic markers and traits. The strategy followed for an association study depends on its efficiency and cost. The efficiency is based on the assumed characteristics of the polymorphisms' allele frequencies and linkage disequilibrium for putative casual alleles. Statistically significant markers (single mutations or haplotypes) that cause a human disorder should be validated and their biological function elucidated. The aim of this chapter is to present a subset of bioinformatics tools for haplotype inference, tag SNP selection, and genome-wide association studies using a high-throughput generated SNP data set.

Key words: SNP genotyping, bioinformatics, complex diseases.

1. Introduction

The human genome, estimated to contain approximately 3 billion base pairs, differs between individuals by a single nucleotide every 100–300 base pairs (1). This variation (0.1%) is mainly due to the presence of about 9–11 million common *single-nucleotide polymorphisms* (SNPs) (*see* **Note 1**) (1, 2). For nucleotide variation to be considered an SNP, it must occur at a frequency of 1% or more in a particular population. The lowest allele frequency at a locus is termed a *minor allele frequency* (MAF). Almost all common SNPs are biallelic, and most genotyping platforms only consider two alleles. The vast majority of SNPs apparently do not have phenotypic effects, but recent association and linkage studies have

R. Matthiesen (ed.), *Bioinformatics Methods in Clinical Research*, Methods in Molecular Biology 593,
DOI 10.1007/978-1-60327-194-3_3, © Humana Press, a part of Springer Science+Business Media, LLC 2010

begun to identify a growing number of SNP variants, which significantly change the orthography or expression of known genes (genetics of global gene expression), altering individual susceptibility to complex diseases or the individual response to drugs (pharmacogenetics). The advances reached in the aforementioned disciplines (accurate molecular diagnosis and pharmacogenetics) will facilitate the development of so-called personalized medicine.

The need to understand the distribution of SNPs in the human genome and to develop novel efficient strategies to identify risk variants of complex diseases encouraged the creation of the International HapMap Project in 2002 [http://www.hapmap.org/; (2–4)]. The objectives of this project were twofold: (1) to generate a catalog of common genetic variants to describe their nature, location, and distribution, and (2) to determine the **haplotype** (combination of marker alleles on a single chromosome, Haploid Geno*type*) structure and diversity of the human genome. The definition of these haplotype blocks allows researchers to genotype in a cost-effective way a significant proportion of the total genomic variation among individuals by analyzing only a few hundred thousand *haplotype tagging SNPs* (*htSNPs*), each one representing the variation of its corresponding block. This htSNP strategy assumes that this kind of point mutation has occurred only once during human evolutionary history and, therefore, SNPs are considered unique event polymorphisms (UEPs) (3).

The information included in the HapMap Project is based mostly on SNP technology, and the consortium has developed several bioinformatics tools for the management and analysis of the generated data. During the first stage of the international project, populations of African, Asian, and European ancestry were characterized: 30 adult-and-both-parents trios from Ibadan, Nigeria (YRI); 30 trios of U.S. residents of northern and western European ancestry (CEU); 44 unrelated individuals from Tokyo, Japan (JPT); and 45 unrelated Han Chinese individuals from Beijing, China (CHB). In a later phase, the project is being completed through pilot studies of other populations in an effort to maximize the human diversity analyzed. We should take into account that the proper characterization of the reference population is crucial for association studies, especially for those projects based on candidate genes.

2. Materials

2.1. Genomic DNA

In this chapter, we provide an overview, focusing on genotype analysis from human genomic DNA (gDNA). The higher the quality of the gDNA, the better the results will be. DNA consumption is important in studies with irreplaceable clinical samples, and, therefore, gDNA obtained should be amplified (at the whole genome level) by any of the methodologies such as Repli-G® Mini Kit (Cat.# 150023, QIAGEN), based on the ability of the Phi29 polymerase. However, unamplified gDNA is preferred.

2.2. Genotyping Technologies

Several platforms are available for high-throughput genotyping. The most frequently used are whole genome genotyping arrays standardized by Affymetrix Inc. and Illumina Inc. Each company has developed several unique arrays and their corresponding specific protocols. To date, the standardized arrays allow the characterization from 96 to more than 2 million SNPs. Some of these designs have also included probes (oligonucleotides) to detect deletions and/or duplications at the chromosomal level, thereby increasing the cost-efficiency of these methodologies.

3. Genotyping Characterization and Association Studies

3.1. Prospects About Strategies to Face Genetic Association Studies

Genetic association studies (GAS) are a powerful method for identifying susceptibility genes for common diseases, offering the promise of novel targets for diagnosis and/or therapeutic intervention that act on the root cause of the disease.

There are two main strategies to achieve association studies:

- Genome-wide association studies (GWAS) involve scanning a huge number of samples, either as case-control cohorts or in family trios, utilizing hundreds of thousands of SNP markers located throughout the human genome. Statistical algorithms are then applied to compare the frequencies of alleles, genotypes, or multimarker haplotypes between disease and control cohorts.
- Candidate gene approaches imply characterizing some polymorphisms that are previously selected in candidate genomic regions in a large number of samples. This strategy requires a priori knowledge of the genetics behind the studied disease or can be based on results from preliminary transcriptomic, proteomic, and/or metabolomic experiments.

Any of these methodologies identify regions (*loci*) with statistically significant differences according to allele or genotype frequencies of cases and controls, suggesting their possible role in the disease or strong linkage disequilibrium with causative *loci*.

When designing the strategy of any association project, one should take into account the *statistical power* of the genotyping possibilities that can be carried out. Statistical power depends on the prevalence of the studied disease, the disease causal allele frequency (if known), the number of *loci* and samples that will be characterized, the linkage disequilibrium (LD) of the markers, the Type I (the error of rejecting a "correct" null hypothesis) and Type II (the error of not rejecting a "false" null hypothesis) error rates, and the genetic model or hypothetical relationships between alleles and disease (e.g., multiplicative, additive, dominant, or recessive model) (5, 6). Thousands of samples should be analyzed to have a significant statistical power (e.g., 95%), which means facing extremely expensive projects.

In order to reduce genotyping costs, it is recommended to perform GAS following a **two-stage** or **multistage** design (7). Both strategies involve analyzing a large number of SNPs in a subset of individuals for possible associations with a disease phenotype (first phase), test all *loci* quality criteria [Hardy–Weinberg equilibrium (HWE) (8), minor allele frequency (MAF), etc.], and only those polymorphisms that exhibit association are further tested in an independent study (second phase). This helps to minimize costs of genotyping and to maximize statistical power.

The haplotype block structure of the human genome allows a current strategy that has proved to be very efficient: to design SNP panels based on *htSNPs* of the regions of interest or all along the genome. When SNPs are in LD with each other and form haplotypes, redundant information is contained within the haplotype. By knowing the marker at one locus, one can make a prediction about the marker that will occur at the linked *loci* nearby. By genotyping htSNPs, it is sufficient to capture most of the haplotype structure of the human genome. The accuracy with which one can make this prediction is dependent upon the strength of LD between the *loci* and the allele frequencies. The premise of this approach is that disease mutations will have occurred on a background of mutations that are already present, and over small distances. The rate at which this background pattern is disrupted will be fairly low. Thus, in theory, one can capture the majority of the diversity within a region by typing its htSNPs.

The aforementioned method is highly efficient in terms of cost and time commitment, but one should be aware that for proper selection of htSNPs, researchers should know very well the population in which they will apply the SNP panel in order to define the haplotypes, and therefore the htSNPs, for the very same population. If previous association studies for a particular population are not available, it is recommended to test the ***portability* or *transferability*** (9, 10) of the htSNPs defined for the HapMap populations, at least for the one that will be used to select the htSNPs. This test could be done by analyzing a preliminarily few SNPs that are not in LD in any of the populations (independent variants), and measuring the genetic divergence based on the F_{ST} value (11) for those markers. If this test results in null divergence, the use of the selected HapMap population should be a suitable approach.

For the selection of htSNPs within or nearby *candidate regions*, we recommend the following steps:

1. Search for information on the genes considered to be related to a certain disease, using, for example,
   - http://www.ncbi.nlm.nih.gov/entrez/query.fcgi?DB=pubmed (PubMed)
   - http://www.ncbi.nlm.nih.gov/entrez/query.fcgi?db=OMIM (OMIM, Online Mendelian Inheritance in Man)
   - http://www.ncbi.nlm.nih.gov/entrez/query.fcgi?CMD=search&DB=nucleotide
2. Prioritize the genes in the order that you want to characterize them.
3. Select the parameters under which the search of SNPs will be done: linkage disequilibrium [LD, (12, 13)], htSNPs, synonymous coding SNPs (cSNPs), nonsynonymous cSNPs, MAF, heterozygosity, etc.

   An example is outlined below:

   * htSNPs selection

   > Copy annotation (e.g., NM_005891, TP53) and paste it in HapMap Genome Browser (http://www.hapmap.org/cgi-perl/gbrowse/hapmap_B36/), with the proper formatting as defined in the web tutorial, into the "Reference point or region" box, and press "Search."

   > Copy chromosome positions as they appear in the header of the refresh page, and delete – add manually 200–300 bp up and downstream in order to increase the region of search (e.g.. convert *Chr9:660,000..760,000* into *Chr9:659,800..760,200*).

   > Select in "Reports and Analysis" the option "Download SNP Genotype Data"; click "Configure."

   > When the configuration window is open, select population as required (e.g., CEU).

   > Select "Save to Disk"; then click "Go."

   > Save the file for further analysis.

   * Download and open HAPLOVIEW v. 4.1 software [http://www.broad.mit.edu/mpg/haploview/index.php (14)].

     > Select "HapMap Format" to open the saved file.

     > Select the file saved from the browser and leave the values that appear in the window by default.

> When the data are open, write the selected HWE *p*-value cut-off (e.g., 0.01).

> Write the selected minimum minor allele frequency (MAF) (e.g., 0.01).

> Click "Rescore Markers."

> Then, go to the "Tagger" tab.

> Select "Pairwise tagging only."

> Click "Run Tagger."

> Copy and paste table in Excel containing htSNPs and Proxys (*see* **Note 2**).

* To be sure that Exonic SNPs (cSNPs) are included in the analysis, one should search for them specifically as follows:

> Go to dbSNP: http://www.ncbi.nlm.nih.gov/SNP/

> Search Entrez SNPs for the same annotation as used in HapMap Browser (e.g., NM_005891, TP53) and click "Go."

> When the list of SNPs related to human is processed, click "GeneView" in one of them.

> Select the cSNP view option. Then copy and paste the table containing the list of cSNPs in Excel together with all the htSNPs retrieved from HapMap.

4. Design plexes/arrays/beads/chips according to each technology, protocol, and chemistry (generally, this step is evaluated together with the company that will elaborate the assay).

3.2. Statistics for Association Studies

The challenge of the emerging genome association studies is to identify patterns of polymorphisms that vary systematically between individuals with different disease states and could therefore represent the effects of risk-enhancing or -protective alleles. This seems to be straightforward; however, the genome is so large that patterns that are suggestive of a causal polymorphism could well arise by chance. To aid in distinguishing causal from spurious signals, robust standards for statistical significance need to be established. Another method is to consider only patterns of polymorphisms that could plausibly have been generated by causative genetic variants, given our current understanding of human genetic history and evolutionary processes such as mutation and recombination (15).

Data quality is very important for preliminary analysis, and, accordingly, results should be checked thoroughly. In most of the GAS, researchers have tested for **HWE** primarily as a data quality check and have discarded *loci* that, for example, deviate from

HWE among controls at significance level $\alpha = 10^{-3}$ or 10^{-4}, using this criterion as a manner of detecting low-quality genotyping or null alleles (16). Departure from HWE can be due to inbreeding, population stratification, or selection. However, it can also be a symptom of disease association (17), the implications of which are often underexploited (18). In addition, the possibility that a deviation from HWE is due to a deletion polymorphism (19) or a segmental duplication (20) that could be important in disease causation should also be considered before discarding *loci*.

Tests of association can be carried out based on *single-marker tests*. These analyses take into account the significant differences in allele or genotype frequency between the case and control populations. To improve the power to detect additive risks, it is recommended to count alleles rather than genotypes so that each individual contributes twice to a 2×2 table and a Pearson 1-df test can be applied. However, this procedure should be used with caution since it requires an assumption of HWE in cases and controls combined and does not lead to interpretable risk estimates. Such analysis can be run in Haploview (14) or PLINK (21) software. The Cochran–Armitage test, also known as the Armitage test (22), is similar to the allele-count method but is more conservative and does not rely on an assumption of HWE. The idea is to test the hypothesis of zero slope for a line that fits the three genotypic risk estimates best. Performing the Armitage test implies sacrificing power if the genotypic risks are far from being additive. Nevertheless, there is no widely accepted answer to the question of which single-SNP test to use. Decisions regarding what test to choose are difficult, but the test to use would be the one that fits best.

Another factor that scientists have to deal with is the possibility of getting fake association values due to the *stratification* of their case and control populations (**Fig. 3.1**). One way of testing this effect is to genotype about 100 widely spaced SNPs in both cases and controls in addition to the candidate association study

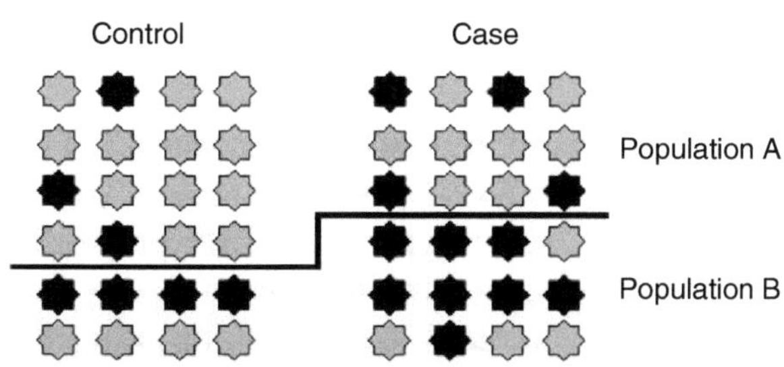

Fig. 3.1. Fake association due to population structure.

SNPs, and test those for HWE. Devlin and Roeder (23) proposed a method, termed "genomic control" (GC), which obviates many of the concerns about population substructure by using the features of the genomes present in the sample to correct for stratification. The goal of GC is to achieve control in population-based designs in the same way that is obtained for a family-based study. The GC approach exploits the fact that population substructure generates an "overdispersion" of statistics used to assess association. The degree of overdispersion generated by population substructure can be estimated and taken into account by testing multiple polymorphisms throughout the genome. GC can be calculated with PLINK software (21).

Another model for testing population stratification, unlike GC, does take into account the possible differences in allele frequency among the studied populations. This clustering method using multilocus genotype data, assigns individuals to populations, and can be processed by STRUCTURE software (24). A similar strategy to STRUCTURE is analyzing the data by principal component analysis (PCA). This could be done using one's own developed bioinformatics tools or using EIGENSTRAT (25). EIGENSTRAT detects and corrects for population stratification even in genome-wide association studies that include hundreds of thousands of markers. The method unambiguously models ancestry differences between cases and controls along continuous axes of variation. The resulting correction is specific to a candidate marker's variation in frequency across ancestral populations, minimizing spurious associations while maximizing the power to detect true associations. EIGENSTRAT is implemented as part of the EIGENSOFT package. Source code, documentation, and executables for the EIGENSOFT package are freely available at http://genepath.med.harvard.edu/~reich/Software.htm. (Note: It runs in a Linux platform.)

Advice for Using Eigensoft Package

Eigensoft includes these four tools:

```
convertf → smartpca → eigenstrat → gc.perl
```

- ➢ *convertf*: converts among five different file formats (*see* below)
- ➢ *smartpca.perl*: runs PCA on input genotype data
- ➢ *eigenstrat*: computes association statistics between genotype and phenotype, both uncorrected and corrected for stratification
- ➢ *gc.perl*: applies Genomic Control (23) to the uncorrected and EIGENSTRAT-corrected statistics

Convertf

Note that "file format" simultaneously refers to the formats of three distinct files:

➢ *genotype file*: contains genotype data for every individual at each SNP

➢ *snp file*: contains information about each SNP

➢ *indiv file*: contains information about each individual

There are five formats:

- ANCESTRYMAP
- EIGENSTRAT
- PED
- PACKEDPED
- PACKEDANCESTRYMAP

PED and PACKEDPED can be used within the PLINK package [(21), http://pngu.mgh.harvard.edu/~purcell/plink/].

The example below takes input in PED format and outputs in EIGENSTRAT format to be used as an input file for the *smartpca.perl* program.

PED format

- The *genotype file* must be saved with the extension **.ped* and contains one line per individual. Each line contains six or seven columns of information about the individual, plus two genotype columns for each SNP in the order the SNPs are specified in the *snp file*.

  For example,

```
1 SAMPLE0 0 0 2 2 1 2 3 3 1 1 1 1 3 3 1 1 3 3
2 SAMPLE1 0 0 1 2 1 2 1 3 1 4 1 1 1 3 1 1 3 3
3 SAMPLE2 0 0 2 1 1 2 1 1 1 4 1 2 1 3 1 4 3 4
4 SAMPLE3 0 0 1 1 2 2 1 3 4 4 2 2 1 1 1 4 3 4
5 SAMPLE4 0 0 2 1 2 2 1 1 1 4 2 2 1 1 1 4 4 4
```

The genotype format must be either 0ACGT or 01234, where 0 means missing data.

The first six or seven columns of the *genotype file* are the following:

1st column = family ID

2nd column = sample ID

3rd and 4th columns = sample ID of parents

5th column = gender (male is 1, female is 2)

6th column = case/control status (1 is control, 2 is case), quantitative trait value, or population group label

7th column (this column is optional): always set to 1.

Convertf does not support pedigree information, so the first, third, and fourth columns are ignored in *convertf* input and set to arbitrary values in *convertf* output.

- The *snp file* must be saved with the extension ***.pedsnp** although *convertf* also supports the ***.map** suffix for this input filename.

  For example,

```
11 rs0000 0.000000      0 A C
11 rs1111 0.001000 100000 A G
11 rs2222 0.002000 200000 A T
11 rs3333 0.003000 300000 C A
11 rs4444 0.004000 400000 G A
11 rs5555 0.005000 500000 T A
11 rs6666 0.006000 600000 G T
```

The *snp file* contains one line per SNP and six columns (last two are optional):

1st column = chromosome. Use X for X chromosome (Note: SNPs with illegal chromosome values, such as 0, will be removed.)

2nd column = SNP name.

3rd column = genetic position (in Morgans).

4th column = physical position (in base pairs).

Optional 5th and 6th columns are reference and variant alleles. (Note: For monomorphic SNPs, the variant allele can be encoded as X.)

- The *indiv file* must be saved with the extension ***.pedind.** *Convertf* also supports the full **.ped** file for this input file.

  For example,

```
1 SAMPLE0 0 0 2 2
2 SAMPLE1 0 0 1 2
3 SAMPLE2 0 0 2 1
4 SAMPLE3 0 0 1 1
5 SAMPLE4 0 0 2 1
```

The *indiv file* contains the same first six or seven columns of the *genotype file*.

The syntax to run *convertf* is "../bin/convertf -p parfile".

Parfiles:

par.ANCESTRYMAP.EIGENSTRAT > converts ANCESTRYMAP to EIGENSTRAT format.

par.EIGENSTRAT.PED > converts EIGENSTRAT to PED format.

par.PED.EIGENSTRAT > converts PED to EIGENSTRAT format (used to estimate possible population structure with EIGENSTRAT).

par.PED.PACKEDPED > converts PED to PACKEDPED format.

par.PACKEDPED.PACKEDANCESTRYMAP > converts PACKEDPED to PACKEDANCESTRYMAP.

par.PACKEDANCESTRYMAP.ANCESTRYMAP > converts PACKEDANCESTRYMAP to ANCESTRYMAP.

Below is the description of each parameter in parfile for *Convertf* (par.PED.EIGENSTRAT):

genotypename: input *genotype file*

snpname: input *snp file*

indivname: input *indiv file*

outputformat: ANCESTRYMAP, EIGENSTRAT, PED, PACKEDPED, or PACKEDANCESTRYMAP

genotypeoutname: output *genotype file*

snpoutname: output *snp file*

indivoutname: output *indiv file*

Smartpca

Smartpca runs principal components analysis (PCA) on input genotype data and outputs principal components (eigenvectors) and eigenvalues.

The method assumes that samples are unrelated. However, having a small number of cryptically related individuals is usually not a problem in practice since they will typically be discarded as outliers. The following example takes input in EIGENSTRAT format.

The syntax to run *smartpca* is "../bin/smartpca.perl" followed by

– i example.geno: *genotype file* in EIGENSTRAT format.

– a example.snp: *snp file*.

– b example.ind: *indiv file*.

– k k: (default is 10) the number of principal components to output.

– o example.pca: output file of principal components. Individuals removed as outliers will have all values set to 0.0 in this file.

– p example.plot: prefix of output plot files of top two principal components (labeling individuals according to labels in *indiv file*).

– e example.eval: output file of all eigenvalues.

– l example.log: output logfile.

- m maxiter: (default is 5) the maximum number of outlier removal iterations. To turn off outlier removal, set –m 0.
- t topk: (default is 10) the number of principal components along which the software takes away outliers during each outlier removal iteration.
- s sigma: (default is 6.0) the number of standard deviations an individual must exceed, along one of the topk top principal components, in order to be removed as an outlier.

Eigenstrat

The syntax to run *eigenstrat* is "../bin/eigenstrat" followed by

- i example.geno: *genotype file* in EIGENSTRAT format.
- j example.pheno: input file of phenotypes. File contains one line, which encloses one character per individual: 0 means control, 1 means case, 9 means missing phenotype. (Note: ../CONVERTF/ind2pheno.perl will convert from *indiv file* to *.pheno file.)
- p example.pca: input file of principal components (output of *smartpca.perl*).
- l l: (default is 10) the number of principal components along which to correct for stratification. Note that l must be less than or equal to the number of principal components reported in the file example.pca.
- o example.chisq: chi square (schisq) association statistics. File contains log of flags to eigenstrat program, followed by one line per SNP.
  - The first entry of each line is Armitage chisq statistic (22). If the set of individuals with both a valid genotype and phenotype is monomorphic for either genotype or phenotype, then NA is reported.
  - The second entry of each line is the EIGENSTRAT chisq statistic. If the set of individuals with both a valid genotype and phenotype is monomorphic for either genotype or phenotype, then NA is reported.

  Note: Even if $l = 0$, there is a tiny difference between the two statistics because Armitage uses NSAMPLES while this program uses NSAMPLES-1.

Gc.perl

The syntax to run *gc.perl* is "../bin/gc.perl infile outfile":

- *infile* is an input file of chisq statistics produced by the EIGENSTRAT program. It contains both uncorrected and EIGENSTRAT statistics for each SNP.
- *outfile* is an output file that lists lambda inflation values (for both uncorrected and EIGENSTRAT) chisq statistics

after scaling by lambda (uncorrected and EIGENSTRAT-corrected).

The computation of lambda is as described in Devlin and Roeder (23): A lambda value above 1 indicates inflation in chisq statistics. By definition, lambda is not allowed to be less than 1.

Association tests for multiple SNPs (*multimarker tests*) are another controversial point in GAS. A popular strategy is to use *haplotypes* (12, 13, 26), estimated by the LD among adjacent SNPs (9, 10), to try to capture the correlation structure of SNPs in regions of low recombination. This approach can lead to analyses with fewer degrees of freedom, but this benefit is minimized when SNPs are ascertained through a tagging strategy. For these tests, Bonferroni correction (27) is too conservative; thus, a nonparametric permutation approach is recommended since it offers asymptotically exact control over the false-positive rate. Permutations are theoretically a simple method, but their estimation demands powerful computational resources. This procedure keeps genotype or haplotype data constant and the phenotypes are randomized over individuals in order to generate several data sets that conserve the LD structure but do not associate the structure with a phenotype.

The inclusion of oligo-probes to detect deletions and/or duplications into the SNP genotyping technologies together with the adjustments carried out for the proper interpretation of SNP fluorescence intensity and heterozygosity data allow the *copy number variation* (*CNV*) to be measured at a chromosomal level as another parameter of association. CNVs may account for a considerable proportion of the normal human phenotypic variation (28, 29) but can also be the cause of several genomic disorders (29), especially those CNVs that interrupt genes, since their presence may alter transcription levels (dosage-sensitive genes). Several statistical methods have been developed to detect associations of CNVs to diseases (30–32), but more accurate algorithms should be considered, since there is still a lack of reproducibility among experiments and analyses carried out in different laboratories (33). In addition, special attention should be focused on selecting the reference population to be used in this kind of analysis (33).

It is worth mentioning that although most analyses in GAS data focus on the effect of individual variants, some algorithms to estimate both *gene–gene* (epistatic) and *gene-environment* interactions are already incorporated into SNP- or haplotype-based regression models and related tests (34, 35). However, since these estimates require enormous calculation resources for data integration, it is still a newly developing field that will yield very promising results for the understanding of complex diseases (e.g., polygenic and multifactorial disorders).

Markers that exhibit association must be exhaustively studied and their role in the studied disease should be elucidated using *in vivo* models (*Drosophila* spp., mice strains, and/or cell cultures).

4. Notes

1. An SNP is defined as a DNA sequence variation of a single nucleotide (A, T, C, G) in the genome between species or chromosomes within a species. SNPs can be located in genes (promoter, exon, intron, or UTRs) or intergenic regions. Those SNPs in coding regions can be divided into nonsynonymous (result in an amino acid change that is called SAP, single amino acid change) or synonymous (silent mutation codes for identical amino acids) SNPs. The nonsynonymous SNPs can be further divided into missense (results in a different amino acid) and nonsense (results in a stop codon) types. The Human Genome Variation Society provides recommendations for the description of SNPs and sequence variants (http://www.hgvs.org/mutnomen/recs.html).
2. Proxies are flanking markers in linkage disequilibrium with the reference SNP.
3. Linkage disequilibrium (LD) measures the nonrandom association of alleles. It is the deviation of the observed haplotype frequency from the expected haplotype frequency. Note that LD can be calculated in different ways (36).

Glossary

Allele – One of the variant forms of a gene or a genetic *locus.*

Causative SNPs – Changes in a single nucleotide that cause a disease or trait.

Coding SNPs (cSNPs) – SNPs that occur in regions of a gene that are transcribed into RNA (i.e., an exon) and eventually translated into protein. cSNPs include synonymous SNPs (i.e., confer identical amino acid) and nonsynonymous SNPs (i.e., confer different amino acid).

Genetic map – Also known as a **linkage map**. A genetic map shows the position of genes and/or markers on chromosomes relative to each other, based on genetic distance (rather than physical distance). The distance between any two markers is represented as a function of recombination.

Genetic marker – A DNA sequence whose presence or absence can be reliably measured. Because DNA segments that are in close proximity tend to be inherited together, markers can be used to indirectly track the inheritance pattern of a gene or region known to be nearby.

Genotype – The combination of alleles carried by an individual at a particular genetic *locus.*

Haplotype – Haplotypes are an ordered set of alleles located on one chromosome. They reveal whether a chromosomal segment was maternally or paternally inherited and can be used to delineate the boundary of a possible disease-linked locus.

Haplotype tagging SNPs (htSNPs) – Those SNPs that represent the variation in each block based on the linkage disequilibrium among the markers considered within a block.

Hardy–Weinberg equilibrium (HWE) – The equilibrium between the frequencies of alleles and the genotype of a population. The occurrence of a genotype stays constant unless mating is nonrandom or inappropriate, or mutations accumulate. Therefore, the frequency of genotypes and the frequency of alleles are said to be at "genetic equilibrium." Genetic equilibrium is a basic principle of population genetics.

Intronic SNPs– Single-nucleotide polymorphisms that occur in noncoding regions of a gene that separate the exons (i.e., introns).

Linkage disequilibrium (LD) – Phenomenon by which the alleles that are close together in the genome tend to be inherited together (haplotype).

Linkage map – See *genetic map.*

Mendelian pattern of inheritance – Refers to the predictable way in which single genes or traits can be passed from parents to children, such as in autosomal dominant, autosomal recessive, or sex-linked patterns.

Minor allele frequency (MAF) – Given an SNP, its minor allele frequency is the frequency of the SNP's less frequent allele in a given population.

Mutation – A change in the DNA sequence. A mutation can be a change from one base to another, a deletion of bases, or an addition of bases. Typically, the term "mutation" is used to refer to a disease-causing change, but technically any change, whether or not it causes a different phenotype, is a mutation.

Penetrance – Penetrance describes the likelihood that a mutation will cause a phenotype. Some mutations have a high penetrance, almost always causing a phenotype, whereas others have a low penetrance, perhaps only causing a phenotype when other genetic or environmental conditions are present. The best way to measure penetrance is phenotypic concordance in monozygotic twins.

Phenotype – Visible or detectable traits caused by underlying genetic or environmental factors. Examples include height, weight, blood pressure, and the presence or absence of disease.

Polygenic disorders – Disorders that are caused by the combined effect of multiple genes, rather than by just one single gene. Most common disorders are polygenic. Because the genes involved are often not located near each other, their inheritance does not usually follow Mendelian patterns in families.

Surrogate SNPs – Single-nucleotide polymorphisms that do not cause a phenotype but can be used to track one because of their strong physical association (linkage) to an SNP that does cause a phenotype.

Susceptibility – The likelihood of developing a disease or condition.

References

1. Hinds DA, Stuve LL, Nilsen GB, Halperin E, Eskin E, Ballinger DG, Frazer KA, Cox DR. (2005) Whole-genome patterns of common DNA variation in three human populations. *Science* 307:1072–1079.
2. The International Haplotype Consortium. (2007) A second generation human haplotype map of over 3.1 million SNPs. *Nature* 449:851–862.
3. The International Haplotype Consortium. (2003) The International HapMap Project. *Nature* 426:789–796.
4. The International Haplotype Consortium. (2005) A haplotype map of the human genome. *Nature* 437:1299–1320.
5. Gordon D, Finch SJ, Nothnagel M, Ott J. (2002) Power and sample size calculations for case-control genetic association tests when errors are present: application to single nucleotide polymorphisms. *Hum Hered* 54:22–33.
6. Zhang K, Calabrese P, Nordborg M, Sun F. (2002) Haplotype block structure and its applications to association studies: power and study designs. *Am J Hum Genet* 71: 1386–1394.
7. Thomas D, Xie R, Gebregziabher M. (2004) Two-stage sampling designs for gene association studies. *Genet Epidemiol* 27: 401–414.
8. Hartl DL, Clark AG. (1997) *Principle of Population Genetics,* 3rd ed., Sinauer Associates, Inc., Sunderland, MA.
9. Ribas G, Gonzalez-Neira A, Salas A, Milne RL, Vega A, Carracedo B, Gonzalez E, Barroso E, Fernandez LP, Yankilevich P, et al. (2006) Evaluating HapMap SNP data transferability in a large-scale genotyping project involving 175 cancer-associated genes. *Hum Genet* 118:669–679.
10. Huang W, He Y, Wang H, Wang Y, Liu Y, Wang Y, Chu X, Wang Y, Xu L, Shen Y, et al. (2006) Linkage disequilibrium sharing and haplotype-tagged SNP portability between populations. *Proc Natl Acad Sci USA* 103:1418–1421.
11. Reynolds J, Weir BS, Cockerham CC. (1983) Estimation of the coancestry coefficient: basis for a short-term genetic distance. *Genetics* 105:767–779.
12. Lewontin RC. (1988) On measures of gametic disequilibrium. *Genetics* 120: 849–852.
13. Pritchard JK, Przeworski M. (2001) Linkage disequilibrium in humans: models and data. *Am J Hum Genet* 69: 1–14.

14. Barrett JC, Fry B, Maller J, Daly MJ. (2005) Haploview: analysis and visualization of LD and haplotype maps. *Bioinformatics* 21: 263–265.
15. Cavalli-Sforza LL, Menozzi P, Piazza A. (1994) *The History and Geography of Human Genes*, Princeton University Press, Princeton, NJ.
16. Carlson CS, Smith JD, Stanaway IB, Rieder MJ Nickerson DA. (2006) Direct detection of null alleles in SNP genotyping data. *Hum Mol Genet* 15:1931–1937.
17. Nielsen DM, Ehm MG, Weir BS. (1998) Detecting marker-disease association by testing for Hardy-Weinberg disequilibrium at a marker locus. *Am J Hum Genet* 63:1531–1540.
18. Wittke-Thompson JK, Pluzhnikov A, Cox NJ. (2005) Rational inferences about departures from Hardy-Weinberg equilibrium. *Am J Hum Genet* 76:967–986.
19. Conrad DF, Andrews TD, Carter NP, Hurles ME, Pritchard JK. (2006) A high-resolution survey of deletion polymorphism in the human genome. *Nat Genet* 38: 75–81.
20. Bailey JA, Eichler EE. (2006) Primate segmental duplications: crucibles of evolution, diversity and disease. *Nat Rev Genet* 7:552–564.
21. Purcell S, Neale B, Todd-Brown K, Thomas L, Ferreira MA, Bender D, Maller J, Sklar P, de Bakker PI, Daly MJ, et al. (2007) PLINK: a tool set for whole-genome association and population-based linkage analyses. *Am J Hum Genet* 81: 559–575.
22. Armitage P. (1955) Tests for linear trends in proportions and frequencies. *Biometrics* 11:375–386.
23. Devlin B, Roeder K. (1999) Genomic control for association studies. *Biometrics* 55:997–1004.
24. Pritchard JK, Stephens M, Donnelly P. (2000) Inference of population structure using multilocus genotype data. *Genetics* 155:945–959.
25. Price AL, Patterson NJ, Plenge RM, Weinblatt ME, Shadick NA, Reich D. (2006) Principal component analysis corrects for stratification in genome-wide association studies. *Nat Genet* 38:904–909.
26. Gabriel SB, Schaffner SF, Nguyen H, Moore JM, Roy J, Blumenstiel B, Higgins J, DeFelice M, Lochner A, Faggart M, et al. (2002) The structure of haplotype blocks in the human genome. *Science* 296:2225–2229.
27. Bonferroni CE. (1936) Teoria statistica delle classi e calcolo delle probabilità [in Italian]. *Pubblicazioni del R Istituto Superiore di Scienze Economiche e Commerciali di Firenze* 8:3–62.
28. Iafrate AJ, Feuk L, Rivera MN, Listewnik ML, Donahoe PK, Qi Y, Scherer SW, Lee C. (2004) Detection of large-scale variation in the human genome. *Nat Genet* 36:949–951.
29. Sebat J, Lakshmi B, Troge J, Alexander J, Young J, Lundin P, Maner S, Massa H, Walker M, Chi M, et al. (2004) Large-scale copy number polymorphism in the human genome. *Science* 305:525–528.
30. Peiffer DA, Le JM, Steemers FJ, Chang W, Jenniges T, Garcia F, Haden K, Li J, Shaw CA, Belmont J, et al. (2006) High-resolution genomic profiling of chromosomal aberrations using Infinium whole-genome genotyping. *Genome Res* 16:1136–1148.
31. Colella S, Yau C, Taylor JM, Mirza G, Butler H, Clouston P, Bassett AS, Seller A, Holmes CC, Ragoussis J. (2007) QuantiSNP: an objective Bayes hidden-Markov model to detect and accurately map copy number variation using SNP genotyping data. *Nucleic Acids Res* 35:2013–2025.
32. Wang K, Li M, Hadley D, Liu R, Glessner J, Grant SF, Hakonarson H, Bucan M. (2007) PennCNV: an integrated hidden Markov model designed for high-resolution copy number variation detection in whole-genome SNP genotyping data. *Genome Res* 17:1665–1674.
33. Baross A, Delaney AD, Li HI, Nayar T, Flibotte S, Qian H, Chan SY, Asano J, Ally A, Cao M, et al. (2007) Assessment of algorithms for high throughput detection of genomic copy number variation in oligonucleotide microarray data. *BMC Bioinformatics* 8:368.
34. Millstein J, Conti DV, Gilliland FD, Gauderman WJ. (2006) A testing framework for identifying susceptibility genes in the presence of epistasis. *Am J Hum Genet* 78:15–27.
35. Lake SL, Lyon H, Tantisira K, Silverman EK, Weiss ST, Laird NM, Schaid DJ. (2003) Estimation and tests of haplotype-environment interaction when linkage phase is ambiguous. *Hum Hered* 55:56–65.
36. Hedrick P, Sudhir K. (2001) Mutation and linkage disequilibrium in human mtDNA. *Eur J Hum Genet* 9:969–972.

Chapter 4

R Classes and Methods for SNP Array Data

Robert B. Scharpf and Ingo Ruczinski

Abstract

The Bioconductor project is an "open source and open development software project for the analysis and comprehension of genomic data" (1), primarily based on the R programming language. Infrastructure packages, such as Biobase, are maintained by Bioconductor core developers and serve several key roles to the broader community of Bioconductor software developers and users. In particular, Biobase introduces an S4 class, the eSet, for high-dimensional assay data. Encapsulating the assay data as well as meta-data on the samples, features, and experiment in the eSet class definition ensures propagation of the relevant sample and feature meta-data throughout an analysis. Extending the eSet class promotes code reuse through inheritance as well as interoperability with other R packages and is less error-prone. Recently proposed class definitions for high-throughput SNP arrays extend the eSet class. This chapter highlights the advantages of adopting and extending Biobase class definitions through a working example of one implementation of classes for the analysis of high-throughput SNP arrays.

Key words: SNP array, copy number, genotype, S4 classes.

1. Introduction

The Bioconductor project is an "open source and open development software project for the analysis and comprehension of genomic data," primarily based on the R programming language, and provides open source software for researchers in the fields of computational biology and bioinformatics-related disciplines (1). Infrastructure packages such as Biobase settle basic organizational issues for high-throughput data and facilitate the interoperability of R packages that utilize this infrastructure. Transparency and

R. Matthiesen (ed.), *Bioinformatics Methods in Clinical Research*, Methods in Molecular Biology 593,
DOI 10.1007/978-1-60327-194-3_4, © Humana Press, a part of Springer Science+Business Media, LLC 2010

reproducibility are emphasized in Bioconductor through package vignettes.

A key element of infrastructure for high-throughput genomic data is the eSet, a virtual class for organizing high-throughput genomic data defined in Biobase. An instance of an eSet-derived class contains the high-throughput assay data and the corresponding meta-data on the experiment, samples, covariates, and features (e.g., probes) in a single object. While much of the development of the eSet has been in response to high-throughput gene expression experiments that measure RNA (or cDNA) abundance, the generality of the eSet class enables the user to extend the class to accommodate a variety of high-throughput technologies. Here, we focus on single-nucleotide polymorphism (SNP) microarray technology and the eSet-derived classes specific to this technology.

SNP microarrays provide estimates of genotype and copy number at hundreds of thousands of SNPs along the genome, and several recent papers describe approaches for the genotype (2–9). In addition to probes targeting the polymorphic regions of the genome, the latest Affymetrix and Illumina platforms contain a set of nonpolymorphic probes for estimating the copy number.

The S4 classes and methods proposed here are organized around the multiple levels of SNP data. In particular, we refer to the raw samples containing probe intensities as the features-level data and the processed data containing summaries of genotype calls and copy number as the SNP-level data. Finally, there is a third level of analytic data obtained from methods that smooth the SNP-level summaries as a function of the physical position on the chromosome, such as hidden Markov models (HMMs). Algorithms at the third tier are useful for identifying genomic features such as deletions (hemizygous or homozygous), amplifications (more than two copies), and copy-neutral loss of heterozygosity.

This chapter is organized as follows. We begin with a brief overview of S4 classes, illustrating concepts such as inheritance using minimal class definitions for the high-throughput SNP data. With these minimal definitions in place, we discuss their shortcomings and motivate the development of the current class definitions. We conclude with an example that illustrates the following workflow: (i) creating an instance of an SNP-level class from matrices of genotype calls and copy number, (ii) plotting the SNP-level data as a function of physical position along the chromosome, (iii) plotting a hidden Markov model to identify alterations in copy number or genotype, and (iv) plotting the predicted states from the hidden Markov model alongside the genomic data.

2. S4 Classes and Methods

In the statistical environment R, an object can be a value, a function, or a complex data structure. To perform an action on an object, we write a function. For instance, we could write a function to calculate the row means of a matrix. When the object and functions become complex, classes and methods become useful as an organizing principle. An S4 class formally defines the ingredients of an object. A method for a class tells R which function should be performed on the object. A useful property of classes and methods is inheritance. For instance, a matrix is an array with only two dimensions: rows and columns. Using the language of classes, we say that an array is a parent class (or superclass) that is extended by the class matrix. Inheritance refers to the property that any methods defined for the parent class are available to the children of the parent class. In this section, we will discuss two approaches that can be used to construct classes that extend a parent class, illustrate the concept of inheritance by minimally defining S4 classes for storing estimates of genotype and copy number, provide examples of how to construct methods to access and replace elements of an instantiated class, and show how methods that check the validity of an instantiated object can be used to reduce errors. This section provides a very brief overview of S4 classes and methods; *see* Chambers (10) for a detailed description. The classes defined in this section are solely for the purpose of illustration and are not intended to be used for any analytic data.

2.1. Initializing Classes

To construct classes for SNP-level summaries of genotype calls and copy number estimates after preprocessing, we can use the following classes as minimal definitions:

```
> setClass("MinimalCallSet", representation(calls
    = "matrix"))
[1] "MinimalCallSet"
> setClass("MinimalCopyNumberSet", representation(copyNumber
    = "matrix"))
[1] "MinimalCopyNumberSet"
```

An instance of MinimalCallSet contains a slot for the matrix of genotype calls, and an instance of MinimalCopyNumberSet contains a slot for the matrix of copy number estimates.

2.2. Extending Classes

A parent class of MinimalCallSet and MinimalCopyNumberSet, called SuperSet, is created by the function setClassUnion:

```
> setClassUnion("SuperSet", c("MinimalCallSet",
   "MinimalCopyNumberSet"))
[1] "SuperSet"
```

```
> showClass("SuperSet")
Virtual Class "SuperSet"
No Slots, prototype of class "NULL"
Known Subclasses: "MinimalCallSet", "MinimalCopyNumberSet"
> extends("MinimalCallSet", "SuperSet")
[1] TRUE
```

MinimalCallSet and MinimalCopyNumberSet extend SuperSet. Note that SuperSet is a virtual class, and therefore we cannot instantiate an object of class SuperSet. However, instantiating one of the derived classes requires only a matrix of the SNP-level summaries. Using a recent version of R (> 2.7), one may obtain an example data set from the VanillaICE R package.

```
> source("http://www.bioconductor.org/biocLite.R")
> biocLite("VanillaICE", type = "source")
> library(VanillaICE)
> data(sample.snpset)
> gt <- calls(sample.snpset)[1:3, 1:3]
> gt[gt == 1] <- "AA"
> gt[gt == 2] <- "AB"
> gt[gt == 3] <- "BB"
> cn <- copyNumber(sample.snpset)[1:3, 1:3]
> colnames(cn) <- colnames(gt) <- sapply(colnames(gt),
   function(x) strsplit(x, "_")[[1]][1])
> callset <- new("MinimalCallSet", calls = gt)
> cnset <- new("MinimalCopyNumberSet", copyNumber = cn)
> attributes(callset)
$calls
NA17101 NA17102 NA17103


SNP_A-1507972 "AB" "BB" "AB"
SNP_A-1641761 "AB" "AB" "AB"
SNP_A-1641781 "AB" "AA" "AA"
$class
[1] "MinimalCallSet"
attr(,"package")
[1] ".GlobalEnv"
> attributes(cnset)
$copyNumber
NA17101 NA17102 NA17103
SNP_A-1507972 3.176972 2.775924 3.051108
SNP_A-1641761 1.705276 1.793427 1.647903
SNP_A-1641781 2.269756 1.741290 1.806562
$class
[1] "MinimalCopyNumberSet"
attr(,"package")
[1] ".GlobalEnv"
```

As MinimalCallSet and MinimalCopyNumberSet extend SuperSet, methods defined at the level of the parent class are inherited. For instance, we define, show, and call this function on the instantiated objects of MinimalCallSet and MinimalCopyNumberSet.

```
> setMethod("show", "SuperSet", function(object)
   attributes(object))
[1] "show"
```

```
> show(callset)
$calls
NA17101 NA17102 NA17103
SNP_A-1507972 "AB" "BB" "AB"
SNP_A-1641761 "AB" "AB" "AB"
SNP_A-1641781 "AB" "AA" "AA"
$class
[1] "MinimalCallSet"

attr(,"package")
[1] ".GlobalEnv"
> show(cnset)
$copyNumber
NA17101 NA17102 NA17103
SNP_A-1507972 3.176972 2.775924 3.051108
SNP_A-1641761 1.705276 1.793427 1.647903
SNP_A-1641781 2.269756 1.741290 1.806562
$class
[1] "MinimalCopyNumberSet"
attr(,"package")
[1] ".GlobalEnv"
```

The contains argument in the function setClass can be used to extend an existing parent class. For instance,

```
> setClass("MinimalSnpSet", contains = "SuperSet",
    representation(calls = "matrix",
+ copyNumber = "matrix"))

[1] "MinimalSnpSet"
```

By defining methods that access specific elements of a class at the level of the parent class, it is not necessary to define these methods for any of the derived classes.

2.3. Signatures

The signature of a generic function is a named list of classes that determines the method that will be dispatched. Consider the generic function foo in the following code chunk. The method that is dispatched when foo(object) is called depends on the class of object.

```
> setGeneric("foo", function(object) standardGeneric("foo"))
[1] "foo"
> setMethod("foo", signature(object = "ANY"),
    function(object) message("message 1"))
[1] "foo"
> setMethod("foo", signature(object = "matrix"),
    function(object) message("message 2"))

[1] "foo"
> foo(1)
> foo(as.matrix(1))
```

More precisely, the dispatched method depends on the "distance" of the class of the argument to the generic function and the signature of the method. For example, if we define a new class A that extends class matrix, message 2 will be printed, as the distance between the object and class matrix is 1 whereas the distance between A and ANY is greater than 1.

```
> setClass("A", contains = "matrix")
[1] "A"
> x <- as(matrix(1), "A")
> foo(x)
> setMethod("foo", signature(object = "A"), function(object)
   message("message 3"))
[1] "foo"
> foo(x)
[1] "genotypeCalls"
[1] "genotypeCalls"
NA17101 NA17102 NA17103
SNP_A-1507972 "AB" "BB" "AB"
SNP_A-1641761 "AB" "AB" "AB"
SNP_A-1641781 "AB" "AA" "AA"
```

In addition to defining methods that access information from an object, one may define a method that replaces information in an object. An example of such a method follows:

```
> setGeneric("genotypeCalls<-", function(object, value)
   standardGeneric("genotypeCalls<-"))
[1] "genotypeCalls<-"
> setReplaceMethod("genotypeCalls", c("SuperSet", "matrix"),
   function(object, value) {
+ object@calls <- value
+ return(object)
+ })
[1] "genotypeCalls<-"
```

2.4. Validity Methods

Validity methods can be useful to avoid committing errors when instantiating a class that can have unfortunate consequences on downstream analyses. For instance, for objects of class MinimalSnpSet, it is useful to require that the row names and column names of the copy number and genotype matrices be identical. Therefore, we can define a validity method for the class MinimalSnpSet that checks whether the names are identical and, if not, throws an error.

```
> setValidity("MinimalSnpSet", function(object) {
+ valid <- identical(rownames(object@calls),
   rownames(object@copyNumber))
+ if (!valid)
+ stop("rownames are not identical")
+ valid <- identical(colnames(object@calls),
   colnames(object@copyNumber))
+ if (!valid)
+ stop("colnames are not identical")
+ return(msg)
+ })
Class "MinimalSnpSet"
Slots:
Name: calls copyNumber
Class: matrix matrix
Extends: "SuperSet"
```

```
> colnames(gt) <- letters[20:22]
> tryCatch(new("MinimalSnpSet", calls = gt, copyNumber = cn),
   error = function(e) print(e))
<simpleError in·validityMethod(object): colnames are not
   identical>
```

3. SNP-Level Classes and Methods

When constructing S4 classes for the purpose of analyzing high-throughput SNP data, the following considerations are useful:

1. Develop as little new code as possible, reusing code that has been extensively tested and documented in other packages.
2. The SNP-level summaries that are available as assay data may depend on the preprocessing algorithm or the particular SNP microarray technology.
3. Attaching meta-data on the samples, features, and experiment to the object storing the assay data (as is commonly done with eSet-derived classes) is useful for ensuring that the meta-data are attached to the assay data throughout an analysis.
4. Adopting standard data structures defined in widely used packages such as Biobase promotes interoperability of R packages that perform complementary tasks.

The schematic in **Fig. 4.1** illustrates the relationships of our implementation of SNP-level classes in the package oligoClasses. We briefly discuss each of these classes below.

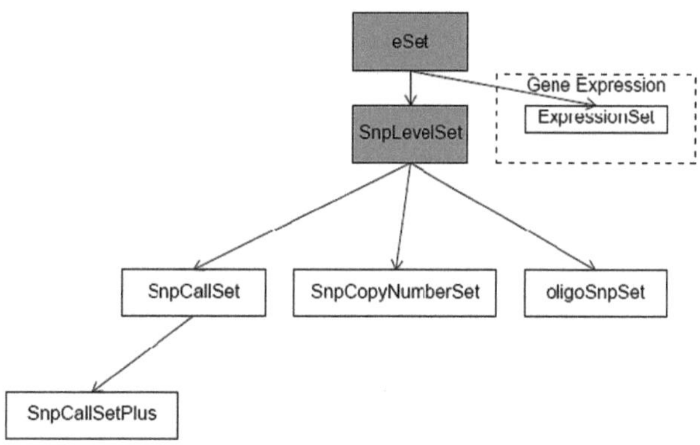

Fig. 4.1. Classes for SNP-level data, as defined in the Bioconductor package oligoClasses. Note that eSet and SnpLevelSet are virtual classes.

eSet: eSet is a virtual class defined in the R package Biobase (1) and provides a basic container for high-throughput genomic data. Slots in eSet are defined for assay data (assayData: e.g., genotype calls), characteristics of the samples (slot phenoData: e.g., phenotype), characteristics of the features (slot featureData: e.g., the name of the feature), and experimental data (slot experimentData: e.g., details of the laboratory and experimental methods). Via inheritance, each of the SNP-derived classes contains these components; accessors and replacement methods defined for the eSet can be readily applied to the eSet-derived classes.

SnpLevelSet: SnpLevelSet is a virtual class that extends eSet directly. Note that all SNP-level classes in **Fig. 4.1** extend SnpLevelSet directly. To understand why we define a virtual superclass for SNP-level data (when eSet is already available), consider that many methods are likely to be applicable to all SNP-derived classes, but perhaps not eSet-derived classes such as the ExpressionSet. For instance, the plotting methods in SNPchip and the hidden Markov model in VanillaICE rely on the chromosome and physical position of the SNP. While this information is critical for statistical methods such as an HMM that smooths SNP-level summaries as a function of physical position on the chromosome, it may be less useful or of no use for gene expression microarrays. Furthermore, because accessors for chromosome and physical position are useful for all of the SNP-derived classes, defining these accessors at the level of SnpLevelSet eliminates the need to define accessors for each of the derived classes. Of course, the flexibility to define methods specific to each of the derived classes remains.

SnpLevelSet progeny: Progeny of SnpLevelSet, including SnpCallSet, SnpCopyNumberSet, and oligoSnpSet, are defined according to the elements in the assayData slot. Elements of the assayData in SnpCallSet include calls (genotype calls) and callsConfidence (confidence scores for the genotype calls), whereas assayData elements in SnpCopyNumberSet are copyNumber and cnConfidence (confidence scores for copy number estimates). The assay data of an oligoSnpSet are the union of the assayData elements in SnpCallSet and SnpCopyNumberSet.

3.1. Example

We suggest the Bioconductor package oligo for preprocessing high-throughput SNP array data for the various Affymetrix platforms (100k, 500k, 5.0, and 6.0). In addition to genotype calls, the crlmm function in oligo provides confidence scores of the genotype calls that can be propagated to higher-level analyses, such as the hidden Markov models discussed in the following section. A method for estimating copy number in oligo is currently under development. In this section, we assume that the user has obtained SNP-level summaries of genotype and copy number by some means. We show how to create an instance of oligoSnpSet from matrices of genotype calls and copy number estimates, plot the SNP-level summaries versus physical position on the genome, and fit an HMM to identify alterations in copy number or genotype.

3.2. Instantiating an oligoSnpSet Object

To create an instance of oligoSnpSet, we take advantage of an example provided with the Bioconductor package VanillaICE, using only the matrices of copy number estimates, cn, and genotype calls, gt. The data we extract from the VanillaICE package are simulated data for a chromosome 1 on the Affymetrix 100k platform. Note that the matrices are organized such that the columns are samples and the rows are SNPs. While the elements of cn can be any positive number, the elements of gt are the integers 1, 2, 3, and 4, corresponding to the genotypes AA, AB, BB, and NA (not available), respectively. The row names (here, Affymetrix identifiers for the SNP) and column names (sample identifiers) of cn and gt must be identical. Confidence scores for the copy number estimates and genotype calls, when available, are stored similarly.

```
> library(VanillaICE)
> data(chromosome1)
> copynumber <- copyNumber(chromosome1)
> calls <- calls(chromosome1)
> cnConf <- callsConf <- matrix(NA, nrow = nrow(copynumber),
   ncol = ncol(copynumber),
+ dimnames = list(rownames(copynumber), colnames(copynumber)))
> snpset <- new("oligoSnpSet", copyNumber = copynumber,
   calls = calls, cnConfidence = cnConf,
+ callsConfidence = callsConf)
> annotation(snpset) <- "pd.mapping50k.hind240,pd.mapping50k.
   xba240"
> validObject(snpset)
[1] TRUE
```

The annotation slot is important for accessing the appropriate annotation package (available at Bioconductor). In this example, the SNPs originate from two Affymetrix platforms – the 50k Xba and 50k Hind chips. The annotation packages can be installed from Bioconductor with the following command:

```
> source("http://www.bioconductor.org/biocLite.R")
> biocLite(c("pd.mapping50k.hind240", "pd.mapping50k.xba240"))
```

Because the plotting methods and the HMM both frequently access the chromosome and physical position of the SNPs in the object, it is generally more convenient to store this information in the featureData slot. The position and chromosome methods first check the variable labels in the featureData and, if not present, retrieve this information from the annotation packages.

```
> featureData(snpset)$position <- position(snpset)
> featureData(snpset)$chromosome <- chromosome(snpset)
```

3.3. Visualizing the Data

The Bioconductor package SNPchip provides several useful methods for visualizing objects instantiated from one of the derived classes of SnpLevelSet (11). Similar to the R package lattice (13), the plotting method does not plot the data; rather, it returns an object of class ParSnpSet that contains all of the default graphical parameters used to plot an instance of oligoSnpSet. The show method called on an object returned by plotSnp produces a plot. The following command plots the snpset object using the default graphical parameters:

```
> show(plotSnp(snpset))
```

The resulting plot, together with the assessment of DNA copy number alterations, is shown in **Fig. 4.2**.

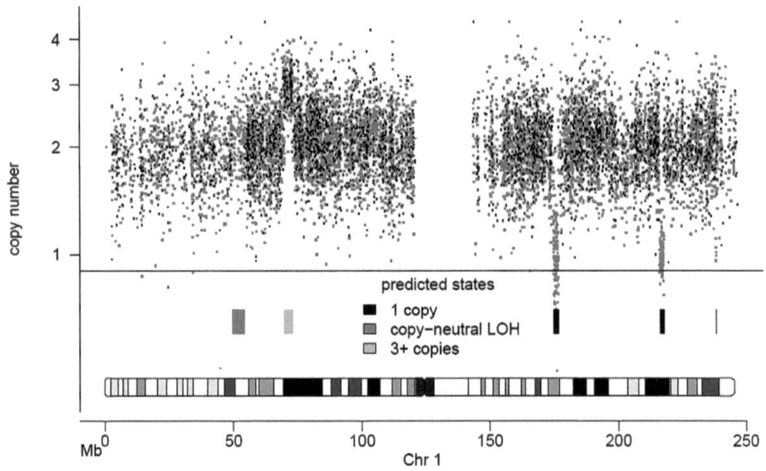

Fig. 4.2. Simulated data for chromosome 1 on the Affymetrix 100k platform. The *x*-axis denotes the loci along the chromosome, and the *y*-axis denotes the copy number estimates. Homozygous genotype calls are plotted in *light gray*, while heterozygous genotype calls are plotted in *dark gray*. Also shown is the inference for DNA copy numbers and alterations, using a hidden Markov model. This HMM captured the DNA alterations we simulated, namely (from *left* to *right*), a region of copy-neutral loss of heterozygosity, an amplification, and three deletions of various sizes on the q-arm.

3.4. Identifying Chromosomal Alterations

The simulated data used in this example contain five alterations that we utilize as benchmarks when testing the HMM model in the VanillaICE package. Details on the simulation and on the HMM model are described elsewhere (12). In order to fit the HMM, we must specify the hidden states and compute the emission and transition probabilities. We assume that the copy number estimates are Gaussian on the log-2 scale. To calculate the emission probabilities for the copy number, we require the location parameter of the Gaussian distribution (on the copy number scale) to be specified for each of the hidden states. If confidence scores for the copy number estimates are not available, the scale parameter is computed using a robust estimate of the log-2 copy number distribution and is assumed to be the same for each state. For genotype calls, one must specify the probability of a homozygous genotype call (AA or BB) for each of the hidden states. The transition probabilities, using an estimate of genomic distance, are SNP-specific.

```
> options <- new("HmmOptions", states = c("D","N","L",
   "A"), snpset = snpset,
+ copyNumber.location = c(1,2,2,3), probHomCall
   = c(0.99, 0.7, 0.99, 0.7))
> params <- new("HmmParameter", states = states(options),
   initialStateProbability = 0.99)
> cn.emission <- copyNumber.emission(options)
> gt.emission <- calls.emission(options)
> emission(params) <- cn.emission + gt.emission
> genomicDistance(params)
   <- exp(-2 * physicalDistance(options)/(100 * 1e+06))
> transitionScale(params)
   <- scaleTransitionProbability(options)
> fit <- hmm(options, params)
> class(fit)
```

The object returned by the hmm method is an instance of the class HmmPredict. HmmPredict extends SnpLevelSet directly. The following code can be used to plot the SNP-level summaries of genotype and copy number alongside the predicted states from the HMM.

```
> gp <- plotSnp(snpset(options), fit)
> gp$col <- c("grey60", "black", "grey60")
> gp$cex <- c(2, 1.5, 2)
> gp$hmm.ycoords <- c(0.6, 0.7)
> gp$ylim <- c(0.4, 4.5)
> gp$xlim[1] <- -10000
> gp$abline <- TRUE
> gp$abline.h <- 0.9
> gp$abline.col <- "black"
> gp$cytoband.ycoords <- c(0.4, 0.45)
> gp$col.predict <- c("black", "white", "grey60", "grey80")
> print(gp)
> legend(95 * 1e+06, 0.9, fill = gp$col.predict[-2],
   legend = c("1 copy", "copy-neutral LOH",
+ "3+ copies"), bty = "n", title = "predicted states")
```

4. Closing Remarks

The Bioconductor project has several infrastructure packages that are useful for organizing and annotating genomic data. In particular, the Biobase package introduces the virtual class eSet, which provides an organization for the high-throughput assay data set and the corresponding meta-data on the samples, features, and experiment. Extensions of the eSet class to a variety of different platforms and architectures are feasible. As our focus is on S4 classes and methods for high-throughput SNP data, we discuss the classes that are currently in place and the considerations that motivated these definitions. We emphasize the importance of using standardized data structures and the ease by which code can be reused through inheritance, both of which are facilitated by utilizing S4 classes and methods. The visualization methods in the SNPchip package and the HMM in the VanillaICE package serve as useful illustrations of how one can build on these definitions.

Acknowledgments

This work was supported by NSF grant DMS034211, NIH training grants 5T32HL007024 and 1K99HG005015 (RBS), and NIH R01 grants GM083084 and HL090577 (IR). The authors also acknowledge the support from a CTSA grant to the Johns Hopkins Medical Institutions.

Appendix

This document was created using Sweave (13).

- R version 2.8.0 Under development (unstable) (2008-06-18 r45949), powerpc-apple-darwin8.11.0
- Locale: C
- Base packages: base, datasets, grDevices, graphics, methods, stats, tools, utils
- Other packages: Biobase 2.1.0, DBI 0.2-4, RSQLite 0.6-4, SNPchip 1.5.2, VanillaICE 1.3.7, oligoClasses 1.1.22, pd.mapping50k.hind240 0.4.1, pd.mapping50k.xba240 0.4.1

References

1. Gentleman RC, Carey VJ, Bates DM, Bolstad B, Dettling M, Dudoit S, Ellis B, Gautier L, Ge Y, Gentry J, Hornik K, Hothorn T, Huber W, Iacus S, Irizarry R, Leisch F, Li C, Maechler M, Rossini AJ, Sawitzki G, Smith C, Smyth G, Tierney L, Yang JYH, Zhang J. (2004) Bioconductor: open software development for computational biology and bioinformatics. *Genome Biol* 5(10):R80.
2. Di X, Matsuzaki H, Webster TA, Hubbell E, Liu G, Dong S, Bartell D, Huang J, Chiles R, Yang G, Mei Shen M, Kulp D, Kennedy GC, Mei R, Jones KW, Cawley S. (2005) Dynamic model based algorithms for screening and genotyping over 100 K SNPs on oligonucleotide microarrays. *Bioinformatics* 21(9):1958–1963.
3. Rabbee N, Speed TP. (2006) A genotype calling algorithm for Affymetrix SNP arrays. *Bioinformatics* 22(1):7–12.
4. Affymetrix. (2006) BRLMM: an improved genotype calling method for the genechip human mapping 500 k array set. Tech. rep., Affymetrix, Inc. White paper, Santa Clara, CA.

5. Carvalho B, Bengtsson H, Speed TP, Irizarry RA. (2007) Exploration, normalization, and genotype calls of high-density oligonucleotide SNP array data. *Biostatistics* 8(2):485–499.
6. Nannya Y, Sanada M, Nakazaki K, Hosoya N, Wang L, Hangaishi A, Kurokawa M, Chiba S, Bailey DK, Kennedy GC, Ogawa S. (2005) A robust algorithm for copy number detection using high-density oligonucleotide single nucleotide polymorphism genotyping arrays. *Cancer Res* 65(14): 6071–6079.
7. Huang J, Wei W, Chen J, Zhang J, Liu G, Di X, Mei R, Ishikawa S, Aburatani H, Jones KW, Shapero MH. (2006) CARAT: a novel method for allelic detection of DNA copy number changes using high density oligonucleotide arrays. *BMC Bioinformatics* 7:83.
8. Laframboise T, Harrington D, Weir BA. (2006) PLASQ: a generalized linear model-based procedure to determine allelic dosage in cancer cells from SNP array data. *Biostatistics* 8(2):323–336.
9. Carter NP. (2007) Methods and strategies for analyzing copy number variation using DNA microarrays. *Nat Genet* 39(7 Suppl):S16–S21.
10. Chambers JM. (1998) *Programming with Data: A Guide to the S Language*, Springer-Verlag, New York.
11. Scharpf RB, Ting JC, Pevsner J, Ruczinski I. (2007) SNPchip: R classes and methods for SNP array data. *Bioinformatics* 23(5): 627–628.
12. Scharpf RB, Parmigiani G, Pevsner J, Ruczinski I. (2008) Hidden Markov models for the assessment of chromosomal alterations using high-throughput SNP arrays. *Ann Appl Stat* 2(2):687–713.
13. Leisch F. (2003) Sweave and beyond: Computations on text documents. In Kurt Hornik, Friedrich Leisch, and Achim Zeileis (eds). Proceedings of the 3rd International Workshop on Distributed Statistical Computing, Vienna, Austria, 2003.
Sarkar D. (2008) Lattice: Multivariate Data Visualization with R. Springer, New York.

Chapter 5

Overview on Techniques in Cluster Analysis

Itziar Frades and Rune Matthiesen

Abstract

Clustering is the unsupervised, semisupervised, and supervised classification of patterns into groups. The clustering problem has been addressed in many contexts and disciplines. Cluster analysis encompasses different methods and algorithms for grouping objects of similar kinds into respective categories. In this chapter, we describe a number of methods and algorithms for cluster analysis in a stepwise framework. The steps of a typical clustering analysis process include sequentially pattern representation, the choice of the similarity measure, the choice of the clustering algorithm, the assessment of the output, and the representation of the clusters.

Key words: Clustering algorithm, feature selection, feature extraction, similarity measure, cluster tendency, cluster validity, cluster stability, relevance networks, dendrogram.

1. Introduction

1.1. The Importance of Clustering

Clustering is one of the most useful tasks in the data mining process for discovering groups and identifying new interesting patterns in the underlying data. Clustering algorithms partition data objects into subsets (clusters) based on similarity or dissimilarity. Patterns within a valid cluster are more similar to each other than they are to a pattern belonging to a different cluster. The clustering process is an unsupervised, semisupervised, or supervised method. Since unsupervised cluster algorithms do not use predefined class labels or examples that would indicate grouping properties in the data set, it is the ideal method for identifying new patterns in data. Unsupervised clustering is also frequently used in combination with other supervised classification algorithms since it has the potential to detect incorrect class labels, outliers, errors, bias, and bad experimental designs.

R. Matthiesen (ed.), *Bioinformatics Methods in Clinical Research*, Methods in Molecular Biology 593,
DOI 10.1007/978-1-60327-194-3_5, © Humana Press, a part of Springer Science+Business Media, LLC 2010

Table 5.1
Overview of the methods discussed in the chapter

Pattern representation	Similarity measure	Clustering algorithm	Assessment of the output	Representation of clusters
Feature selection (1)	Euclidean distance (2)	*Hierarchical* BIRCH (3), CURE (4), ROCK (5), DIANA (6), MONA (6),	*Clustering tendency* (7, 8)	*Graphs* Relevance networks (9)
	Manhattan distance (10)	*Partitional* *k*-means (11), ISODATA (12), PAM (6), CLARA (6), CLARANS (13), nearest neighbor (14)		*Partitions* (15)
	Pearson's correlation coefficient (16)	*Density-based* DBSCAN(17), DENCLUE (18)	*Cluster validity* External, internal, and relative criteria. Validity (19) indices (Dunn's) (20)	*Classification trees* (15)
	Vector angle distance (21)	*Grid-based* WaveCluster (22) and STING (23)		
Feature extraction (Principal component analysis) (24)	Squared Pearson's correlation (16)	*Fuzzy clustering* FCM (25)	*Cluster stability* Bagging (26)	*Dendrogram* Displaying the assessment of the uncertainty in hierarchical cluster analysis (27)
	Inner product (28)	Artificial neural networks for clustering. SOM (29), SOTA (30, 31)		
	Spearman's rank correlation (32) and Kendall's Tau (33)	*Evolutionary approaches for clustering* Genetic algorithms (34)		
	Mutual information (35)	Biclustering (36)		

There are an overwhelming number of different strategies that can be combined in cluster analysis. In the following sections, an attempt to discuss a subset of the available clustering methods is provided (*see* **Table 5.1**).

1.2. Computational Steps in the Clustering Process

A typical clustering analysis can be subdivided into the steps outlined below (9). The actual choices made in each step are data-dependent and in many cases are optimized by trial and error.

1. **Pattern representation** (optionally including feature extraction and/or selection): The goal is to select the features on which the clustering is to be performed. The features should encode as much information as possible.
2. **Similarity measure:** Definition of a pattern proximity measure appropriate to the data domain. The similarity measure quantifies how similar two data points or patterns are. In most cases, it is important to check that all selected features contribute equally to the computation of the proximity measure and that no features dominate others.
3. **Clustering algorithm**: This step refers to the choice of the clustering algorithm. It should result in the definition of a good clustering scheme for the data set under analysis. The clustering algorithm is also critical for computational speed.
4. **Assessment of the output:** Clustering algorithms define clusters that are unknown a priori; therefore, the final partition of data requires some kind of evaluation. To verify whether the result of a clustering algorithm is correct, appropriate criteria and techniques must be used. These techniques aim at the quantitative evaluation of the results of the clustering algorithms and are known under the general term of *cluster validity* methods. They answer questions like "how many clusters are there in the data set?", "does the resulting clustering scheme fits our data set?", "is there a better partitioning for our data set?", and "how consistent or robust are the clusters when re-sampling the data?"
5. **Graphical representation:** The cluster results need to be represented with some sort of data abstraction in a graphical display for easy interpretation.

Each of the above steps is discussed in more detail in the sections below.

2. Pattern Representation

A *pattern* (or *feature vector*, *observation*, *object*, or *data point*) (*see* **Note 1**) is a single data item used by the clustering algorithm. It typically consists of a vector of d measurements:

x = (*x*1, . . .,*xd*). The individual scalar components *xi* of a pattern are called *features* (or *attributes*). Pattern representation is concerned with the preprocessing of feature extraction and feature selection and is used to define the number of features, type, weight, and scale of the features available to the clustering algorithm. For example, in mass spectrometry, one can choose to use all data points in a spectrum or only the data points corresponding to peak tops that will affect the number of features. The peak tops can be represented as a signal-to-noise ratio, integrated intensity, or maximum intensity.

The practice of assigning different weights to features and/or scaling of their values is widespread. Translating the importance of each feature using weights allows clusters to be constructed of better shapes and can lead to better classification results (37). Attribute scaling is the application of a mathematical transformation to each of the individual components of the attributes so that all of the attributes make a comparable contribution to the measurement of similarity.

2.1. Feature Selection

Feature selection is the process of identifying the most effective subset of the original features to use in the clustering. It implies selecting a subset of the existing features without any transformation. It is important to distinguish feature selection from dimension reduction. In feature selection, some features are completely removed, whereas in dimension reduction, new features are defined as functions of all features. The problem of feature selection is defined as follows: Given a set of features, select a subset that leads to the smallest clustering error performance. Using all features in a data set often introduces noise, which can confuse the clustering algorithm. Feature selection removes noisy features, improves the performance of the clustering algorithms, increases the speed of the clustering algorithms, and yields a more compact, more easily interpretable representation of the target concept (38).

Search and evaluation of subsets of features are the two main steps in the feature selection process. Search methods can be exhaustive, heuristic, random, or some hybrid of these techniques. Their efficiency is measured by optimality, defined as the best subset of features according to a specified criterion. Exhaustive methods guarantee optimality but are impractical due to their exponential time complexity in a number of features. Random methods generate subsets randomly and return the best subset at any point of time, approaching optimality only asymptotically. A variation of pure random methods is the probabilistic method where the probability of generating a subset varies by some rules. Examples of such rules are genetic algorithms and simulated annealing. Forward and backward selection are examples of heuristic methods. A forward selection method first finds the best feature among all features and stepwise accumulates the features until the performance drops. A backward selection algorithm is the opposite of the forward selection algorithm (39).

When selecting a good attribute subset, there are two fundamentally different approaches. One is to make an independent assessment based on general characteristics of the data; the other is to evaluate the subset using the machine learning algorithm that will ultimately be employed for learning. The latter approach requires that the class labels are known. The first is called a *filter* method, because the attribute set is filtered to produce the most promising subset before the learning commences (39, 40). The second is known as a *wrapper* method, because the learning algorithm is wrapped into the selection procedure (41, 42). Wrapper methods typically require extensive computation to search the best features (38).

The relevance of the features can be evaluated either individually (univariate approaches) or in a multivariate manner. Univariate approaches are simple and fast; however, possible correlation and dependencies between the features are not considered (1).

2.2. Feature Extraction

Feature extraction is the process of using transformations of the input features to produce a new salient feature set. It involves transforming the existing features into a lower-dimensional space. Linear transforms, such as principal component analysis (24), factor analysis (43), linear discriminant analysis (44), partial least-squares regression (45), and projection pursuit (46), have been widely used in feature extraction and dimensionality reduction. The most commonly used linear feature extractor is the principal components analysis.

2.2.1. Principal Component Analysis

Principal component analysis (PCA) (24) is a technique used to reduce multidimensional data sets to lower dimensions for analysis. PCA is an unsupervised technique and as such does not include label or category information given in the data.

In general, PCA changes the original variables into new independent and uncorrelated variables called *principal components* that explain the observed variability. PCA performs an orthogonal linear transformation. This transforms the data into a new coordinate system such that the greatest variance by any projection of the data comes to lie on the first coordinate, the second-greatest variance on the second coordinate, and so on. These coordinates are the principal components. In other words, given m observations on n variables, the goal of PCA is to reduce the dimensionality of the data matrix by finding $r \leq n$ new variables. These r principal components account together for as much of the variance in the original n variables as possible while remaining mutually uncorrelated and orthogonal. For each component, it is possible to find one eigenvalue with an associated variance value. The eigenvalues and their corresponding eigenvectors originate from the covariance matrix obtained from the original data.

The result is a lower-dimensional representation of the data that removes some of the "noisy" directions, making the data more accessible for visualization and analysis. The problem with using PCA is that it assumes that the large variance between groups is the most important, which is not necessarily always true. For example, small changes in the concentration of electrolytes in the blood can have a profound effect on the health of an individual. In such cases, an alternative to PCA is partial least-squares regression (PLS regression) (47). In PLS regression, the overall goal is to extract some factors to predict responses in the population and to describe the common structure underlying the dependent and independent variables. PLS regression searches for a set of components (called *latent vectors*; *see* **Note 2**) that performs a simultaneous decomposition of X (data matrix from independent variables) and Y (data matrix from dependent variables) with the constraint that these components explain as much as possible of the covariance between X and Y. This first step is followed by a regression step where the decomposition of X (latent variables) is used to predict Y.

3. Similarity Measure

A metric system is a decimalized system of measurement (*see* **Note 3**). The similarity measure quantifies how similar two data points (or two objects) are and will provide an indication of proximity, likeness, affinity, or association. Because of the variety of feature types and scales, the similarity measure must be chosen carefully.

There are two types of similarity measures: metric and probability distribution–based similarity measures.

A metric system is a decimalized system of measurement (*see* **Note 3**). A number of the more common metrics are discussed below (15, 48); in each case, X_i and Y_i are two n-dimensional vectors (patterns) being compared.

Euclidean distance: The most popular metric for continuous features is the Euclidean distance (2) d_e:

$$d_e = \sqrt{\sum_{i=1}^{n} (X_i - Y_i)^2} \qquad [1]$$

where the index i iterates over all values in the vectors. The Euclidian distance metric (2) is a measure of the geometric distance between two components. It is purely based on magnitude. So the case of two vectors whose values are clearly highly correlated would not be well represented by the Euclidian distance d_e, where only the distance between them is taken into account.

Manhattan distance: Manhattan distance (10) d_M is similar to the Euclidian distance. It is calculated as the sum of the absolute distances of two vector values (10):

$$d_M = \sum_{i=1}^{n} |X_i - Y_i| \qquad [2]$$

Pearson's correlation: Pearson's correlation coefficient (16) r is calculated by

$$r = \frac{\sum_{i=1}^{n} (X_i - \overline{X})(Y_i - \overline{Y})}{\sqrt{\sum_{i=1}^{n} (X_i - \overline{X})^2}\sqrt{\sum_{i=1}^{n} (Y_i - \overline{Y})^2}} \qquad [3]$$

where $\overline{X}$ is the mean of vector X and $\overline{Y}$ is the mean of vector Y_i. Pearson's correlation coefficient is a measure of the tendency of the vectors to increase or decrease together; in other words, it measures the association between two variables. It ranges from –1 (negatively correlated), to 0 (no correlation), to 1 (positively correlated), with 1 meaning that the two series are identical, 0 meaning they are completely independent, and −1 meaning they are perfect opposites. It measures only linear correlations. The correlation coefficient is invariant under a scalar transformation of the data (adding, subtracting, or multiplying the vectors with a constant factor), meaning that it is independent of both origin and scale. If two patterns have a common peak or valley at a single feature, the correlation will be dominated by this feature, although the patterns at the remaining features may be completely dissimilar. As a consequence, it is very sensitive to outliers. Another drawback of Pearson's correlation coefficient is that it assumes an approximate Gaussian distribution of the points and may not be robust for non-Gaussian distributions (49).

Vector angle distance: Cos(α) (21), where 1 – Cos(α) is a measure of the angle between two vectors, is also known as the uncentered Pearson's correlation distance. It captures a similarity that does not change if scales are multiplied by a common factor. Another strong property is that the cosine similarity does not depend on the length of the vectors. The cosine measure assigns a high similarity to points that are in the same direction from the origin, zero similarity to points that are perpendicular to one another, and negative similarity for those that are pointing in opposing directions to one another. This is basically the same formula as above, except the mean is expected to be 0.

$$\cos\alpha = \frac{\sum_{i=1}^{n} X_i Y_i}{\sqrt{\sum_{i=1}^{n} X_i^2 \sqrt{\sum_{i=1}^{n} Y_i^2}}} \qquad [4]$$

Squared Pearson's correlation coefficient: The squared Pearson correlation coefficient (16) r_{sq} calculates the square of the Pearson correlation coefficient, so that negative values become positive. It is a measure of how well the regression line approximates the real data points. An r_{sq} of 1 indicates that the regression line perfectly fits the data.

$$r_{sq} = \left(\frac{\sum_{i=1}^{n} (X_i - \overline{X})(Y_i - \overline{Y})}{\sqrt{\sum_{i=1}^{n} (X_i - \overline{X})^2} \sqrt{\sum_{i=1}^{n} (Y_i - \overline{Y})^2}} \right)^2 \qquad [5]$$

Inner product: The simplest measurement of association between two vectors is the inner product, also referred as the scalar or the dot product (28). It is a value expressing the angular relationship between two vectors. When the two vectors are perpendicular, the result of the inner product will be zero, because the cosine of 90° will be zero. If the angle between the two vectors is less than 90°, the dot product will be positive as the cosine will be positive, and the vector lengths are always positive values. If the angle between the two vectors is greater than 90°, the dot product will be negative, as the cosine will be negative, and the vector lengths are always positive values. The inner product between X and Y is defined as the sum of products of components and can be modified by defining an adjusted or averaged dot product d:

$$d = \frac{1}{n} \sum_{i=1}^{n} X_i Y_i = \frac{1}{n} \sum_{i=1}^{n} |X_i|\,|Y_i| \cos\alpha \qquad [6]$$

Rank-based metrics: Spearman's rank correlation (32) and Kendall's Tau (33) are nonparametric or rank correlations. They are techniques for determining the correlation between two ordinal variables (*see* **Note 4**) or metric variables reduced to an ordinal scale, and they are used to measure the degree of correspondence between the resulting two rankings. When replacing the value of each X_i and Y_i by the value of its rank among all the other, that is, 1, 2, 3, . . ., n, the resulting list of numbers will be drawn from a perfectly known distribution function, from the integers from 1 to n. Spearman's correlation coefficient and Kendall's Tau do not require the assumption of Gaussian distribution and therefore are more robust against outliers than Pearson's correlation coefficient. However, as a consequence of ranking, a significant amount of information present in the data is lost.

Mutual information (35) and the Kullback–Leibler divergence (50) are examples of probability distribution–based similarity measures. Mutual information is a special case of Kullback–Leibler divergence. In fact, many of the quantities in information theory can be considered as special cases of Kullback–Leibler divergence.

Kullback–Leibler divergence: For two probability functions P and Q of discrete random variables, the Kullback–Leibler divergence is defined by

$$D_{KL}(P||Q) = \sum_i P(i) \log \frac{P(I)}{Q(I)} \qquad [7]$$

Mutual information: Mutual information (35) is a quantity that measures the mutual or statistical dependence of the two variables. It quantifies the reduction in the uncertainty of one random variable given knowledge about another random variable. It takes into account nonlinear correlations. The mutual information of two discrete random variables X and Y, $I(X; Y)$, can be defined as

$$I(X;Y) = \sum_{y \in Y} \sum_{x \in X} p(x,y) \log \left(\frac{p(x,y)}{p(x)p(y)} \right) \qquad [8]$$

where $p(x)$, $p(y)$, and $p(x, y)$ are thge probabilities of a given x- and y-value, and the co-occurrence of the x- and y-values. Mutual information can be equivalently expressed as

$$\begin{aligned} I(X;Y) &= H(X) - H(X/Y) = H(Y) - H(Y/X) \\ &= H(X) + H(Y) - H(X,Y) \end{aligned} \qquad [9]$$

$$H(X) = -\sum_{x \in X} P(x) \log(P(x)) \qquad [10]$$

$$H(Y) = -\sum_{y \in Y} P(y) \log(P(y)) \qquad [11]$$

$$H(X;Y) = -\sum_{x \in X, y \in Y} P(x,y) \log(P(x,y)) \qquad [12]$$

where $H(X)$ and $H(Y)$ are the marginal entropies, $H(X|Y)$ and $H(Y_i|X)$ are the conditional entropies, and $H(X,Y)$ is the joint entropy of X and Y. Recall that the entropy quantifies uncertainty (*see* **Note 5**).

4. Clustering Algorithm Categories

According to the method adopted to define clusters, the algorithms can be broadly classified into the categories described below (15, 19, 51):

- *Hierarchical clustering* proceeds iteratively by either merging smaller clusters into larger ones, or by splitting larger clusters, resulting in a tree of clusters, called a *dendrogram*, that shows how the clusters are related. Typical examples for hierarchical clustering include BIRCH (3), CURE (4), and ROCK (5).
- *Partitional clustering* decomposes the data set into a set of disjoint clusters. More specifically, it attempts to determine an integer number of partitions that optimize in an iterative way a certain criterion function that may represent the local or global structure of the data. The simplest and most commonly used partitional algorithm is k-means (11).
- *Density-based clustering* proceeds by grouping neighboring patterns of a data set into clusters based on density conditions. Some examples include DBSCAN (17) and DENCLUE (18).
- *Grid-based clustering* operates with the assumption that the space is divided into a finite number of cells and all of the operations are done in the divided space. STING (23) and WaveCluster (22) are examples of programs that implement this type of clustering.

  For each of above categories, there are a number of subtypes and different algorithms for finding the clusters (15).

Another classification criterion is the way clustering handles uncertainty in terms of cluster overlap. The output clustering can be hard (crisp) or fuzzy (soft): A hard clustering algorithm allocates each pattern to a single cluster. In a fuzzy clustering method, each pattern has a variable degree of membership in each of the output clusters. If an instance belongs to a group with a certain probability, the clustering is *probabilistic*; if not, it is said to be *deterministic*. Probabilistic clustering algorithms assume an underlying probability model with parameters that describe the probability that an instance belongs to a certain cluster.

Two schemes are related to the sequential or simultaneous use of features in the clustering process. In a monothetic scheme, cluster membership is based on the presence or absence of a single feature. Polythetic schemes use more than one feature.

5. Clustering Algorithms

5.1. Hierarchical Clustering Algorithms

Hierarchical clustering (52) transforms a distance matrix of pairwise similarity measurements between all items into a hierarchy of nested groupings. The hierarchy is represented with a binary tree-like dendrogram that shows the nested grouping of patterns and the similarity levels at which groupings change. According to the algorithmic structure and operation, hierarchical algorithms can be further categorized into two procedures:

1. Agglomerative procedures: This procedure begins with each pattern in a distinct cluster and successively merges clusters together until a stopping criterion is satisfied.
2. Divisive procedures: A divisive method begins with all patterns in a single cluster and iteratively splits the cluster until a stopping criterion is met.

The basic process of agglomerative hierarchical clustering has the following steps:

1. A similarity distance matrix is constructed by calculating the pairwise distance between all patterns. Each pattern is assigned to a single cluster, so each pattern represents one cluster.
2. The two clusters r and s with the minimum distance to each other are found.
3. The clusters r and s are merged and r is replaced with the new cluster. s is deleted and distances (similarities) between the new cluster and each of the old clusters are computed.
4. Repeat steps 2 and 3 until the total number of clusters is one.

In the first step, when each item represents its own cluster, the distances between those items are defined by the chosen similarity measure. Then a linkage or amalgamation rule is needed to determine if two clusters are sufficiently similar to be linked together (*see* **Note 6**). Step 3 can be done in different ways, which is what distinguishes *single-linkage* from *complete-linkage*, *average-linkage*, and *Ward's method* clustering:

In *single-linkage* clustering, the distance between one cluster and another is considered to be equal to the shortest distance from any member of one cluster to any member of the other cluster.

In *complete-linkage* clustering, the distances between clusters are determined by the greatest distance between any two members in the different clusters.

In *average-linkage* clustering, the distance between two clusters is calculated as the average distance between all pairs of members in the two different clusters.

Ward's method clustering (53) uses an analysis of a variance approach to evaluate the distances between clusters. This method attempts to minimize the sum of squares of any two clusters that can be formed at each step. In general, this method is very efficient; however, it tends to create clusters that are small in size.

Hierarchical clustering algorithms only require a matrix with the pairwise similarities based on a predefined distance. However, when the number of data points to be clustered, m, is large, such a matrix requires a lot of storage space, of the order $o(m^2)$. In recent years, because of the requirement for handling large-scale data sets, many new hierarchical clustering techniques have appeared and have greatly improved the clustering performance. Typical examples for agglomerative hierarchical clustering include BIRCH (3), CURE (4), and ROCK (5). Examples of divisive hierarchical clustering include DIANA (6) and MONA (6).

Hierarchical clustering is the most commonly used clustering strategy for gene expression analysis at the moment (48). The biggest advantage is that the only parameters that need to be specified are the type of similarity distance measurement and the choice of the amalgamation rule (*see* **Note 6**). However, if the feature selection procedure is used, many more parameters must be chosen and the choice of such parameters indeed becomes a parameter optimization problem. Hierarchical clustering algorithms are applicable to any attribute type, and they do not make any assumptions about the underlying data distribution. They do not require the number of clusters to be known in advance, while they provide a complete hierarchy of clusters, and a "flat" partition can be derived using a cut through the dendrogram. However, both divisive and agglomerative procedures suffer from their inability to perform adjustments once the splitting or merging decision is made. The deterministic nature of this procedure and the impossibility of reevaluating the results can cause the pattern to be clustered based on local decisions rather than a global one (54).

When clusters overlap or vary considerably in shape, density, or size, this class of methods is known to perform poorly (55). Hierarchical clustering presents drawbacks when dealing with data containing a high amount of noise and is dependent on the data order (30).

5.2. Partitional Clustering Algorithms

The result of a partitional clustering algorithm is a single partition of the data into a set of disjoint clusters. Partitional methods are used when the analysis involves very large data sets for which the construction of a dendrogram is computationally prohibitive. A drawback of partitional algorithms is that the number of clusters must be specified. In partitional techniques, the clusters produced optimize a criterion function defined either locally, that is, on a subset of patterns, or globally, so defined over all of the patterns. As the combinatorial search of the set of possible labeling for an optimum value of a criterion is computationally very expensive, in practice, the algorithm is run multiple times with different starting points, using as the output the best configuration obtained from all of the runs.

The most commonly used strategy in partitional clustering is based on the squared-error criterion. It tries to minimize the total intracluster variance, or the squared-error function:

$$V = \sum_{i=1}^{k} \sum_{x_j \in S_i} |x_j - \mu_i|^2 \qquad [13]$$

where there are k clusters S_i, $i = 1, 2, \ldots, k$, and μ_i is the centroid or mean point of all the points.

The strategy based on the squared-error criterion works well with isolated and compact clusters, but when outliers (*see* **Note 7**) are present, they can influence the clusters that are found because the resulting cluster centroids may not be representative.

5.2.1. k-Means

The simplest and most commonly used algorithm employing the squared-error criterion is the *k*-means (11) partitional clustering algorithm. The parameter *k*, which accounts for the number of clusters, must be specified. Then *k* points are chosen at random as cluster centers. All items are assigned to their closest cluster center according to the distance measure being used. Next the centroid, or mean, of the items in each cluster is calculated. These centroids are taken to be new center values for their respective clusters. The whole process is repeated with the new cluster centers. Iteration continues until the same or similar points are assigned to each cluster in consecutive rounds or there is a minimal decrease in the squared error. Using centroids has the advantage of a clear geometric and statistical meaning while keeping the algorithm insensitive to data ordering, but means of points within a cluster only work with numerical attributes and can be negatively affected by a single outlier (56).

The *k*-means is easy to implement and its time complexity is $O(l^* k^* n)$, where n is the number of patterns, l is the number of iterations, and k is the number of clusters.

The major drawback of this algorithm is that it is sensitive to the selection of the initial partition and may converge to a local minimum of the criterion function value if the initial partition is not properly chosen. To find the global optimum, techniques such as simulated annealing and generic algorithms can be incorporated with the *k*-means algorithm.

The basic steps of the *k*-means algorithm are as follows:

1. An initial partition with *k* clusters containing randomly chosen samples is selected, and the centroids of the clusters are computed.
2. By assigning each sample to the closest cluster center, a new partition is generated.
3. New cluster centers as the centroids of the clusters are recomputed.
4. Steps 2 and 3 are repeated until the optimal value of the criterion function is found or until the cluster membership stabilizes.

Some variants like the ISODATA (12) algorithm include a procedure to search for the best *k* and therefore overcome the problem of estimating *k*. ISODATA can dynamically adjust the number of clusters by merging and splitting clusters according to some predefined thresholds. In this sense, the problem of identifying the initial number of clusters becomes a matter of tweaking parameters.

5.2.2. Partitioning Around Medoids

Partitioning around medoids (PAM) (6) is a partitioning method that operates on a distance matrix and requires a prespecified number of clusters *k*. The PAM procedure is based on the search for *k* representative patterns, or medoids, among the observations. The medoids minimize the sum of the distances of the observations to their closest medoid. Therefore, a medoid can be defined as an object of a cluster, whose average dissimilarity to all the objects in the cluster is minimal, resulting in the medoid's being the most centrally located data point in the data set. After finding a set of *k* medoids, *k* clusters are constructed by assigning each observation to the nearest medoid. Representation by *k*-medoids has the advantages that it presents no limitations on attribute types and it is not very sensitive to the presence of outliers, as the choice of medoids is dictated by the location of a predominant fraction of points inside a cluster (56). PAM works well for small data sets, but does not scale well with large data sets.

CLARA (*clustering large applications*) (6) is an implementation of PAM in which PAM works on different subsets of the data set. First, multiple samples of the data set are drawn, then PAM is applied on the samples, and, finally, the best clustering output of these samples is used as the output.

CLARANS (*clustering large applications based on randomized search*) (13) combines the sampling techniques with PAM. The clustering process can be presented as searching a graph where every node is a potential solution, that is, a set of *k* medoids. The *neighbor* of the current clustering is the clustering obtained after replacing a medoid. CLARANS selects a node and searches for a local minimum among a specified number of neighbors. If a better neighbor is found, CLARANS moves to the neighbor's node and repeats the process; otherwise, the current clustering is a local optimum. When the local optimum is found, CLARANS starts with a new randomly selected node in search of a new local optimum.

5.2.3. k-Nearest-Neighbor Clustering

k-nearest neighbors (14) is a supervised clustering algorithm. It classifies new, unlabeled patterns based on training samples. The parameter *k*, which accounts for the number of patterns or (training points) closest to the query point, must be specified. The classification is performed using majority vote among the classification of the *k* patterns. It works based on the minimum distance from the query instance to the training samples to determine the *k*-nearest neighbors. The *k*-nearest neighbors, each already assigned to a class, is used to make a majority vote. The class labeled that is assigned to the highest number of these *k*-nearest neighbors is assigned to the query instance. A major drawback is that one needs to maintain the training set for each classification procedure.

5.3. Density-Based Clustering

The key idea of this type of clustering is to group neighboring objects of a data set into clusters based on density conditions measured in terms of the local distribution of nearest neighbors. Density-based algorithms typically assign clusters in dense regions of objects in the data space that are separated by regions of low density. Density-based algorithms are capable of discovering clusters of arbitrary shapes, providing a natural protection against outliers. Some examples include DBSCAN (17) and DENCLUE (18).

5.4. Grid-Based Clustering

This type of algorithm is mainly proposed for spatial data mining and it inherits the topology from the underlying attribute space. These algorithms divide the spatial area into a finite number of rectangular cells, generating several levels of cells corresponding to different levels of resolution, and then perform all operations on the quantized spatial area, which has the advantage of limiting the search combinations. STING (23) and WaveCluster (22) are some examples of this kind of algorithm.

5.5. Fuzzy Clustering

The issue of uncertainty support in the clustering task leads to the introduction of algorithms that use fuzzy logic in their procedure. Fuzzy clustering associates each pattern with every cluster using a membership function. In fuzzy clustering, each cluster is a fuzzy set of all the patterns. So they consider that a pattern can be classified as more than one cluster, as the pattern can belong to all clusters with a degree of membership. This is particularly useful when boundaries among the clusters are not well separated and are ambiguous. Moreover, the memberships may help to discover more sophisticated relationships between a given object and the disclosed clusters.

The most popular fuzzy clustering algorithm is fuzzy *c*-means (FCM) (25), which is an extension of the classical *k*-means algorithm for fuzzy applications. Fuzzy *c*-means is better than *k*-means at avoiding local minima.

5.6. Artificial Neural Networks for Clustering

Artificial neural networks (ANNs) (57) are mathematical or computational models motivated by biological neural networks. Specifically, they attempt to mimic biological neural systems' capacity to learn. ANNs consist of an interconnected group of artificial neurons (nodes) that build an architecture capable of processing the information.

Each artificial neuron receives a number of inputs either from original data or from the output of other neurons in the network. Each input comes via a connection that has a strength or weight. Each neuron has a single threshold value. The activation of the neuron is the integrated signal obtained by weighting the sum of the inputs and then subtracting the threshold.

The architecture defining how neurons are connected together models the relationship between inputs and outputs. ANNs are adaptive systems because they learn the input–output relationship through training. Two types of training are used in ANNs: supervised learning and unsupervised learning. In supervised learning, the training data contain examples of inputs together with the corresponding outputs, and the network learns to infer the relationship between the two. In unsupervised learning, however, the training algorithms adjusts the weights between the input nodes and the output nodes in the neural network by reference to a training data set that includes input variables only.

An unsupervised learning kind of ANN called *competitive neural networks* has been recognized as a powerful tool for pattern analysis, feature extraction, and cluster analysis. This kind of ANN is single-layered. Patterns are presented at the input and are associated with the output nodes. The network based on data correlations groups similar input patterns, which represent a single output neuron, which is indeed a pattern, an extracted feature, or a cluster, respectively.

ANNs are applicable to multivariate, nonlinear problems and have the advantage that there is no need to assume an underlying data distribution, which is usually done in statistical modeling.

5.6.1. Self-Organizing Maps (SOM)

One of the most popular competitive unsupervised neural network models today is the principle of a *self-organizing map* (*SOM*) (29). The SOM network has input and output nodes. For each attribute of the record, the input layer (input nodes) has a node, each one connected to every output node (output layer). The self-organizing map describes a mapping that seeks to preserve the topological properties from the higher-dimensional input space to a lower, discrete-dimensional map space (typically two-dimensional). SOM can be considered a nonlinear generalization of principal component analysis. Each connection is associated with a weight, which determines the position of the corresponding output node. The algorithm initially populates its nodes by randomly sampling the data and then, during a training process, changes the weights in a systematic fashion, adjusting the nodes in a way that captures the distribution of the data set's variability. At the end of the training process, each output node represents the average pattern of the data that map into it and move to form a cluster. This reduction of the dimensionality in the data space makes a very interesting property when dealing with large data sets.

The fact that SOM is based on neural networks confers a series of advantages that makes it suitable to the clustering of large amounts of noisy data with outliers (30). However, in this approach, the training of the network – and therefore the clusters – depends on the number of nodes, and the number of clusters must be arbitrarily fixed from the beginning, making the recovery of the cluster structure a very complex and subjective job (30). SOM does not perform well with invariant profiles (30). Additionally, when a particular kind of profile is abundant, SOM will populate the majority of the clusters with this profile; the most interesting profiles will map in a few clusters and their resolution might be low (30). These problems and the lack of a tree structure to detect the relationship between the clusters have motivated the appearance of neuro-hierarchical approaches like the Self-Organizing Tree Algorithm (SOTA) (31) discussed below that combine the advantages of hierarchical clustering techniques and SOM.

5.6.2. Self-Organizing Tree Algorithm (SOTA)

SOTA (31) is a hierarchical neural network that grows into a binary tree topology. SOTA is based on SOM and growing cell structures (58). It offers a criterion to stop the growing of the tree based on the approximate distribution of the probability obtained by randomization of the original data set and therefore provides a statistical support for the cluster definition.

SOTA's run times are approximately linear with the number of items to be classified, making it suitable for large data sets. Also, because SOTA follows a top-to-bottom hierarchical approach, it forms higher clusters in the hierarchy before forming the lower clusters, with the ability to stop the algorithm at any level of hierarchy to obtain meaningful intermediate results.

SOTA was originally designed for phylogenetic reconstruction (31). It has since been applied to microarray expression data analysis (30), where it has been widely used to discover gene expression patterns in time-course microarray experiments.

5.7. Evolutionary Approaches for Clustering

The basic objective of search techniques is to find the global or approximate global optimum for combinatorial optimization problems. Combinatorial optimization problems usually have NP-hard complexity and need to search the solution space exponentially. Clustering algorithms organize a set of data points in *k* subsets by optimizing some criterion function, which is why clustering can be regarded as an optimization problem. Simple local search techniques, like hill-climbing algorithms, are used to find the partitions, but they cannot guarantee optimality, as they easily get stuck in the local optimum. More complex stochastic methods like evolutionary algorithms, genetic algorithms, simulated annealing, and Tabu search, or deterministic methods like deterministic annealing, can explore the solution space more efficiently (51).

Evolutionary approaches are inspired by natural evolution. They make use of evolutionary operators and a population of solutions (also called *individuals*) to obtain the globally optimal partition of the data. Candidate solutions are encoded as chromosomes. Selection, recombination, and mutation are the most commonly used evolutionary operators. An optimization function, called the *fitness function*, is used to evaluate the optimizing degree of the population, in which each individual has its corresponding fitness value. After an initial population of solutions is generated randomly, for example, thenselection, crossover, and mutation are iteratively applied to the population until the stop condition is satisfied (15).

A more detailed description of an evolutionary approach is described below (15):

1. A random population of solutions is chosen where each solution corresponds to a valid *k*-partition of the data. A fitness value, typically inversely proportional to the squared-error value, is associated with each solution.
2. The evolutionary operators' selection, recombination, and mutation are used on a subpopulation of solutions with the highest fitness value to generate the next population of solutions, and their fitness values are calculated.
3. Step 2 is repeated until some termination condition is satisfied.

The best-known evolutionary techniques are genetic algorithms (GAs) (59, 60), evolution strategies (ESs) (61), and evolutionary programming (EP) (62). Of these three approaches, GAs have been most frequently used in clustering.

Typically, solutions are binary strings in GAs. In GAs, solutions are propagated from the current generation to the next generation based on their fitness by a selection operator. This selection operator uses a probabilistic scheme to assign a higher probability of getting reproduced to the solutions with higher fitness, so that by favoring the best individuals in the next generation, the selection operator ensures their continuity in the population.

The recombination and mutation operators are responsible for introducing diversity in the population by performing perturbations in the individuals. From the variety of recombination operators in use, crossover is the most popular one. Crossover takes as input a pair of chromosomes (called *parents*) and outputs a new pair of chromosomes (called *children* or *offspring*) where parts of the parents' parameters have been interexchanged. *Mutation* takes as input an existing chromosome and complements a bit value at a randomly selected location, generating a new chromosome.

The GA algorithm proposed by Hall, Özyurt, and Bezdek can be regarded as a general scheme (34).

5.8. Biclustering

Biclustering (36) is a data mining technique that allows simultaneous clustering of the rows and columns of a matrix. It has acquired a lot of relevance in gene expression analysis, where the results of the application of standard clustering methods to genes are limited (63). Gene expression matrices have been extensively analyzed in two dimensions separately: the gene dimension and the condition (or sample) dimension, where the goal when analyzing gene expression data with ordinal cluster analysis is either grouping of genes according to their expression under multiple conditions or grouping of samples according to the expression of multiple genes. Biclustering seeks to find submatrices, that is, subgroups of genes and subgroups of conditions, where the genes exhibit highly correlated activities for each condition in the subgroup.

Biclustering algorithms usually define a priori the number of biclusters, and they assume that either (i) there is one bicluster in the data matrix, or (ii) the data matrix contains K biclusters, where K is the number of biclusters. In the latter scenario, the following biclusters may overlap.

Some approaches attempt to identify *one bicluster at a time*, others discover *one set of biclusters at a time*, and there are also algorithms that find all the biclusters at the same time.

From its simplest form, the problem of biclustering is NP-complete, requiring either a large computational effort or the use of some sort of heuristic approach to short-circuit the calculation. A number of different heuristic approaches have been used to address this problem:

- Iterative row and column clustering combination. The approach consists of separately applying the clustering algorithms to the rows and columns of the data matrix, combining the results afterwards using some sort of iterative procedure to combine the two cluster arrangements.

- Divide and conquer. The algorithm breaks the problem into several subproblems that are similar to the original problem but smaller in size. Then it solves the problems recursively, combining the solutions to get the solution to the original problem.
- Greedy iterative search. Makes a locally optimal choice, hoping that such a choice will lead to a globally good solution.
- Exhaustive bicluster enumeration applies a search restriction on the size of the biclusters to speed up the search.
- Distribution parameter identification. The biclusters are generated using a given statistical model. The aim is to identify the distribution parameters that best fit the available data, by minimizing a certain criterion through an iterative approach.

6. Assessment of the Output

6.1. Clustering Tendency

The majority of the clustering algorithms impose a clustering structure on the data set even if it does not possess a structure. Indeed, clustering applied to a data set with no naturally occurring clusters will impose an artificial and meaningless structure. Therefore, it is important to verify whether the data set has a structure before applying any clustering algorithm. The problem of verifying whether or not clusters actually exist in data is known as *clustering tendency determination* (7).

Cluster tendency has mainly focused on the problem of determining the optimal number of clusters present in the data. If the optimal clustering contains only one group, then a null tendency must be concluded (8).

6.2. Cluster Validity

Cluster validity is the assessment of a clustering procedure's output. The clustering process has no predefined classes; therefore, it is difficult to find an appropriate metric for measuring if the found cluster configuration is acceptable or not. A clustering structure is valid if it cannot have occurred either by chance or as an artifact of the clustering algorithm.

The objective of the clustering methods is to discover significant groups present in a data set. In general, they should search for clusters whose members have a high degree of similarity with each other and are well separated from the members of the other clusters. In cluster analysis, we face the problem that when different algorithms are applied to the same data set, they give different results. Moreover, most of the clustering algorithms are very sensitive to their input parameters and we must thus decide the optimal number of clusters that fits the data set.

There are three types of validation studies (19):

1. An *external* assessment of validity compares the recovered structure to an a priori structure. The results of a clustering algorithm are evaluated based on a prespecified structure imposed on a data set and reflects the intuition about the data set's clustering structure.
2. An *internal* examination of validity determines whether the clustering structure is intrinsically appropriate for the data. The results are evaluated in terms of quantities that involve the vectors of the data themselves.
3. A *relative* test compares two structures and measures their relative merit. Here the clustering structure is evaluated by comparing it to other clustering schemes, resulting in the same algorithm but with different parameter values.

These validation assessments are carried out using some validity indexes that provide a quantitative evaluation of the clustering results based on two criteria:

Heterogeneity of the clusters, also known as the cluster *cohesion* or *compactness*: The members of each cluster should be as close to each other as possible.

Separation or *intercluster distances*: The clusters themselves should be widely separated. There are three common approaches measuring the distance between two different clusters: distance between the closest member of the clusters, distance between the most distant members, and distance between the centers of the clusters.

Thus, a basic clustering approach may aim to search for a partition that minimizes intracluster distances and maximizes intercluster distances.

6.2.1. Validity Indices

These indices are used for measuring the "goodness" of a clustering result compared to other ones that were created by other clustering algorithms, or by the same algorithms but using different parameter values. They are based on geometric properties.

Hard clustering indices are often based on some geometric motivation to estimate how compact and well separated clusters are (e.g., Dunnś index) (20). Others are statistically motivated, for example, by comparing the within-cluster scattering with the between-cluster separation (64).

6.2.1.1. The Dunn Index

The Dunn index is defined as

$$D_m = \min_{i=1,..,m} \left\{ \min_{j=i+1,...,m} \left(\frac{d(C_i, C_j)}{\max_{k=1,...,m} \mathrm{diam}(C_k)} \right) \right\} \qquad [14]$$

where the dissimilarity function between two clusters C_i and C_j is

$$d(C_i, C_j) = \min_{x \in C_i,\ y \in C_j} d(x, y) \qquad [15]$$

and the diameter of a cluster C is defined as

$$\mathrm{diam}(C) = \max_{x, y \in C} d(x, y) \qquad [16]$$

If the data set contains compact and well-separated clusters, Dunn's index will be large, since the distance between clusters is expected to be large and the diameter of the cluster is expected to be small. The main disadvantages of the Dunn index are that the calculation of the index is time-consuming and the index is very sensitive to noise (as the maximum cluster diameter can be large in a noisy environment).

6.3. Cluster Stability

Cluster stability research is involved with the validity of clusters generated by a clustering algorithm. It answers whether generated clusters are true clusters or are due to chance. Recent work on cluster validity research has concentrated on a kind of *relative index* called *cluster stability.*

Cluster stability exploits the fact that when multiple data sources are sampled from the same distribution, the clustering algorithms should behave in the same way and produce similar structures.

Bagging (26), or *bootstrap aggregating*, can be used to assess stability and improve classification in terms of stability. In this ensemble method, a partitioning clustering procedure is applied to bootstrap learning sets and the resulting multiple partitions are combined by voting (65).

7. Representation of Clusters

A partition of the data set is the end product where the number of clusters and their structure are discovered. This partition shows the separability of the data points into the clusters. The notion of cluster representation was introduced in Duran and Odell (66) and subsequently studied by Diday and Simon (67) and Michalski et al. (68). They suggested the following representation schemes:

1. Represent a cluster of points by its centroid or by a set of distant points in the cluster.
2. Represent clusters using nodes in a classification tree.
3. Represent clusters by using conjunctive logical expressions, for example, the expression $[X1 > 3]\,[X2 < 2]$.

Examples of cluster representations are shown in **Figs. 5.1 and 5.2.** Apart from these representation schemes, nowadays more sophisticated and informative representations of the clusters have been proposed. For example, these include relevance network and hierarchical clustering, including probability values (p-values) for each cluster using bootstrap resampling techniques (*see* **Fig. 5.3**).

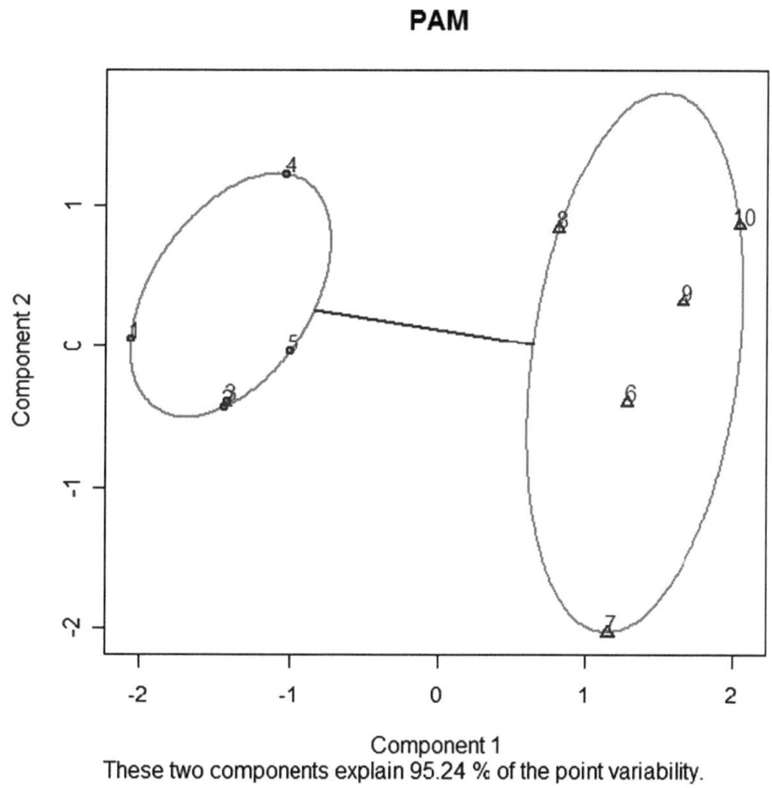

Fig. 5.1. Representation of a clusters by points (15). The data are expression values from two genes on a DNA microarray. (*Left*) From nontumor tissue; (*right*) from tumor tissues.

7.1. Relevance Networks

To explore the most relevant associations of the features, Butte and Kohane (9) proposed performing pairwise calculations of all features using a chosen similarity metric, in this case mutual information. An association with a high mutual information

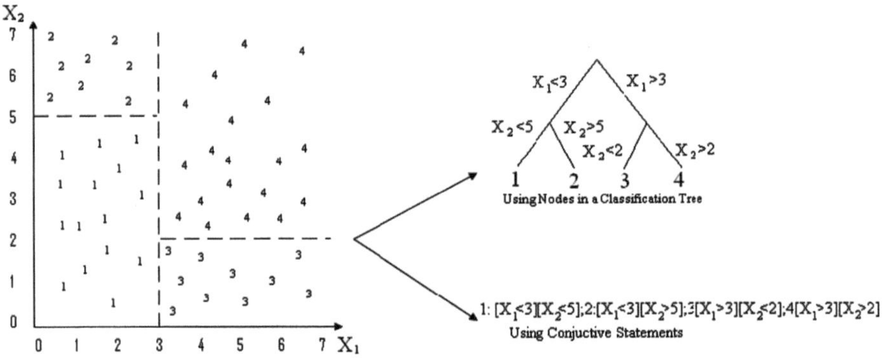

Fig. 5.2. Representation of a cluster using nodes in a classification tree and representation clusters by using conjunctive logical expressions (15).

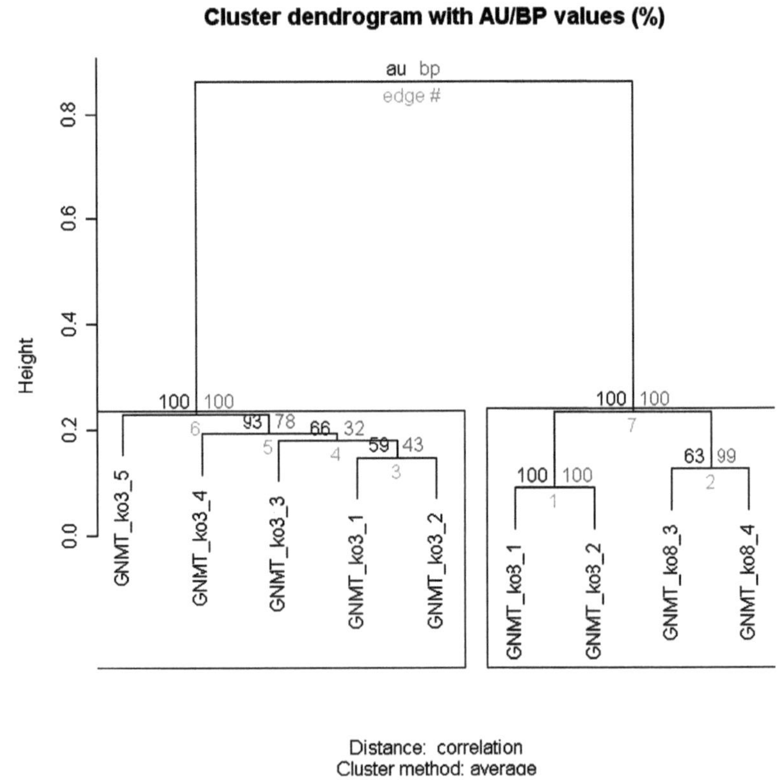

Fig. 5.3. A Pvclust output example. The example shows the hierarchical clustering of microarray data from 3- and 8-month GNMT knockout model vs. wild-type. The clustering algorithm distinguishes between the 3- and 8-month GNMT knockout models.

means that one feature is nonrandomly associated with another. They construct relevance networks by picking threshold mutual information and displaying only associations at or above the threshold.

7.2. Assessment of the Uncertainty in Hierarchical Cluster Analysis

Pvclust (27) is a package for the R statistical software that assesses the uncertainty in hierarchical cluster analysis. It calculates probability values (p-values) for each cluster in the dendrogram using bootstrap resampling techniques and highlights clusters with significant p-values. The p-value represents the possibility that the cluster is the true cluster. Two types of p-values are available: the bootstrap probability (BP) value and the approximately unbiased (AU) p-value. In both cases, thousands of bootstrap samples are generated by randomly sampling with replacement elements of the data, and bootstrap replicates of the dendrogram are obtained by repeatedly applying cluster analysis to them. The BP value of a cluster is the frequency it appears in the bootstrap replicates. Although the BP test is very useful in practice, it is biased (69–73). Multiscale bootstrap resampling is used for the calculation of the AU p-value (72, 74–76), which has superiority in bias over the BP value calculated by ordinary bootstrap resampling.

8. Notes

1. An object can be any thing, entity, or being. For example, it can be a datum, a vector, DNA, RNA, or a protein sequence.
2. Latent variables are variables inferred through a mathematical model from other variables that are observed and directly measured.
3. A metric is a nonnegative geometric function $g(x, y)$ that describes the distances between pairs of points in space.

   A metric satisfies the triangle inequality: $g(x, y) + g(y, z) \geq g(x, z)$.

   A metric should be symmetric: $g(x, y) = g(y, x)$.

   A metric also satisfies $g(x, x) = 0$.

   And lastly, it should fulfill the condition that $g(x, y) = 0$ implies $x = y$.
4. Ordinal variables do not establish the numeric difference between data points. They indicate only that one data point is ranked higher or lower than another.
5. Shannon entropy: "Information" and "uncertainty" are technical terms used to describe any process that selects one or more objects from a set of objects. If we have a device that can produce three symbols, A, B, or C, while we wait for a symbol, we are *uncertain* as to which symbol it will produce. Once a symbol appears, our uncertainty *decreases*, and we say that we have received some *information*. That is, the information is a decrease in uncertainty. The Shannon entropy is

a measure of the uncertainty, and it provides a way to estimate the minimum average message length, in bits, needed to encode a string of symbols, based on the frequency of the symbols.

6. Linkage or amalgamation rules determine how the distance between two clusters can be measured. Those include *single-linkage*, *complete-linkage*, *average-linkage*, and *minimum-variance* or *Ward's methods*. Clusters are linked sequentially to form new clusters. At each stage of this clustering process, the clusters with the shortest distance between them are combined, and the distances between the resulting set of clusters recomputed.
7. An *outlier* in statistics is a data point that does not fit a probability distribution.

References

1. Saeys Y, Inza I, Larrañaga P. (2007) *Bioinformatics* 23:2507–2517.
2. Densmore D, Heath TL. (2002) *Euclid's Elements,* Green Lion Press, Santa Fe, NM.
3. Zhang T, Ramakrishnman R, Linvy M. (1996) In *ACM SIGMOD International Conference on Management of Data.*
4. Guha S, Rastogi R, Shim K. (1998) In *ACM SIGMOD International Conference on Management of Data.*
5. Guha S, Rastogi R, Shim K. (1999) In *IEEE Conference on Data Engineering.*
6. Kaufman L, Rousseeuw P. (1990) *Finding Groups in Data: An Introduction to Cluster Analysis,* John Wiley & Sons, New York.
7. Gonzalez MD. (2005) In *Mathematics,* University of Puerto Rico, Puerto Rico.
8. Massey L. (2002) In *Recent Advances in Soft-Computing (RASC02),* Nottingham, UK.
9. Butte AJ, Kohane IS. (2000) In *Pacific Symposium on Biocomputing.*
10. Krause EF. (1987) *Taxicab Geometry,* Dover Publications, Dover, UK.
11. MacQueen JB. (1967) In *5th Berkeley Symposium on Mathematical Statistics and Probability,* Vol. 1, University of California Press, Berkeley.
12. Ball G, Hall D. (1967) *Behav Sci* 12:153–155.
13. Ng R, Han J. (1994) In *Proceedings of 20th VLDB Conference,* Santiago, Chile.
14. Lu SY, Fu KS. (1978) *IEEE Trans Syst Man Cybern* 8:381–389.
15. Jain A K. (1999) *ACM Comp Surv* 31:264–323.
16. Pearson K. (1896) *Philos Trans Roy Soc* 187:253–318.
17. Ester M, Kriegel H, Sander J, Xu X. (1996) In *2nd International Conference On Knowledge Discovery and Data Mining (KDD'96),* pp. 226–231.
18. Hinneburg A, Keim D. (1998) In *4th International Conference On Knowledge Discovery and Data Mining (KDD'98),* pp. 58–65.
19. Halkidi M, Batistakis Y, Vazirgiannis M. (2001) *J. Intell Inform Syst* 17: 107–145.
20. Dunn J. (1974) *J Cybern* 4:95–104.
21. Knudsen S. (2002) *A Biologist's Guide to Analysis of DNA Microarray Data,* John Wiley & Sons, New York.
22. Sheikholeslami G, Chatterjee S, Zhang A. (1998) In *Proceedings of 24th VLDB Conference,* pp. 428–439.
23. Wang W, Yang J, Muntz R. (1997) In *Proceedings of 23rd VLDB Conference.*
24. Pearson K. (1901) *Philos Mag* 2:559–572.
25. Bezdeck JC, Ehrlich R, Full W. (1984) *Comput Geosci* 10:191–203.
26. Breiman L. (1996) *Mach Learn* 24:123–140.
27. Suzuki R, Shimodaira H. (2006) *Bioinformatics* 22:1540–1542.
28. Arfken G. (1985) In *Mathematical Methods for Physicists,* Academic Press, Orlando, FL, pp. 13–18.
29. Kohonen T. (1995) *Self-Organizing Maps,* Springer-Verlag, Heidelberg, Germany.
30. Herrero J, Valencia A, Dopazo J. (2001) *Bioinformatics* 17:126–136.
31. Dopazo J, Carazo JM. (1997) *J Mol Evol* 44:226–233.

32. Spearman C. (1906) *Br J Psychol* 2:89–108.
33. Kendall M. (1938) *Biometrika* 30:81–89.
34. Hall L, Özyurt I, Bezdek J. (1999) *IEEE Trans Evol Comput* 3:103–112.
35. Shannon CE. (1948) *Bell Syst Tech J* 27:379–423 and 623–656.
36. Mirkin B. (1996) *Mathematical Classification and Clustering*, Kluwer Academic Publishers, Dordrecht, the Netherlands.
37. Bandeira LPC, Sousa JMC, Kaymak U. (2003) In *Fuzzy Sets and Systems – IFSA 2003*, Vol. 2715. Springer, Berlin.
38. Witten IH, Frank E. (2005) *Data Mining: Practical Machine Learning Tools and Techniques*, Elsevier, San Francisco.
39. Dash M, Choi K, Scheuermann P, Liu H. (2002) In *IEEE International Conference on Data Mining (ICDM'02)*.
40. Yu L, Liu H. (2003) in *Proceedings ICML*, Washington, DC.
41. Xiong M, Fang X, Zhao J. (2001) *Genome Res* 11:1878–1887.
42. Blanco R, Larrañaga P, Inza I, Sierra B. (2004) *Int J Patt Recog. Artif Intell* 18:1373–1390.
43. Subbarao C, Subbarao NV, Chandu SN. (1995) *Environ Geol* 28:175–180.
44. Fisher RA. (1936) *Ann Eugen* 7:179–188.
45. Frank I, Friedman J. (1993) *Technometrics* 35:109–148.
46. Friedman JH, Tukey JW. (1974) *IEEE Trans Comput* 23:881–890.
47. Wold H. (1966) In *Multivariate Analysis* (Krishnaiaah PR, Ed.), Academic Press, New York, pp. 391–420.
48. Sturn A. (2000) The Institute for Genomic Research, Rockville, MD.
49. Jiang D, Tang C, Zhang A. (2004) *Trans Knowl Data Eng* 16:1370–1386.
50. Kullback S, Leibler RA. (1951) *Ann Math Stat* 22:79–86.
51. Xu R. (2005) *IEEE Trans Neural Netw* 16:645–678.
52. Johnson SC. (1967) *Psychometrika* 2:241–254.
53. Ward JH. (1963) *J Am Stat Assoc* 58:236–244.
54. Tamayo P, Slonim D, Mesirov J, Zhu Q, Kitareewan S, Dmitrousky E, Lander ES, Golub TR. (1999) *Proc Natl Acad Sci* 96:2907–2912.
55. Fung, G. (2001) A Comprehensive Overview of Basic Clustering Algorithms. Available at http://pages.cs.wisc.edu/~gfung/
56. Berkhin, P. (2002) Survey of clustering data mining techniques. Technical report, Accrue.
57. Hertz J, Krogh A, Palmer RG. (1991) *Introduction to the Theory of Neural Computation*, Addison-Wesley, Reading, MA.
58. Fritzke B. (1994) *Neural Netw* 7:1441–1460.
59. Goldberg DE. (1989) *Genetic Algorithms in Search, Optimization and Machine Learning*, Addison-Wesley, Redwood City, CA.
60. Holland JH. (1975) *Adaption in Natural and Artificial Systems*, University of Michigan Press, Ann Arbor.
61. Schewefel HP. (1981) *Numerical Optimization of Computer Models*, John Wiley and Sons, New York.
62. Fogel LJ, Owens AJ, Wals MJ. (1965) *Artificial Intelligence Through Simulated Evolution*, John Wiley and Sons, New York.
63. Madeira SC, Oliveira AL. (2004) *IEEE/ACM Trans Comput Biol Bioinform* 1:24–45.
64. Davies DL, Bouldin DW. (1979) *IEEE Trans Patt Recog Mach Intell* 1:224–227.
65. Dudoit S, Fridlyand J. (2003) *Bioinformatics* 19:1090–1099.
66. Duran BS, Odell PL. (1974) *Cluster Analysis: A Survey*, Springer-Verlag, New York.
67. Diday E, Simon JC. (1976) Clustering analysis. In *Digital Pattern Recognition*, Springer-Verlag, Secaucus, NJ.
68. Michalski R, Stepp RE, Diday E. (1981) In *Progress in Pattern Recognition* (Kanal L, Rosenfeld A, Eds.), Vol. 1, Springer-Verlag, North-Holland, New York, pp. 33–55.
69. Hillis D, Bull J. (1993) *Syst Biol* 42:182–192.
70. Felsenstein J, Kishino H. (1993) *Syst Biol* 42:193–200.
71. Zharkikh A, Li WH. (1992) *Mol Biol Evol* 9:1119–1147.
72. Efron B, Halloran E, Holmes S. (1996) *Proc Natl Acad Sci* 93:13429–13434.
73. Sanderson MJ, Wojciechwski MF. (2000) *Syst Biol* 49:671–685.
74. Shimodaira H. (2002) *Syst Biol* 51:492–508.
75. Shimodaira H. (2004) *Ann Stat* 32:2616–2641.
76. Suzuki R, Shimodaira H. (2004) In *15th International Conference on Genome Informatics*.

Chapter 6

Nonalcoholic Steatohepatitis, Animal Models, and Biomarkers: What Is New?

Usue Ariz, Jose Maria Mato, Shelly C. Lu, and Maria L. Martínez Chantar

Abstract

Nonalcoholic fatty liver disease (NAFLD) is a clinicopathological term that encompasses a spectrum of abnormalities ranging from simple triglyceride accumulation in the hepatocytes (hepatic steatosis) to hepatic steatosis with inflammation (steatohepatitis, also known as nonalcoholic steatohepatitis or NASH). NASH can also progress to cirrhosis and hepatocellular carcinoma (HCC). Steatohepatitis has been estimated to affect around 5% of the total population and 20% of those who are overweight. The mechanisms leading to NASH and its progression to cirrhosis and HCC remain unclear, but it is a condition typically associated with obesity, insulin resistance, diabetes, and hypertriglyceridemia. This point corroborates the need for animal models and molecular markers that allow us to understand the mechanisms underlying this disease. Nowadays, there are numerous mice models to study abnormal liver function such as steatosis, NASH, and hepatocellular carcinoma. The study of the established animal models has provided many clues in the pathogenesis of steatosis and steatohepatitis, although these remain incompletely understood and no mice model completely fulfills the clinical features observed in humans.

In addition, there is a lack of accurate sensitive diagnostic tests that do not involve invasive procedures. Current laboratory tests include some biochemical analysis, but their utility for diagnosing NASH is still poor. For that reason, a great effort is being made toward the identification and validation of novel biomarkers to assess NASH using high-throughput analysis based on genomics, proteomics, and metabolomics. The most recent discoveries and their validation will be discussed.

Key words: Nonalcoholic fatty liver disease, steatosis, steatohepatitis, animal models, biomarkers.

1. Nonalcoholic Steatohepatitis (NASH)

1.1. Background

Nonalcoholic fatty liver disease (NAFLD) is a clinical-pathological term that includes a spectrum of alterations that go from the simple accumulation of triglycerides in the hepatocytes (steatosis)

R. Matthiesen (ed.), *Bioinformatics Methods in Clinical Research*, Methods in Molecular Biology 593,
DOI 10.1007/978-1-60327-194-3_6, © Humana Press, a part of Springer Science+Business Media, LLC 2010

to hepatic steatosis with inflammation (steatohepatitis, or NASH). NAFLD is one of the most common causes of hepatic damage in developed countries. The prevalence of the disease in the general population varies from 13–15%. Due to the fact that this type of disease is increasing, NASH is considered an emergent disease in the majority of developed countries. Current methods use invasive technologies such as hepatic biopsy in order to differentiate simple steatosis from steatohepatitis. The clinical characteristics are increased levels of transaminases or the changes observed by imaging studies [ultrasonography, computerized tomography (CT) scan, and magnetic resonance imaging (MRI)] have all been used to diagnose NAFLD. However, these techniques do not have sufficient criteria to distinguish between steatosis and steatohepatitis. The distinction in these cases is of great importance, because while NASH is considered to be a disease that can evolve to cirrhosis and even to a cellular hepatocarcinoma, steatosis, or fatty liver, normally does not progress and is considered benign.

The etiology of NASH is not known in depth, but it is normally correlated with obesity, type II diabetes, and hyperlipidemia. This information suggests that both fatty liver and NASH are hepatic manifestations of a metabolic dysfunction known as metabolic steatohepatitis (MESH). The lack of information about the factors involved in the pathogenesis, prognostics, and treatment of NASH points out the necessity for a research project aimed at understanding the mechanisms involved in the development of this disease.

1.2. History of NAFLD

The injuries, principally those corresponding to alcoholic hepatitis, have been considered indicative of alcohol abuse. Nevertheless, for decades it has been known that lesions similar to those induced by alcohol occur in people who do not consume it. For this reason, Thaler proposed replacing the term "alcoholic hepatitis" by "fatty hepatitis" (*Fetterberhepatitis*) (1). At present, it is considered that NASH forms part of a wider spectrum of lesions that include, in addition to NASH, nonalcoholic fatty liver, fatty liver with inflammation, and probably a great number of cryptogenic cirrhoses (2).

1.3. Nomenclature

A proper diagnosis of fatty liver disease should include the stage of the disease as well as its etiology (alcohol, type II diabetes, lipid disorders, bariatric surgery, etc.). Traditionally, fatty disorders of the liver have been classified as alcoholic or nonalcoholic. NAFLD includes both nonalcoholic fatty liver and NASH. The underlying mechanism as well as prognosis of the disease are completely different. While fatty liver is a stable lesion that evolves to more severe stages in just 3% of cases, NASH evolves to cirrhosis in 15–25% of cases. Using the term "nonalcoholic" to describe fatty liver disease associated with all of the above-mentioned etiologies renders the condition heterogeneous in terms of etiology and, possibly, natural history as well as response to therapy. There is currently no consensus on the best way to classify fatty disorders of the liver.

1.4. Criteria for the Diagnosis of Steatosis and Steatohepatitis

The distinctive morphological features of steatohepatitis, regardless of the clinical background, include some "alcoholic hepatitis-like" findings: steatosis, lobular inflammation, which includes polymorphonuclear leukocytes, and perisinusoidal fibrosis in zone 3 of the acinus. Other common features are hepatocellular ballooning, poorly formed Mallory's hyaline, and glycogenated nuclei (3, 4). The clinical course is indolent in most patients, but cirrhosis is a known complication that has been reported in 7–16% (5) of patients with NASH. NASH might be a cause for some cases of cryptogenic cirrhosis (6).

To date, there are no clinical or histopathological "markers" that predict patients at risk for progression to cirrhosis. It is now appreciated that, in a given patient, only some of these features may be present. Mallory bodies are less frequently seen in NASH compared with alcoholic steatohepatitis and may even be absent (7, 8). There are no criteria to diagnose steatohepatitis, and there is considerable interobserver variation in the assessment of inflammation. This has led to the definition of variables commonly described in NASH, most of which were significant for necroinflammatory activity as follows: *intra-acinar (lobular) inflammation,* the cellular components of inflammation (polymorphonuclear leukocytes, lymphocytes, and other mononuclear cells;, eosinophils; and microgranulomas) and location (sinusoidal, surrounding Mallory's hyaline, or hepatocellular necrosis), portal tract inflammation, acidophil bodies, PAS-D Kupffer cells, lipogranulomas (intra-acinar lipogranulomas), and, finally, hepatocellular iron (9).

1.5. Symptoms

As with many other types of chronic liver disease, most patients with NAFLD are asymptomatic. The liver disease is normally discovered during routine laboratory examination or associated with conditions such as hypertension, diabetes, or morbid obesity. NAFLD is the most common cause for unexplained persistent elevation of ALT levels once hepatitis C and other known causes of chronic liver disease have been excluded. Only limited are data available in the literature. Fatigue and upper right quadrant discomfort that is typically vague are the most common complaints (10). Then there is a small percentage of the population that develops pruritus, anorexia, and nausea, symptoms indicative of more serious liver disease.

1.6. Physical Signs

There are no signs directly associated with NASH. Obesity is the most common abnormality on physical examination and is present in 30–100% of patients. The most common physical sign of liver disease is hepatomegaly, which has been reported in up to 50% of subjects in different studies (11, 12). A smaller percentage of patients have stigmata of chronic liver disease.

1.7. Hepatic Profile

The percentage of NASH patients with abnormal aminotransferase activity varies from 50–90% (13). The degree of abnormal activity changes and is usually between one and four times the upper limit of normal values. Depending on the etiology of the disease, ALT is higher or lower than AST. Although gamma-S glutamyltransferase levels may be elevated, there are little data on the frequency and degree of elevation. The alkaline phosphatase level may also be variably elevated up to twice the upper limit of normal (14). The hepatic functional capacity is not compromised until cirrhosis appears and liver failure has set in. In diabetic subjects with NASH, isolated hypoalbuminemia may also occur due to proteinuria related to diabetic nephropathy. Hematological parameters are usually normal unless cirrhosis and portal hypertension lead to hypersplenism (15). Patients with NASH (30–50%) have either diabetes or glucose intolerance (16).

There are three criteria for the diagnosis of NASH as described by Powell et al.: first, the characteristic of the histological samples; second, minimal or no alcohol consumption ($\leq$40 g/wk); and, finally, the absence of serological evidence of viral hepatitis. Although these criteria are widely used in clinical practice, each criterion has specific limitations (17). The variability in the histological expression of NASH has been in discussion and has avoided the development of a universally accepted set of diagnostic criteria for steatohepatitis.

1.7.1. Imaging Diagnosis

There are no very accurate noninvasive methods for the diagnosis of NASH. The presence of fat in the liver can be diagnosed in many cases using various imaging modalities. Ultrasonography, computerized tomography (CT) scan, and magnetic resonance imaging (MRI) have all been used to diagnose NAFLD. The total content of liver fat can be estimated semiquantitatively by CT and also by MRI (18). The distribution of fat seen via CT imaging is not equal. There is an imbalance between the right lobe of the liver compared to the left lobe. There is considerable variability between multiple examinations in the content of liver fat using CT imaging (19). This is due to nonidentical calibration of different machines, different types of CT scanners, and differing regions of interest during multiple examinations and changes in hepatic fat content. Accurate measurement of fatty liver and changes in hepatic fat content require careful calibration during each examination.

Differences in the precession frequency (3.7 ppm) between water and fat protons can be used in which the fat signal is subtracted from the water signal to diagnose fatty liver using MRI (20). Several newer modifications in MRI techniques like fast gradient echo techniques (modified Dixon method) in the assessment of hepatic fat content have resulted in considerable improvement in the ability to diagnose a fatty liver by MRI (21). Compared to CT scan, sonography is more sensitive in detecting changes in the total amount of fatty content (22). CT scan and MRI are superior to sonography when the changes in fatty content are focal. Despite the utility of these imaging modalities, none of these modalities can distinguish between fatty liver versus steatohepatitis. Thus, the only way to diagnose steatohepatitis is liver biopsy.

1.7.2. Liver Biopsy

There is little controversy about whether liver biopsy is the only accurate method for the diagnosis of NASH (11). The decision to perform a liver biopsy in routine clinical practice should be taken into consideration depending on the clinical question and also the risk and time consumption of the practice. The patient should be included in this decision-making process.

1.8. NAFLD in Different Populations

1.8.1. NAFLD in Children

Most of the cases already known in children developed around puberty, although there are some studies done in younger people (7–8 years of age) (23). In addition, several studies that were performed in obese children have shown evidence of fatty liver disease, documented by sonography or increased ALT levels, in up to 50–60% of affected children (24). Taking these data into account, the situation of NASH in the pediatric population is similar to that in adults. Many children are asymptomatic. The most common physical findings are obesity, hepatomegaly, and elevated ALT (25). The duration of obesity may also be a determinant of the likelihood of progression to cirrhosis (26). NAFLD is considered a chronic disease, so many issues, including the disease's mechanism, have to be clarified in the etiology in the pediatric population.

1.8.2. NAFLD in Obese People

Obesity is one the main causes of developing NAFLD. As early as 1973, Kern et al. published data about the incidence of patients with fatty liver (92 cases) in 151 obese subjects (27). These were followed by other reports of a high incidence of cirrhosis and diabetes in morbidly obese individuals (28). One of the situations associated with morbid obesity is the development of very severe steatohepatitis after a dramatic weight loss produced by the surgical procedure jejunoileal bypass (29). The prevalence of hepatic steatosis increased from 66 to 95%

in the first year after jejunoileal bypass but returned to baseline values by 5–7 years in one study (30). During the first 18 months, some patients also show a marked increase in inflammation and hepatocellular injury that may manifest as subacute liver failure. The reason for this accurate progression normally is based on the metabolic disorder that encompasses this type of patient.

1.8.3. NAFLD and Insulin Resistance

There are several types of insulin resistance that have already been associated with NAFLD. Diabetes mellitus associated with lipoatrophy is the prototypical example of such a condition. In some cases, the disease can progress to fibrosis, portal hypertension, and splenomegaly (31). To date, the development of cirrhosis has not been characterized.

1.8.4. NAFLD and Hepatitis C

NAFLD and hepatitis C can coexist in the same individual. It has to be taken into consideration that the presence of hepatitis C produces portal and lobular inflammation, so these parameters cannot be used to assess the presence of fatty liver disease. Also, minor degrees of hepatic steatosis are often seen in those with hepatitis C (32). Additionally, it is already known that interferon therapy, widely used in hepatitis C, increases the serum levels of triglycerides (33).

1.8.5. NAFLD and Liver Transplantation

The accumulation of lipid droplets in the liver affects the availability of promising results in liver transplantation. Recent data indicate that up to 20% of potential donors have hepatic steatosis (34). Hepatic macrovesicular steatosis is associated with primary nonfunction of the graft. Several mechanisms are related to this dysfunction, such as a generation of toxic metabolites and a decrease in adenosine triphosphate production (35). In addition, there is a risk of developing NASH (approximately 20%) in those people who get a transplant due to alcoholic liver disease, hepatitis B, or primary biliary cirrhosis.

1.9. Treatment of NAFLD

The treatment of any condition requires consideration of the natural history of the condition, the relative efficacy and safety of the therapeutic options, and the cost. As previously reported, NASH can also progress to cirrhosis, and it has to be taken in consideration that (i) fat and ballooning degeneration or (ii) fat, ballooning degeneration, and Mallory bodies, or (iii) perisinusoidal fibrosis may be at greater risk for progression (36). There are no published controlled trials of treatment modalities for NAFLD. In the absence of treatment, the therapy selected is directed toward correction of the risk factors or the etiology that produced NASH (i.e., insulin resistance, decreasing delivery of fatty acids to the liver, and use of drugs with potentially hepatoprotective effects).

1.10. Risk Factors

1.10.1. Weight Management

There are several reports on the role of weight loss in the development of NASH. However, there are no clinical trials where weight control has been used as a treatment for NAFLD (37). In overweight individuals with elevated aminotransferase levels, weight reduction by 10% or more has been shown to correct aminotransferase activities and decrease hepatomegaly and also the cardiovascular risk profile. In this special risk factor, the type of diet is an important component in the weight loss. For example, saturated fats in a diet did not improve insulin resistance, whereas a diet rich in fiber can improve insulin resistance (38). There are no controlled studies of the value of diet in the management of NAFLD. Also, the effects of polyunsaturated fatty acids and the specific fiber supplements designed to decrease insulin resistance or dietary fat have not been evaluated (39). Thus, the reasonable recommendation in this point is for overweight people to consume a heart-healthy diet. Also, in this picture we should include the value of exercise. Exercise has been shown to increase the oxidative capacity of muscle cells and the utilization of fatty acids for oxidation (40). For example, the degree of improvement in insulin sensitivity is related to the intensity of the exercise. Finally, the role of drugs for weight reduction in NAFLD has to be considered. At present, there are three approved drugs for weight reduction: phentermine, sibutramine, and orlistat (41). Although the value of these drugs in achieving weight loss is established, their value in the management of NAFLD remains to be shown.

1.10.2. Pharmacologic Treatment of Insulin Resistance

Insulin resistance seems to be the common denominator in many cases of NASH. NASH is associated with decreased insulin-mediated suppression of lipolysis (41). As a consequence of this decrease, NASH patients show high levels of free fatty acid in the serum. This increase in the fatty acid concentration can regulate the insulin action in the hepatocytes in the way molecules are used for mitochondrial oxidation and the proapoptotic mitochondrial uncoupling protein 2 expression (42). These considerations, along with the well-known association of NASH with obesity and diabetes, have led to attempts to treat NASH by treating insulin resistance. There are no data on this issue. Two major types of compounds are used to improve insulin resistance: biguanides (e.g., metformin) and thiazolidinediones (e.g., rosiglitazone and pioglitazone). It has already been shown that treatment with metformin improved inflammation and hepatic steatosis in steatotic animal models (43).

Thiazolidinediones are drugs that act via peroxisome proliferator-activated receptor and improve insulin sensitivity. In one small study, NASH patients were treated for six months with troglitazone (44). The treatment resulted in decreased ALT values as well as inflammatory scores.

Finally, the effects of different drugs on the insulin sensitivity of various metabolic pathways are variable (45). Despite these limitations, these promising results form the basis for using these drugs in future trials.

1.10.3. Lipid-Lowering Agents

Hypertriglyceridemia is often associated with NASH, but lipid-lowering agents are not normally used for treatment of patients with NASH (46). This is an interesting field to be explored with this type of drug.

1.10.4. Drugs That Protect Hepatocytes

Several drugs that can be used as protectors of the function of the hepatocyte have already been used in patients with NASH. The most important are betaine, vitamin E, lecithin, and beta-carotene. There are publications about the effects of vitamin E (DL–tocopherol) in the treatment of NASH. It can be observed that the levels of ALT follow-up with 5.2 months of treatment either improved markedly or normalized in all cases (47). Although these data are promising, there is no histological confirmation of this benefit. There are two additional studies where betaine supplements were given for the treatment of patients with NASH. Betaine is a very important metabolite of the methionine cycle, a precursor of *S*-adenosyl methionine, a hepatoprotective factor. In one study, a group of 10 people decreased aminotransferase activity as well as liver histology (48). Similarly, a 25% improvement in hepatic steatosis was reported in a randomized controlled study in which betaine was administered along with diethanolamine glucuronate and nicotinamide ascorbate for eight weeks (49). These findings now require confirmation in large, long-term prospective trials.

2. Mouse Models of NASH

The literature contains numerous different rodent models to study abnormal liver function such as steatosis, NASH, and hepatocellular carcinoma. These animal models are extremely useful, as there are still many events to be elucidated in the pathology of NASH. The study of the established animal models has provided many clues in the pathogenesis of steatosis and steatohepatitis, but these remain incompletely understood (50, 51).

The different mouse models can be classified in two big groups. The first one includes genetically modified (transgenic or knockout) mice that spontaneously develop liver disease, and the second one includes the mice that acquire the disease after dietary or pharmacological manipulation. Ideal animal models of disease should closely resemble the pathological characteristics observed in humans. For the study of NASH, they should show ballooning hepatocyte degeneration in addition to simply fatty liver change and an inflammatory infiltrate (52), along with other biochemical parameters. To date, no animal model has completely fulfilled the clinical features observed in humans (53).

2.1. Genetically Modified Mice

2.1.1. ob/ob Mouse

ob/ob mice have a naturally occurring mutation that prevents the synthesis of leptin, a satiety hormone that inhibits feeding behavior and increases energy expenditure (54). Leptin is synthesized predominately by white adipose tissue and exerts its major anorexigenic effect by acting on neurons in the hypothalamus. These mice have hyperphagia and become obese (55), which is accompanied by hyperinsulinemia and hyperglycemia, as well as hyperlipidemia and fatty liver. Although the ob/ob mice do not spontaneously progress from steatosis to steatohepatitis, after a "second hit," they progress to NASH. The livers of these mice are predisposed to injury after various types of stimulus: LPS, ischemia-reperfusion injury, methionine-/choline-deficient diet, and ethanol feeding, which evolve to steatohepatitis and acute mortality (56–58). Regeneration of these livers after partial hepatectomy is also diminished (50). The mechanism for hepatic steatosis in ob/ob mice is not well understood, but the evidence points to the increased hepatic lipogenesis. The expression of TNF-α by adipose tissue is increased as well as serum free fatty acid concentration, suggesting that the delivery of fatty acids to the liver is also increased (55). The expression of uncoupling protein (UCP)-2 mRNA and protein is induced in ob/ob livers. UCP-2 is a mitochondrial transmembrane protein that promotes the accumulation of protonated fatty acid anions in the mitochondrial matrix (59) and is related to oxidative stress. There is conflicting evidence regarding the activity of hepatocyte fatty acid beta-oxidation pathways, but the expression of some CYP4A and CYP2E1 microsomal enzymes involved in ω-oxidation is also increased (60), generating ROS, which might be a compensatory response to the increased rate of fatty acid production.

2.1.2. db/db Mouse

Mutations in the diabetes (db) gene result in an autosomal recessive diabetic, obese phenotype similar to the ob/ob mouse (61). db/db mice have normal or elevated levels of leptin but are resistant to its effects. Studies have shown that the db gene encodes the leptin receptor, which is structurally similar to a class I cytokine receptor (62). The lack of activity of the leptin hormone resembles the phenotype of the leptin-deficient mice, and the comments made about ob/ob mice can be applied to this model, too.

2.1.3. MAT1A −/− Mouse

As will be explained later in more detail, it is known that when mice are fed a diet deficient in lipotropes (methionine, choline, folate, and vitamin B12), the liver develops steatosis within a few days, and if the diet continues, it evolves to NASH, fibrosis, cirrhosis, and HCC in some cases (63, 64). This deficiency entails a decrease in S-adenosyl methionine (SAM) (65, 66) that may be related the pathogenesis of NASH, as all observed cirrhotic patients had an important decrease in SAM hepatic synthesis (67). SAM is a metabolite that actively participates in many physiological processes, where it donates its methyl group to different molecules (DNA, RNA, phospholipids, and proteins), its sulfur atom to cellular antioxidants, its propylamine group to polyamines necessary for cellular growth, and its MTA to the "methionine recovery pathway" for the synthesis of this amino acid. These reactions can affect a great spectrum of biological processes, including gene expression, proliferation, differentiation, and apoptosis, among others (68). Mice deficient in methionine adenosyl transferase (MAT) 1A (the enzyme responsible for SAM synthesis in the adult liver) have a decrease in hepatic SAM levels and spontaneously develop steatosis, NASH, and HCC (69). By three months of age, these mice have hepatomegaly with macrovesicular steatosis. These mice also have increased mRNA levels of CYP2E1 and UCP2, and the levels of GSH (glutathione), the most important antioxidant, are reduced. On the other hand, several key enzymes involved in cystein and GSH synthesis are increased, suggesting that there is an oxidative stress that is trying to be compensated. Also, these mice have changes in the expression of genes involved in proliferation and lipid and carbohydrate metabolism (70), are predisposed to liver injury, and have impaired liver regeneration after partial hepatectomy (71).

2.1.4. PTEN −/− Mouse

PTEN (phosphatase and tensin homologue) is a multifunctional phosphatase whose substrate is phosphatidylinositol-3,4,5-triphosphate (PIP3) and acts as a tumor suppressor gene that downregulates phosphatidyl inositol kinases (PI3K) (72, 73). Hepatocyte-specific PTEN-deficient mice spontaneously develop steatosis, steatohepatitis, and hepatocellular carcinoma (74). By 10 weeks of age, these mice have increased concentrations of triglyceride and cholesterol esters, and, after histological analysis, micro- and macrovesicular lipid vacuoles can be observed. After 40 weeks of age, they have macrovesicular steatosis, Mallory bodies, ballooning degeneration, and sinusoidal fibrosis (75): NASH events that also occur in human NASH (76). The pathogenesis is probably mediated by an increase in PPARγ (peroxisome proliferator-activated receptor γ) and its downstream adipogenic targets, an increase in SREBP1c (sterol regulatory element-binding protein 1c), a transcriptional activator of lipogenesis, and an increase in β-oxidation–related genes. An upregulation in PPARγ and SREBP1c increases adipogenesis and lipogenesis, leading to liver steatosis, enhancing beta-oxidation, which leads to oxidative stress (75).

2.1.5. SREBP1c Transgenic Mouse

SREBP1c is a transcription factor involved in adipocyte differentiation (77, 78). Transgenic mice that overexpress nuclear SREBP1c in adipose tissue at 30 weeks of age exhibited intralobular inflammation with ballooning degeneration, Mallory hyaline bodies, and pericellular fibrosis characteristic of NASH (79). These mice also showed lipodystrophy, insulin resistance, and hyperglycemia, accompanied by hyperlipidemia and marked fatty liver. In humans, lipodystrophy and obesity share common metabolic disorders, including insulin resistance and liver steatosis, potentially leading to NASH. At 20 weeks of age, these mice showed serum levels of AST, cholesterol, and triglycerides higher than those of wild-types. There were no significant differences in body weight, but the levels of serum leptin and adiponectin were significantly lower in transgenic mice (79). At 30 weeks of age, there was an increase in oxidative DNA damage that may also be involved in the development of the NASH-like lesions. The molecular mechanism involved in the development of NASH in the nSREBP1c mice remains unknown; however, insulin resistance is likely associated with the development of fatty liver and the progression to NASH (80–83).

2.2. Environmental

2.2.1. Methionine-/Choline-Deficient (MCD) Diet

Choline is an essential nutrient with roles in cell membrane integrity, transmembrane signaling, phosphatidyl choline synthesis, neurotransmission, and methyl metabolism. Mice fed a diet that is deficient in both choline and methionine develop inflammation and hepatic fibrosis in addition to simple steatosis (84). Evidence suggests that a methionine-/choline-deficient diet impairs mitochondrial β-oxidation and leads to the induction of CYP2E1 expression, an event that was confirmed in NASH patients (85). This situation promotes oxidative stress and induces steatohepatitis along with elevated plasma TNF-α levels. In summary, a methionine-/choline-deficient diet induces ROS production, mitochondrial DNA damage, and apoptotic cell death (86), making this probably the best-established model to study inflammation and fibrosis in NAFLD, although some features do not resemble those in humans, such as cachexia, low plasma trygliceride levels, and reduced liver-to-body weight ratio, histological distribution of steatosis (50), plasma ALT levels, and insulin resistance (87).

2.2.2. High-Fat Diet

It is relatively difficult to induce obesity in normal rats and mice (88). Certain diets that have been shown to cause obesity and fatty livers in mice might be important in the development of obesity-related fatty liver disease. Complex traits such as obesity and fatty liver disease are influenced by genetic variation and by the type of diet the mice are fed. When normal, lean C57BL6/J male mice are fed diets that contain 45% fat, at four weeks of age they exhibit about a 15% increase in body weight that is due to a gradual accumulation of body fat (89) accompanied with increases in circulating leptin levels. This persistent hyperleptinemia induces leptin resistance and hyperphagia. The expression of hepatic lipid synthesis (SREBP1c, SREBP-2, and Stearoyl-coA desaturase 1) was increased by HFD in two different strains of mice: the C57BL6 and the 129S6/SvEvTac (a more resistant strain to HFD-induced obesity) (90, 91). These environmentally induced forms of leptin-resistant steatosis resemble the same observed phenotype as the ob/ob and db/db mice.

3. Biomarkers in NASH

3.1. Current Biomarkers in NASH

Nonalcoholic fatty liver disease (NAFLD) includes a broad spectrum of liver abnormalities, from simple steatosis to nonalcoholic steatohepatits (NASH) with various degrees of inflammation and fibrosis (92), which can eventually develop into cirrhosis and hepatocellular carcinoma (93). Simple steatosis has a benign course and only a small percentage of these patients will develop NASH, which is a potentially serious condition associated with a significant increase in overall and liver-related morbidity and mortality (94, 95). It is very important then to distinguish between simple steatosis and NASH. To date, the noninvasive tests available have limited utility in general, and liver biopsy remains the gold standard for diagnosing NASH. As NAFLD has increased to 24% in industrialized countries (96), it is obvious that this invasive procedure is not suitable as a screening test. There is a need to obtain a noninvasive test that is able to diagnose NASH, follow disease progression, and monitor the response to therapy in these patients. Current laboratory tests include some biochemical analysis such as ALT, AST, GGT, ALP, prothrombin time, and complete blood count, but the utility for diagnosing NASH is still poor.

For that reason, a great effort is now being made toward the identification and validation of novel biomarkers to assess NASH. An ideal biomarker should be simple, reproducible, inexpensive, readily available, and accurate. Finding biomarkers that fulfill all the requirements is a major challenge today. Looking at the pathological processes that take place in the disease, several biomarkers are now under investigation. Markers for oxidative stress, inflammation, apoptosis, and fibrosis – events occurring in

NASH – are currently under study, but none of them has yet been validated (97). It is widely known that oxidative stress is a key mechanism in liver damage and disease progression in NAFLD. Several oxidation pathways take place in this process, and quantification of the products of these reactions could be a valuable diagnostic tool. But the main problem is that these reactive oxygen species (ROS) react rapidly and in situ in the environment where they are produced, and their measurement in blood might not reflect the situation in the liver. Oxidized low-density lipoprotein and thiobarbituric acid-reacting substance (TBARS), although found to be significantly higher in the blood of NASH patients, did not pass stepwise regression analysis (98). Another study measured total antioxidant response (TAR) and total lipid peroxide levels and were found to be significantly lower and higher, respectively, in NASH plasma, although the size of the experiment was small and the time between biopsy and blood tests was not exactly the same (99). Other parameters also measured but not found of usefulness were TBARS, plasma vitamin E levels, glutathione peroxidase activity, erythrocyte glutathione peroxidase activity, Cu-to-Zn superoxide dismutase activities, and breath ethane (100, 101). Therefore, although the existence of oxidative stress in the liver is well known, the results obtained so far are mixed and a deeper study of NASH oxidative stress biomarkers should be undertaken to establish good noninvasive biomarkers for diagnosis. Inflammation is a central process in NASH. Several cytokines and other proteins involved in inflammation have been proposed as potential biomarkers of NASH, although currently there is no clinical evidence of the usefulness of any of these markers. There is an open discussion regarding TNF-α serum levels in NASH patients compared to simple steatosis and controls, as different results have been obtained from different research groups (102–105). As TNF-α has a relative short life and low circulating levels, it may not reflect the changes occurring in the liver tissue. Also, differences in experimental designs may explain the different results obtained. Serum adiponectin was quantified and found to be significantly lower in patients with NASH compared to controls and steatosis in two different studies (102, 103). IL-6 was increased in NASH and steatosis compared to controls but was not able to distinguish between NASH and steatosis (106). No association was found between C-reactive protein and any histological feature of NASH (102, 106). CC-chemokine ligand-2 levels were increased in patients with NASH compared to simple steatosis (106). Some of these observations look promising although they need larger studies to be confirmed.

Apoptosis is also a characteristic of hepatocytes in NASH, which is absent in simple steatosis (107), making it an ideal process to distinguish between simple steatosis and NASH. In apoptosis, the effector caspases cleave a number of different substrates, including cytokeratin 18, which has been measured in the liver and plasma (108), obtaining good results in distinguishing between NASH and controls. A large multicenter validation study is currently under way.

The presence of fibrosis in the liver suggests a more severe progressive liver damage, and quantifying the degree of fibrosis is essential for patients with NAFLD. It is important to detect early stages of fibrosis for adequate pharmacological intervention, but many of the noninvasive markers studied to date are valid to detect advanced, severe fibrosis but have low utility for the presence of mild to moderate fibrosis (109). Most of them use a combination of clinical and/or biochemical parameters such as age, body mass index, transaminases, triglycerides, platelet count, hepatic proteins, and extracellular matrix proteins (110–115).

Although some of these potential biomarkers look promising, a new and promising approach that uses high-throughput techniques is now being used to identify useful biomarkers. These techniques include large-scale studies in the fields of genomics, proteomics, and metabolomics.

3.2. The Use of High-Throughput Technologies for the Development of Novel Biomarkers in NASH

3.2.1. Genomics

Mutations, insertions, and deletions in the sequences of the genes can affect the structure and activity of proteins and change the physiopathological state of a cell. Also, the level of expression of each gene in the cell determines the cellular processes that are taking place in it. Large-scale genomics analysis with microarrays can be performed at the level of both genotyping and gene expression and can provide a tremendous amount of information about a pathological process.

3.2.1.1. Genotyping

Single-nucleotide polymorphisms (SNPs) are a type of punctual mutation that affects one nucleotide of the DNA sequence. The majority of these mutations do not have any changes in the activity of the protein, but some others are important because they affect the function of the protein and can potentially predispose to a certain disease. Several family clustering studies in NAFLD show that there is a link between genetics and the development of the disease (116, 117). Identification of these mutations related to NASH can serve as biomarkers for the diagnosis or prognosis of the disease.

To date, studies to discover mutations have been carried out based on candidate genes related to lipid metabolism, oxidative stress, cytokines, bacterial receptors, and extracellular matrix synthesis and degradation (118). But the possibility of performing high-throughput analysis of SNPs gives a new perspective to investigate potential links between a certain mutation and NASH that could help in its diagnosis.

3.2.1.2. Gene Expression

Large-scale gene expression analysis is a very powerful technique to measure the differences between two or more samples. This technique is able to measure the mRNAs that are being expressed in the tissue analyzed and identify those that are relevant in the disease, but one major problem is that it is not possible to directly propose them as biomarkers to use in clinics because the samples are obtained directly from the liver and some of the important characteristics of an ideal biomarker are that they should be simple and inexpensive. In a strict sense, this technique is not suitable for the direct identification of biomarkers but is very valuable for identifying new pathways involved in the disease, which could serve as a starting point toward the discovery of novel therapeutic targets.

As in any other type of assay, the key for success in large-scale gene expression analysis is to start with a very well-planned experiment. As the amount of information from a relatively small number of samples that are going to be obtained is huge, it is crucial that the samples are well selected and will answer the questions proposed. For obvious reasons, human tissue samples are often very difficult to obtain and are unique most of the time, so it is essential to get the maximum amount of information from these precious samples. Samples should be very well diagnosed, including the patient's clinical parameters, biochemical tests, and histology at the time of biopsy. Also, they should be matched to ethnic group, gender, age, and other features that may alter the pattern of gene expression. It is also a good idea to include samples of related diseases that could have common features and similar expression patterns to specifically identify gene expression differences unique to the disease that could serve to rule out any other disease. Good-quality mRNA is crucial to obtaining reliable results, and this should be checked before its utilization. Gene expression analysis with microarrays can give us information of up to 47,000 transcripts in one set of experiments. Even when considering a very small false-positive rate, it is very risky to assess the result without any previous validation. The information that microarrays generate should be taken as a global view of the transcriptome of the tissue that provides clues on the altered pathways rather than information on single genes unless validated with a different technique such as qPCR or Northern blot. Taking these premises into account, the amount of information about a pathological process that can be obtained with the use of these techniques is enormous.

The use of this technology opens a broad spectrum of possibilities to elucidate molecular mechanisms related to NASH. A brief review of some examples of the use of high-throughput gene expression analysis in the field of NASH pathology is detailed below. We have selected three different approaches to the disease to show some of the possibilities of this technology. There is one study that uses human liver samples to identify differentially expressed genes in NASH (119), another study that uses one of the described mouse models to evaluate the therapeutic effect of a compound (120), and one study that combines the use of human samples and a mouse model to identify early markers of NASH (121). The aim of this chapter is to provide some insights for the application of the high-throughput gene expression technologies and have been reviewed under a clinical and biological point of view. No details are provided about the data analysis processes in these studies. Bioinformatics experts in other chapters of this book review data analysis processes in detail.

Younossi et al.'s study (119)

Objective: To elucidate steps in the complex pathogenesis of NASH through a genomic approach using microarray technology.

The selected samples were liver biopsies from patients with NAFLD and controls. The control samples were obtained from potential liver transplant organ donors or from patients undergoing hepatic resection for liver mass and no evidence of chronic liver disease. Histological analysis was done and NASH was defined when, in addition to steatosis, at least one unequivocal Mallory body was identified and/or some degree of zone 3 pericellular fibrosis or bridging fibrosis was identified on the trichrome stain (122). Biochemical analyses were also performed. Samples were classified into four groups: nonobese controls ($n=6$), obese controls ($n=7$), steatosis ($n=12$), and NASH ($n=29$). The microarrays used were custom-spotted and had 5,220 human cDNA clones with genes involved in inflammation pathways, genes related to liver diseases, 329 ESTs, and approximately 1,000 cDNA of unknown function.

Gene expression analysis was performed comparing NASH expression profiles to nonobese control and to obese control profiles. Also, obese controls were compared to nonobese controls. Genes differentially expressed in NASH and/or obese samples vs. nonobese samples were considered as obesity-related genes. The gene expression pattern of samples with simple steatosis was also analyzed. And genes differentially expressed in NASH vs. obese and nonobese controls were considered to be directly related to NASH-specific gene expression. These genes encode key enzymes of lipid metabolism, extracellular matrix remodeling, liver regeneration, apoptosis, and the detoxification process. Four genes were selected for real-time quantitative polymerase chain reaction (RT-qPCR) confirmation, and this verification confirmed that all the PCR results were in agreement with the microarray data.

This is a well-designed study of human NASH gene expression. Different types of control samples were also included in the study to analyze the expression profile exclusively related to NASH. Obesity and steatosis are risk factors for NASH, although not all obese patients or patients with steatosis will evolve to NASH. In order to differentiate between them, those samples were included in the study and only the genes differentially expressed in NASH vs. different controls were considered as related to the disease. When clustering all the samples with the differentially expressed genes in NASH vs. nonobese controls, NASH and obese samples did not separate as expected, suggesting overlapping gene expression profiles involved in the pathogenesis of NASH and its main risk factor, obesity. These results indicate the need to include obese controls in this type of study. The availability of human samples is frequently a very limiting factor; for that reason, it is important to obtain the maximum amount of precise information from them. Ideally, samples should be matched to age, gender, body mass index (BMI), or other parameters that could affect the gene expression profile in the liver, but usually this is not a very easy question to address. Although the samples were not matched, extensive clinical and laboratory data were collected from the patients, ensuring the suitability of the samples for the study.

In conclusion, we can say that this is a good approach to evaluate gene expression differences in NASH exclusively due to the disease and points to the necessity of including obese non-NAFLD samples in the studies to obtain more precise information. The identification of these genes promotes the understanding of the progressive form of NAFLD and the potential targets for future therapy.

de Oliveira et al.'s study (120)

Objective: Understand the molecular mechanisms underlying NASH prevention by S-nitroso-N acetylcysteine (SNAC). The study examines hepatically differentially expressed genes between ob/ob mice receiving or not receiving SNAC treatment concomitantly with an MCD diet.

Animals for gene expression analysis were divided into three groups: The first one included ob/ob mice fed a standard diet ($n = 6$), the second one included ob/ob mice fed an MCD diet ($n = 6$), and the third one consisted of ob/ob mice fed an MCD diet and an oral solution of SNAC ($n = 6$). The microarrays used in this study were CodeLink UniSet Mouse 20 K I Bioarray (GE, Healthcare Bio-Sciences), which contain 19,801 probes. Parallel to gene expression analysis, biochemical and histopathological analysis were performed on the same samples.

This study combines the use of a genetically modified mouse model (ob/ob) with an environmentally acquired NASH through a methionine-/choline-deficient diet to evaluate the potential therapeutic effect of SNAC. This mixed model could be a simulation of the two-hit hypothesis of NASH development in humans proposed by Day and James (123), because ob/ob mouse spontaneously develop steatosis (not NASH), but a second hit like the MCD diet promotes oxidative stress inducing NASH. In this study, SNAC was tested as a therapeutic agent based on the principle that the oxidative stress produced by the MCD diet could be prevented by the concomitant administration of SNAC, a potent inhibitor of lipid peroxidation (123). The effectiveness of the use of SNAC to prevent NASH was previously demonstrated (124), but in this study a gene expression profile was analyzed to elucidate the mechanism responsible for the observed effect. First, the biochemical and histopathological analyses were performed to assess the effect of the MCD diet on the livers of the ob/ob mice and its prevention by SNAC. Before analyzing any sample for gene expression, it is important to check that it fully meets the characteristics of the expected phenotype. In this case, biochemical parameters (AST, ALT, triglycerides, cholesterol, and weight change) were measured as well as histology for the presence of steatosis, hepatocellular ballooning, and lobular inflammation. Second, the gene expression analysis of the liver compared the ob/ob mice vs. the ob/ob mice fed an MCD diet, and the ob/ob mice fed an MCD diet vs. the ob/ob mice fed an MCD diet and SNAC. This second comparison yielded a set of downregulated genes belonging to the pathways related to fatty acid metabolism, especially oxidative phosphorylation, which presumably may be preventing mitochondrial overload by downregulating genes participating in the electron transport chain. Although the gene expression results obtained were very conclusive, there is no validation by any other type of technique like RT-qPCR for the differentially expressed genes.

This is a very interesting approach to identify novel therapeutic strategies for NAFLD treatment. Biochemical analysis, liver histology, and liver gene expression profiles were performed to characterize the mechanism underlying the effect of SNAC. Analyzing the mechanism of this effect provides important clues for the development of this and other potential therapeutic targets. Although these are very interesting results, they should be taken as preliminary results, because even if it is a good animal model of NASH, it does not exactly resemble the human phenotype of NASH. Also, the extrahepatic effects of SNAC were not analyzed, and this is an important point if proposing SNAC as a pharmacological treatment. This issue will be addressed in future studies.

Nevertheless, this promising result could be considered a starting point for the development of novel therapeutic agents for NAFLD treatment, and the use of high-throughput expression analysis for this issue has been proven very valuable in this study.

Rubio et al.'s study (121)

Objective: Identify a gene pathway associated with NASH, comparing the gene expression profile of human liver samples and a well-established mouse model of NASH.

Two separate gene expression analysis were made for the human and mouse samples. Human liver samples from controls ($n=6$), steatosis ($n=6$), and NASH ($n=9$) were obtained from patients undergoing bariatric surgery or cholecystectomy in the case of controls. Mouse samples were obtained from livers of MAT1A −/− mice and WT controls of 15 days, and 1, 3, 5, and 8 months of age. HG-U133A Plus 2 (Affymetrix) was used for the human samples and MOE430A (Affymetrix) for the mouse samples. Parallel to gene expression, the same samples were also processed for routine histology and biochemical analysis in these patients and mice.

This study combines the use of a mouse NASH model (MAT1A-deficient mice, described above) and human samples to study the pathogenesis of the disease. In MAT1A-deficient mice, steatosis develops at three months of age, leading to NASH at eight months of age. One of the main handicaps of working with human samples is that when the first clinical changes are observed in patients, many events have already taken place related to the pathogenesis of the disease. Working with mouse models gives us the possibility of studying the molecular events happening before the onset of the disease. In this study, mouse samples were analyzed and some genes were found to be differentially expressed. Then the same type of analysis was performed with the human samples comparing just the NASH vs. the controls. Matching these two sets of genes revealed a set of common gene markers (218) of NASH to human and mouse. From this list, 81 genes that were differentially expressed at early stages of the disease in mice and whose expression remained differentially expressed through the disease were selected as early markers of the disease. Also, the gene expression pattern of the samples of simple steatosis lay between the expression pattern of controls and NASH, with some samples resembling more healthy controls and others resembling more NASH patients. Some of these genes were validated by qPCR. Further study of the promoters of these 81 genes revealed that many of these genes have binding sequences for the transcriptional factor Sp1 (and validated by Chromatin Immunoprecipitation Assay) and that it is phosphorylated in steatosis and steatohepatitis, suggesting it is potentially involved in the development of the disease.

Combining a mouse NASH model and human samples is a very interesting approach to identify early markers of the disease. The possibility of examining gene expression before any clinical feature is observed, and comparing it to the expression in human samples for the identification of markers or potential therapeutic targets of the disease, is very valuable. Although the steatosis and NASH samples were not matched to age, gender, or BMI to the control samples, the histological analysis and biochemical parameters assessed the suitability of the samples for the gene expression analysis. Only a few genes were selected for validation by RT-qPCR, but a subsequent study of binding the Sp1 transcription factor to some of the differentially expressed genes and its phosphorylation in disease strengthened the overall results.

This study has been able to use the advantages of working with a mouse model and applying it to human clinical samples. The relationship of Sp1 in the development of the disease is an important discovery and has to be studied further in detail.

3.3. Proteomics: Biomarkers in Serum

Proteomic technologies have the potential to help clarify the complex pathogenic mechanisms involved in the progression of NAFLD. Also, this type of approach allows us to distinguish between the molecular pathways that protect from those that may contribute to the progression of NAFLD (125–127). Differences in protein expression are the principal targets looking for biomarkers and drugs. The comparison of two-dimensional (2D) gel images is a well-known method to analyze changes in protein expression (128). The problems of reproducibility have been solved in the two-dimensional difference gel electrophoresis by using the difference gel electrophoresis (DIGE) technique, which enables different samples together with a pooled internal standard to be separated in the same gel (129). Two-dimensional electrophoresis has been used to study different NAFLD pathologies, from steatosis (130) to hepatocarcinoma (131). To date, there is just one well-documented report that describes the profiling of hepatic gene expression and serum protein content in patients with different subtypes of NAFLD (132). Liver biopsy specimens from 98 bariatric surgery patients were classified as NAFLD (91 patients). In those patients, 12 were steatosis alone, 52 were steatosis with nonspecific inflammation, 27 were NASH, and 7 patients without NAFLD served as obese controls. Each group of NAFLD patients was compared to the obese controls, and 22 genes with more than twofold differences in expression levels were revealed. Proteomic analyses were performed for the same samples and revealed 12 significantly different protein peaks. In conclusion, this genomic/proteomic analysis suggests differential expression of several genes and protein peaks in patients within and across the forms of NAFLD. This type of finding may help clarify the pathogenesis of NAFLD and identify potential targets for therapeutic intervention. In addition, future studies involving a large number of patients with sequential liver biopsies and serum specimens will be able to make important contributions in the progression of different subtypes of NAFLD.

3.4. Metabolomics

As mentioned earlier, NAFLD is an increasingly recognized cause of morbidity and mortality. Although the majority of patients do not develop complications, 28% may develop serious liver sequelae, including end-stage liver disease and hepatocellular carcinoma (133, 134). There are several negative points in using liver biopsy for this purpose (135). It is an invasive and costly procedure, and in some cases, patients suffer minor complications like pain, although there is a risk of death (0.01%) (136). Most importantly, the number of patients at risk for NAFLD is high enough that liver biopsy is not a practical and efficient tool for identifying those at risk of NASH and advanced fibrosis. Indeed, an estimated 15–20% of the western European population has steatosis (137), while more than half of Americans are overweight or obese. Because liver biopsy is impossible to perform in such large cohorts of individuals, some investigators have tried to identify simple noninvasive markers of liver injury in patients with NAFLD. A high-throughput analysis that compares controls to NAFLD has yet to be available. The approaches that have been reported combined several parameters in order to develop a predictive test, or the focus has been on specific targets selected by their functional biology. For example, Poynard et al. developed a NashTest (NT) using patented algorithms combining 13 parameters: age, sex, height, weight, and serum levels of triglycerides, cholesterol, alpha2macroglobulin, apolipoprotein A1, haptoglobin, gamma-glutamyltranspeptidase, transaminases ALT, AST, and total bilirubin (138). Among patients with suspected NAFLD, the new generation of biomarkers such as NT will allow better identification of those at risk and reassurance for patients without fibrosis or NASH. Biomarkers as a first-line estimate of injury in chronic liver diseases should reduce the need for liver biopsy.

Another kind of approach to be considered compromises specific molecules selected by their functional biological characteristics. Tarantino et al. focused their research on structural proteins, specifically keratin (K) 18, a component of Mallory bodies (MB) (139). Deregulated expression of K18 may thus be an important determinant of MB formation, which compromises the function of centrosomes and the microtubule network and leads to cell death. The tissue polypeptide-specific antigen (TPS), a serological mirror of K18, is widely used as a marker for various cancers. It can be abundantly released into the extracellular space during the intermediate stage of epithelial cell apoptosis (140). It is conceivable that the instability of hepatocytes could be reflected at a serum level by altered TPS concentrations. Hepatocyte apoptosis is significantly increased in patients with NASH (107). So in this study, Tarantino et al. were able to demonstrate that TPS is a better marker than alanine aminotransferaseactivity, ultrasonography, or a combination of both parameters in differentiating NASH from fatty liver. Finally, adiponectin had also been pointed out as a marker in the NASH diagnosis. Shimada et al. (141) analyzed 19 patients with simple steatosis and 66 patients with early-stage NASH (stages 1 and 2). Approximately 90% of patients with early-stage NASH can be predicted by a combined evaluation of the serum adiponectin level, HOMA-IR, and serum type IV collagen 7S level.

In summary, NASH is a disease that in some cases can evolve to cirrhosis and even to hepatocellular carcinoma. For that reason, it is essential to distinguish it from simple steatosis. To date, the only possible method to discriminate between NASH and simple steatosis is liver biopsy. A great effort is being made using high-throughput techniques in the field of genomics, proteomics, and metabolomics to identify potential biomarkers specific for NASH to be used in diagnosis. Large-scale genomic and proteomic studies performed in liver tissue provide clues of the molecular mechanisms of the pathology of the disease, and large-scale proteomics and metabolomics studies performed in the serum of patients will hopefully identify novel biomarkers for clinical application. Therefore, it is very important to improve existing techniques and data analysis tools to obtain maximal good-quality results in this field.

References

1. Thaler H. (1975) Relation of steatosis to cirrhosis. *Clin Gastroenterol* 4:273–280.
2. Catlin R. (1976) Liver dysfunction after intestinal bypass. *JAMA* 236:1693–1694.
3. Gross PA, Barrett TL, Dellinger EP, Krause PJ, Martone WJ, McGowan JE, Jr., Sweet RL, Wenzel RP. (1994) Purpose of quality standards for infectious diseases. Infectious Diseases Society of America. *Clin Infect Dis:* 18:421.
4. Zelman S, (1952) The liver in obesity. *Arch Intern Med* 90:141–156.
5. Peters RL, Gay T, Reynolds TB. (1975) Post-jejunoileal-bypass hepatic disease. Its similarity to alcoholic hepatic disease. *Am J Clin Pathol* 63:318–331.
6. Payne JH, Dewind LT, Commons RR. (1963) Metabolic observations in patients with jejunocolic shunts. *Am J Surg* 106: 273–289.
7. Christoffersen P, Petersen P. (1978) Morphological features in non-cirrhotic livers from patients with chronic alcoholism, diabetes mellitus or adipositas. A comparative study. *Acta Pathol Microbiol Scand [A]* 86A:495–498.
8. Gluud C, Christoffersen P, Andersen T, Morton JA, McGee JO. (1984) Occurrence and significance of Mallory bodies in morbidly obese patients. An immunohistochemical study. *Acta Pathol Microbiol Immunol Scand [A]* 92:39–43.
9. Brunt EM, Janney CG, Di Bisceglie AM, Neuschwander-Tetri BA, Bacon BR. (1999) Nonalcoholic steatohepatitis: a proposal for grading and staging the histological lesions. *Am J Gastroenterol.* 94:2467–2474.
10. Sonsuz A, Basaranoglu M, Ozbay G. (2000) Relationship between aminotransferase levels and histopathological findings in patients with nonalcoholic steatohepatitis. *Am J Gastroenterol* 95:1370–1371.
11. Bacon BR, Farahvash MJ, Janney CG, Neuschwander-Tetri BA. (1994) Nonalcoholic steatohepatitis: an expanded clinical entity. *Gastroenterology* 107:1103–1109.
12. Ludwig J, Viggiano TR, McGill DB, Oh BJ. (1980) Nonalcoholic steatohepatitis: Mayo clinic experiences with a hitherto unnamed disease. *Mayo Clin Proc* 55:434–438.
13. Teli MR, James OF, Burt AD, Bennett MK, Day CP. (1995) The natural history of

nonalcoholic fatty liver: a follow-up study. *Hepatology* 22:1714–1719.
14. Diehl AM, Goodman Z, Ishak KG. (1988) Alcohollike liver disease in nonalcoholics. A clinical and histologic comparison with alcohol-induced liver injury. *Gastroenterology* 95:1056–1062.
15. Tajiri K, Takenawa H, Yamaoka K, Yamane M, Marumo F, Sato C. (1997) Nonalcoholic steatohepatitis masquerading as autoimmune hepatitis. *J Clin Gastroenterol* 25:538–540.
16. da Silva PM, Eliseu T, Costa MM, Bastos H, Nobre FL. (1995) [Nonalcoholic steatohepatitis]. *Acta Med Port* 8:323–327.
17. Powell EE, Cooksley WG, Hanson R, Searle J, Halliday JW, Powell LW. (1990) The natural history of nonalcoholic steatohepatitis: a follow-up study of forty-two patients for up to 21 years. *Hepatology* 11:74–80.
18. Piekarski J, Goldberg HI, Royal SA, Axel L, Moss AA. (1980) Difference between liver and spleen CT numbers in the normal adult: its usefulness in predicting the presence of diffuse liver disease. *Radiology* 137:727–729.
19. Nomura F, Ohnishi K. Ochiai T, Okuda K. (1987) Obesity-related nonalcoholic fatty liver: CT features and follow-up studies after low-calorie diet. *Radiology* 162:845–847.
20. Outwater EK, Blasbalg R, Siegelman ES, Vala M. (1998) Detection of lipid in abdominal tissues with opposed-phase gradient-echo images at 1.5 T: techniques and diagnostic importance. *Radiographics* 18:1465–1480.
21. Fishbein MH, Gardner KG, Potter CJ, Schmalbrock P, Smith MA. (1997) Introduction of fast MR imaging in the assessment of hepatic steatosis. *Magn Reson Imaging* 15:287–293.
22. Mendler MH, Bouillet P, Le Sidaner A, Lavoine E, Labrousse F, Sautereau D, Pillegand B. (1998) Dual-energy CT in the diagnosis and quantification of fatty liver: limited clinical value in comparison to ultrasound scan and single-energy CT, with special reference to iron overload. *J Hepatol* 28:785–794.
23. Tazawa Y, Noguchi H, Nishinomiya F, Takada G. (1997) Serum alanine aminotransferase activity in obese children. *Acta Paediatr* 86:238–241.
24. Noguchi H, Tazawa Y, Nishinomiya F, Takada G. (1995) The relationship between serum transaminase activities and fatty liver in children with simple obesity. *Acta Paediatr Jpn* 37:621–625.
25. Baldridge AD, Perez-Atayde AR, Graeme-Cook F, Higgins L, Lavine JE. (1995) Idiopathic steatohepatitis in childhood: a multicenter retrospective study. *J Pediatr* 127:700–704.
26. Kinugasa A, Tsunamoto K, Furukawa N, Sawada T, Kusunoki T, Shimada N. (1984) Fatty liver and its fibrous changes found in simple obesity of children. *J Pediatr Gastroenterol Nutr* 3:408–414.
27. Kern WH, Heger AH, Payne JH, DeWind LT. (1973) Fatty metamorphosis of the liver in morbid obesity. *Arch Pathol* 96:342–346.
28. Adler M, Schaffner F. (1979) Fatty liver hepatitis and cirrhosis in obese patients. *Am J Med* 67:811–816.
29. Drenick EJ, Simmons F, Murphy JF. (1970) Effect on hepatic morphology of treatment of obesity by fasting, reducing diets and small-bowel bypass. *N Engl J Med* 282: 829–834.
30. Salmon PA, Reedyk L. (1975) Fatty metamorphosis in patients with jejunoileal bypass. *Surg Gynecol Obstet* 141:75–84.
31. Moore SJ, Auchterlonie IA, Cole GF, Gray ES, Dean JC. (1999) Partial lipodystrophy presenting with myopathy. *Dev Med Child Neurol* 41:127–131.
32. Rubbia-Brandt L, Quadri R, Abid K, Giostra E, Male PJ, Mentha G, Spahr L, Zarski JP, Borisch B, Hadengue A, Negro F. (2000) Hepatocyte steatosis is a cytopathic effect of hepatitis C virus genotype 3. *J Hepatol* 33:106–115.
33. Fernandez-Miranda C, Castellano G, Guijarro C, Fernandez I, Schoebel N, Larumbe S, Gomez-Izquierdo T, del Palacio A. (1998) Lipoprotein changes in patients with chronic hepatitis C treated with interferon-alpha. *Am J Gastroenterol* 93:1901–1904.
34. Marcos A, Fisher RA, Ham JM, Olzinski AT, Shiffman ML, Sanyal AJ, Luketic VA, Sterling RK, Olbrisch ME, Posner MP. (2000) Selection and outcome of living donors for adult to adult right lobe transplantation. *Transplantation* 69:2410–2415.
35. Contos MJ, Cales W, Sterling RK, Luketic VA, Shiffman ML, Mills AS, Fisher RA, Ham J, Sanyal AJ. (2001) Development of nonalcoholic fatty liver disease after orthotopic liver transplantation for cryptogenic cirrhosis. *Liver Transpl* 7:363–373.
36. Angulo P, Keach JC, Batts KP, Lindor KD. (1999) Independent predictors of liver fibrosis in patients with nonalcoholic steatohepatitis. *Hepatology* 30:1356–1362.
37. Luyckx FH, Desaive C, Thiry A, Dewe W, Scheen AJ, Gielen JE, Lefebvre PJ. (1998) Liver abnormalities in severely obese subjects: effect of drastic weight loss after gastroplasty. *Int J Obes Relat Metab Disord* 22: 222–226.
38. Harris RB, Kor H. (1992) Insulin insensitivity is rapidly reversed in rats by reducing

dietary fat from 40 to 30% of energy. *J Nutr* 122:1811–1822.
39. Hallfrisch J, Facn Behall KM. (2000) Mechanisms of the effects of grains on insulin and glucose responses. *J Am Coll Nutr* 19: 320S–325S.
40. Eriksson S, Eriksson KF, Bondesson L. (1986) Nonalcoholic steatohepatitis in obesity: a reversible condition. *Acta Med Scand* 220:83–88.
41. Sanyal AJ, Campbell-Sargent C, Mirshahi F, Rizzo WB, Contos MJ, Sterling RK, Luketic VA, Shiffman ML, Clore JN. (2001) Nonalcoholic steatohepatitis: association of insulin resistance and mitochondrial abnormalities. *Gastroenterology* 120:1183–1192.
42. Mingrone G, DeGaetano A, Greco AV, Capristo E, Benedetti G, Castagneto M, Gasbarrini G. (1997) Reversibility of insulin resistance in obese diabetic patients: role of plasma lipids. *Diabetologia* 40:599–605.
43. Lin HZ, Yang SQ, Chuckaree C, Kuhajda F, Ronnet G, Diehl AM. (2000) Metformin reverses fatty liver disease in obese, leptin-deficient mice. *Nat Med* 6:998–1003.
44. Caldwell SH, Hespenheide EE, Redick JA, Iezzoni JC, Battle EH, Sheppard BL. (2001) A pilot study of a thiazolidinedione, troglitazone, in nonalcoholic steatohepatitis. *Am J Gastroenterol* 96:519–525.
45. Ide T, Nakazawa T, Mochizuki T, Murakami K. (2000) Tissue-specific actions of antidiabetic thiazolidinediones on the reduced fatty acid oxidation in skeletal muscle and liver of Zucker diabetic fatty rats. *Metabolism* 49:521–525.
46. Laurin J. (2002) Motion – all patients with NASH need to have a liver biopsy: arguments against the motion. *Can J Gastroenterol* 16:722–726.
47. Lavine JE. (2000) Vitamin E treatment of nonalcoholic steatohepatitis in children: a pilot study. *J Pediatr* 136:734–738.
48. Abdelmalek MF, Angulo P, Jorgensen RA, Sylvestre PB, Lindor KD. (2001) Betaine, a promising new agent for patients with non-alcoholic steatohepatitis: results of a pilot study. *Am J Gastroenterol* 96:2711–2717.
49. Miglio F, Rovati LC, Santoro A, Setnikar I. (2000) Efficacy and safety of oral betaine glucuronate in non-alcoholic steatohepatitis. A double-blind, randomized, parallel-group, placebo-controlled prospective clinical study. *Arzneimittelforschung* 50:722–727.
50. Koteish A, Diehl AM. (2001) Animal models of steatosis. *Semin Liver Dis* 21:89–104.
51. Diehl AM. (2005) Lessons from animal models of NASH. *Hepatol Res* 33:138–144.
52. Kleiner DE, Brunt EM, Van Natta M, Behling C, Contos MJ, Cummings OW, Ferrell LD, Liu YC, Torbenson MS, Unalp-Arida A, Yeh M, McCullough AJ, Sanyal AJ. (2005) Design and validation of a histological scoring system for nonalcoholic fatty liver disease. *Hepatology* 41:1313–1321.
53. Anstee QM, Goldin RD. (2006) Mouse models in non-alcoholic fatty liver disease and steatohepatitis research. *Int J Exp Pathol* 87:1–16.
54. Campfield LA, Smith FJ, Burn P. (1996) The OB protein (leptin) pathway – a link between adipose tissue mass and central neural networks. *Horm Metab Res* 28: 619–632.
55. Pelleymounter MA, Cullen MJ, Baker MB, Hecht R, Winters D, Boone T, Collins F. (1995) Effects of the obese gene product on body weight regulation in ob/ob mice. *Science* 269:540–543.
56. Yang SQ, Lin HZ, Lane MD, Clemens M, Diehl AM. (1997) Obesity increases sensitivity to endotoxin liver injury: implications for the pathogenesis of steatohepatitis. *Proc Natl Acad Sci USA* 94:2557–2562.
57. Chavin KD, Yang S, Lin HZ, Chatham J, Chacko VP, Hoek JB, Walajtys-Rode E, Rashid A, Chen CH, Huang CC, Wu TC, Lane MD, Diehl AM. (1999) Obesity induces expression of uncoupling protein-2 in hepatocytes and promotes liver ATP depletion. *J Biol Chem* 274: 5692–5700.
58. Faggioni R, Fantuzzi G, Gabay C, Moser A, Dinarello CA, Feingold KR, Grunfeld C. (1999) Leptin deficiency enhances sensitivity to endotoxin-induced lethality. *Am J Physiol* 276:R136–R142.
59. Boss O, Muzzin P, Giacobino JP. (1998) The uncoupling proteins, a review. *Eur J Endocrinol* 139:1–9.
60. Enriquez A, Leclercq I, Farrell GC, Robertson G. (1999) Altered expression of hepatic CYP2E1 and CYP4A in obese, diabetic ob/ob mice, and fa/fa Zucker rats. *Biochem Biophys Res Commun* 255:300–306.
61. Hummel KP, Dickie MM, Coleman DL. (1966) Diabetes, a new mutation in the mouse. *Science* 153:1127–1128.
62. Tartaglia LA, Dembski M, Weng X, Deng N, Culpepper J, Devos R, Richards GJ, Campfield LA, Clark FT, Deeds J, Muir C, Sanker S, Moriarty A, Moore KJ, Smutko JS, Mays GG, Wool EA, Monroe CA, Tepper RI. (1995) Identification and expression cloning of a leptin receptor, OB-R. *Cell* 83: 1263–1271.
63. Best CH, Hershey JM, Huntsman ME. (1932) The effect of lecithine on fat deposition in the liver of the normal rat. *J Physiol* 75:56–66.

64. Newberne PM. (1986) Lipotropic factors and oncogenesis. *Adv Exp Med Biol* 206:223–251.
65. Shivapurkar N, Poirier LA. (1983) Tissue levels of S-adenosylmethionine and S-adenosylhomocysteine in rats fed methyl-deficient, amino acid-defined diets for one to five weeks. *Carcinogenesis* 4:1051–1057.
66. Cook RJ, Horne DW, Wagner C. (1989) Effect of dietary methyl group deficiency on one-carbon metabolism in rats. *J Nutr* 119:612–617.
67. Avila MA, Berasain C, Torres L, Martin-Duce A, Corrales FJ, Yang H, Prieto J, Lu SC, Caballeria J, Rodes J, Mato JM. (2000) Reduced mRNA abundance of the main enzymes involved in methionine metabolism in human liver cirrhosis and hepatocellular carcinoma. *J Hepatol* 33:907–914.
68. Mato JM, Corrales FJ, Lu SC, Avila MA. (2002) S-adenosylmethionine: a control switch that regulates liver function. *FASEB J* 16:15–26.
69. Martinez-Chantar ML, Corrales FJ, Martinez-Cruz LA, Garcia-Trevijano ER, Huang ZZ, Chen L, Kanel G, Avila MA, Mato JM, Lu SC. (2002) Spontaneous oxidative stress and liver tumors in mice lacking methionine adenosyltransferase 1A. *FASEB J* 16:1292–1294.
70. Lu SC, Alvarez L, Huang ZZ, Chen L, An W, Corrales FJ, Avila MA, Kanel G, Mato JM. (2001) Methionine adenosyltransferase 1A knockout mice are predisposed to liver injury and exhibit increased expression of genes involved in proliferation. *Proc Natl Acad Sci USA* 98:5560–5565.
71. Chen L, Zeng Y, Yang H, Lee TD, French SW, Corrales FJ, Garcia-Trevijano ER, Avila MA, Mato JM, Lu SC. (2004) Impaired liver regeneration in mice lacking methionine adenosyltransferase 1A. *FASEB J* 18: 914–916.
72. Maehama T, Dixon JE. (1998) The tumor suppressor, PTEN/MMAC1, dephosphorylates the lipid second messenger, phosphatidylinositol 3,4,5-trisphosphate. *J Biol Chem* 273:13375–13378.
73. Li J, Yen C, Liaw D, Podsypanina K, Bose S, Wang SI, Puc J, Miliaresis C, Rodgers L, McCombie R, Bigner SH, Giovanella BC, Ittmann M, Tycko B, Hibshoosh H, Wigler MH, Parsons R. (1997) PTEN, a putative protein tyrosine phosphatase gene mutated in human brain, breast, and prostate cancer. *Science* 275:1943–1947.
74. Horie Y, Suzuki A, Kataoka E, Sasaki T, Hamada K, Sasaki J, Mizuno K, Hasegawa G, Kishimoto H, Iizuka M, Naito M, Enomoto K, Watanabe S, Mak TW, Nakano T. (2004) Hepatocyte-specific Pten deficiency results in steatohepatitis and hepatocellular carcinomas. *J Clin Invest* 113:1774–1783.
75. Watanabe S, Horie Y, Kataoka E, Sato W, Dohmen T, Ohshima S, Goto T, Suzuki A. (2007) Non-alcoholic steatohepatitis and hepatocellular carcinoma: lessons from hepatocyte-specific phosphatase and tensin homolog (PTEN)-deficient mice. *J Gastroenterol Hepatol* 22 (Suppl 1): S96–S100.
76. Brunt EM. (1999) Nonalcoholic steatohepatitis (NASH): further expansion of this clinical entity? *Liver* 19:263–264.
77. Yokoyama C, Wang X, Briggs MR, Admon A, Wu J, Hua X, Goldstein JL, Brown MS. (1993) SREBP-1, a basic-helix-loop-helix-leucine zipper protein that controls transcription of the low density lipoprotein receptor gene. *Cell* 75:187–197.
78. Wang X, Sato R, Brown MS, Hua X, Goldstein JL. (1994) SREBP-1, a membrane-bound transcription factor released by sterol-regulated proteolysis. *Cell* 77:53–62.
79. Nakayama H, Otabe S, Ueno T, Hirota N, Yuan X, Fukutani T, Hashinaga T, Wada N, Yamada K. (2007) Transgenic mice expressing nuclear sterol regulatory element-binding protein 1c in adipose tissue exhibit liver histology similar to nonalcoholic steatohepatitis. *Metabolism* 56:470–475.
80. Chitturi S, Abeygunasekera S, Farrell GC, Holmes-Walker J, Hui JM, Fung C, Karim R, Lin R, Samarasinghe D, Liddle C, Weltman M, George J. (2002) NASH and insulin resistance: insulin hypersecretion and specific association with the insulin resistance syndrome. *Hepatology* 35:373–379.
81. Garcia-Monzon C, Martin-Perez E, Iacono OL, Fernandez-Bermejo M, Majano PL, Apolinario A, Larranaga E, Moreno-Otero R. (2000) Characterization of pathogenic and prognostic factors of nonalcoholic steatohepatitis associated with obesity. *J Hepatol* 33:716–724.
82. Dixon JB, Bhathal PS, O'Brien PE. (2001) Nonalcoholic fatty liver disease: predictors of nonalcoholic steatohepatitis and liver fibrosis in the severely obese. *Gastroenterology* 121:91–100.
83. Farrell GC, Larter CZ. (2006) Nonalcoholic fatty liver disease: from steatosis to cirrhosis. *Hepatology* 43:S99–S112.
84. Weltman MD, Farrell GC, Liddle C. (1996) Increased hepatocyte CYP2E1 expression in a rat nutritional model of hepatic steatosis with inflammation. *Gastroenterology* 111:1645–1653.

85. Weltman MD, Farrell GC, Hall P, Ingelman-Sundberg M, Liddle C. (1998) Hepatic cytochrome P450 2E1 is increased in patients with nonalcoholic steatohepatitis. *Hepatology* 27:128–133.
86. Gao D, Wei C, Chen L, Huang J, Yang S, Diehl AM. (2004) Oxidative DNA damage and DNA repair enzyme expression are inversely related in murine models of fatty liver disease. *Am J Physiol Gastrointest Liver Physiol* 287:G1070–1077.
87. Rinella ME, Green RM. (2004) The methionine-choline deficient dietary model of steatohepatitis does not exhibit insulin resistance. *J Hepatol* 40:47–51.
88. Harrold JA, Widdowson PS, Clapham JC, Williams G. (2000) Individual severity of dietary obesity in unselected Wistar rats: relationship with hyperphagia. *Am J Physiol Endocrinol Metab* 279:E340–E347.
89. El-Haschimi K, Pierroz DD, Hileman SM, Bjorbaek C, Flier JS. (2000) Two defects contribute to hypothalamic leptin resistance in mice with diet-induced obesity. *J Clin Invest* 105:1827–1832.
90. Toye AA, Lippiat JD, Proks P, Shimomura K, Bentley L, Hugill A, Mijat V, Goldsworthy M, Moir L, Haynes A, Quarterman J, Freeman HC, Ashcroft FM, Cox RD. (2005) A genetic and physiological study of impaired glucose homeostasis control in C57BL/6 J mice. *Diabetologia* 48:675–686.
91. Biddinger SB, Almind K, Miyazaki M, Kokkotou E, Ntambi JM, Kahn CR. (2005) Effects of diet and genetic background on sterol regulatory element-binding protein-1c, stearoyl-CoA desaturase 1, and the development of the metabolic syndrome. *Diabetes* 54:1314–1323.
92. Angulo P. (2002) Nonalcoholic fatty liver disease. *N Engl J Med* 346:1221–1231.
93. Di Bisceglie AM, Lyra AC, Schwartz M, Reddy RK, Martin P, Gores G, Lok AS, Hussain KB, Gish R, Van Thiel DH, Younossi Z, Tong M, Hassanein T, Balart L, Fleckenstein J, Flamm S, Blei A, Befeler AS. (2003) Hepatitis C-related hepatocellular carcinoma in the United States: influence of ethnic status. *Am J Gastroenterol* 98:2060–2063.
94. Adams LA, Angulo P. (2005) Recent concepts in non-alcoholic fatty liver disease. *Diabet Med* 22:1129–1133.
95. Ekstedt M, Franzen LE, Mathiesen UL, Thorelius L, Holmqvist M, Bodemar G, Kechagias S. (2006) Long-term follow-up of patients with NAFLD and elevated liver enzymes. *Hepatology* 44:865–873.
96. Neuschwander-Tetri BA, Caldwell SH. (2003) Nonalcoholic steatohepatitis: summary of an AASLD Single Topic Conference. *Hepatology* 37:1202–1219.
97. Wieckowska A, McCullough AJ, Feldstein AE. (2007) Noninvasive diagnosis and monitoring of nonalcoholic steatohepatitis: present and future. *Hepatology* 46: 582–589.
98. Chalasani N, Deeg MA, Crabb DW. (2004) Systemic levels of lipid peroxidation and its metabolic and dietary correlates in patients with nonalcoholic steatohepatitis. *Am J Gastroenterol* 99:1497–1502.
99. Horoz M, Bolukbas C, Bolukbas FF, Sabuncu T, Aslan M, Sarifakiogullari S, Gunaydin N, Erel O. (2005) Measurement of the total antioxidant response using a novel automated method in subjects with nonalcoholic steatohepatitis. *BMC Gastroenterol* 5:35.
100. Bonnefont-Rousselot D, Ratziu V, Giral P, Charlotte F, Beucler I, Poynard T. (2006) Blood oxidative stress markers are unreliable markers of hepatic steatosis. *Aliment Pharmacol Ther* 23:91–98.
101. Solga SF, Alkhuraishe A, Cope K, Tabesh A, Clark JM, Torbenson M, Schwartz P, Magnuson T, Diehl AM, Risby TH. (2006) Breath biomarkers and non-alcoholic fatty liver disease: preliminary observations. *Biomarkers* 11:174–183.
102. Hui JM, Hodge A, Farrell GC, Kench JG, Kriketos A, George J. (2004) Beyond insulin resistance in NASH: TNF-alpha or adiponectin? *Hepatology* 40:46–54.
103. Musso G, Gambino R, Durazzo M, Biroli G, Carello M, Faga E, Pacini G, De Michieli F, Rabbione L, Premoli A, Cassader M, Pagano G. (2005) Adipokines in NASH: postprandial lipid metabolism as a link between adiponectin and liver disease. *Hepatology* 42:1175–1183.
104. Abiru S, Migita K, Maeda Y, Daikoku M, Ito M, Ohata K, Nagaoka S, Matsumoto T, Takii Y, Kusumoto K, Nakamura M, Komori A, Yano K, Yatsuhashi H, Eguchi K, Ishibashi H. (2006) Serum cytokine and soluble cytokine receptor levels in patients with non-alcoholic steatohepatitis. *Liver Int* 26:39–45.
105. Kugelmas M, Hill DB, Vivian B, Marsano L, McClain CJ. (2003) Cytokines and NASH: a pilot study of the effects of lifestyle modification and vitamin E. *Hepatology* 38:413–419.
106. Haukeland JW, Damas JK, Konopski Z, Loberg EM, Haaland T, Goverud I, Torjesen PA, Birkeland K, Bjoro K, Aukrust P. (2006) Systemic inflammation in nonalcoholic fatty liver disease is characterized by elevated levels of CCL2. *J Hepatol* 44:1167–1174.

107. Feldstein AE, Canbay A, Angulo P, Taniai M, Burgart LJ, Lindor KD, Gores GJ. (2003) Hepatocyte apoptosis and fas expression are prominent features of human nonalcoholic steatohepatitis. *Gastroenterology* 125: 437–443.
108. Wieckowska A, Zein NN, Yerian LM, Lopez AR, McCullough AJ, Feldstein AE. (2006) In vivo assessment of liver cell apoptosis as a novel biomarker of disease severity in nonalcoholic fatty liver disease. *Hepatology* 44: 27–33.
109. Rockey DC, Bissell DM. (2006) Noninvasive measures of liver fibrosis. *Hepatology* 43:S113–S120.
110. Ratziu V, Giral P, Charlotte F, Bruckert E, Thibault V, Theodorou I, Khalil L, Turpin G, Opolon P, Poynard T. (2000) Liver fibrosis in overweight patients. *Gastroenterology* 118:1117–1123.
111. Ratziu V, Massard J, Charlotte F, Messous D, Imbert-Bismut F, Bonyhay L, Tahiri M, Munteanu M, Thabut D, Cadranel JF, Le Bail B, de Ledinghen V, Poynard T. (2006) Diagnostic value of biochemical markers (FibroTest-FibroSURE) for the prediction of liver fibrosis in patients with non-alcoholic fatty liver disease. *BMC Gastroenterol* 6:6.
112. Angulo P, Hui JM, Marchesini G, Bugianesi E, George J, Farrell GC, Enders F, Saksena S, Burt AD, Bida JP, Lindor K, Sanderson SO, Lenzi M, Adams LA, Kench J, Therneau TM, Day CP. (2007) The NAFLD fibrosis score: a noninvasive system that identifies liver fibrosis in patients with NAFLD. *Hepatology* 45:846–854.
113. Suzuki A, Angulo P, Lymp J, Li D, Satomura S, Lindor K. (2005) Hyaluronic acid, an accurate serum marker for severe hepatic fibrosis in patients with non-alcoholic fatty liver disease. *Liver Int* 25:779–786.
114. Lydatakis H, Hager IP, Kostadelou E, Mpousmpoulas S, Pappas S, Diamantis I. (2006) Non-invasive markers to predict the liver fibrosis in non-alcoholic fatty liver disease. *Liver Int* 26:864–871.
115. Rosenberg WM, Voelker M, Thiel R, Becka M, Burt A, Schuppan D, Hubscher S, Roskams T, Pinzani M, Arthur MJ. (2004) Serum markers detect the presence of liver fibrosis: a cohort study. *Gastroenterology* 127:1704–1713.
116. Willner IR, Waters B, Patil SR, Reuben A, Morelli J, Riely CA. (2001) Ninety patients with nonalcoholic steatohepatitis: insulin resistance, familial tendency, and severity of disease. *Am J Gastroenterol* 96:2957–2961.
117. Struben VM, Hespenheide EE, Caldwell SH. (2000) Nonalcoholic steatohepatitis and cryptogenic cirrhosis within kindreds. *Am J Med* 108:9–13.
118. Osterreicher CH, Brenner DA. (2007) The genetics of nonalcoholic fatty liver disease. *Ann Hepatol* 6:83–88.
119. Younossi ZM, Gorreta F, Ong JP, Schlauch K, Giacco LD, Elariny H, Van Meter A, Younoszai A, Goodman Z, Baranova A, Christensen A, Grant G, Chandhoke V. (2005) Hepatic gene expression in patients with obesity-related non-alcoholic steatohepatitis. *Liver Int* 25:760–771.
120. de Oliveira CP, Simplicio FI, de Lima VM, Yuahasi K, Lopasso FP, Alves VA, Abdalla DS, Carrilho FJ, Laurindo FR, de Oliveira MG. (2006) Oral administration of S-nitroso-N-acetylcysteine prevents the onset of non alcoholic fatty liver disease in rats. *World J Gastroenterol* 12: 1905–1911.
121. Rubio A, Guruceaga E, Vazquez-Chantada M, Sandoval J, Martinez-Cruz LA, Segura V, Sevilla JL, Podhorski A, Corrales FJ, Torres L, Rodriguez M, Aillet F, Ariz U, Arrieta FM, Caballeria J, Martin-Duce A, Lu SC, Martinez-Chantar ML, Mato JM. (2007) Identification of a gene-pathway associated with non-alcoholic steatohepatitis. *J Hepatol* 46:708–718.
122. Gramlich T, Kleiner DE, McCullough AJ, Matteoni CA, Boparai N, Younossi ZM. (2004) Pathologic features associated with fibrosis in nonalcoholic fatty liver disease. *Hum Pathol* 35:196–199.
123. Day CP, James OF. (1998) Steatohepatitis: a tale of two "hits"? *Gastroenterology* 114: 842–845.
124. de Oliveira CP, Stefano JT, de Lima VM, de Sa SV, Simplicio FI, de Mello ES, Correa-Giannella ML, Alves VA, Laurindo FR, de Oliveira MG, Giannella-Neto D, Carrilho FJ. (2006) Hepatic gene expression profile associated with non-alcoholic steatohepatitis protection by S-nitroso-N-acetylcysteine in ob/ob mice. *J Hepatol* 45: 725–733.
125. Honda M, Kaneko S, Kawai H, Shirota Y, Kobayashi K. (2001) Differential gene expression between chronic hepatitis B and C hepatic lesion. *Gastroenterology* 120: 955–966.
126. Paradis V, Degos F, Dargere D, Pham N, Belghiti J, Degott C, Janeau JL, Bezeaud A, Delforge D, Cubizolles M, Laurendeau I, Bedossa P. (2005) Identification of a new marker of hepatocellular carcinoma by serum protein profiling of patients with chronic liver diseases. *Hepatology* 41: 40–47.

127. Collins FS, Guttmacher AE. (2001) Genetics moves into the medical mainstream. *JAMA* 286:2322–2324.
128. Rabilloud T. (2002) Two-dimensional gel electrophoresis in proteomics: old, old fashioned, but it still climbs up the mountains. *Proteomics* 2:3–10.
129. Alban A, David SO, Bjorkesten L, Andersson C, Sloge E, Lewis S, Currie I. (2003) A novel experimental design for comparative two-dimensional gel analysis: two-dimensional difference gel electrophoresis incorporating a pooled internal standard. *Proteomics* 3:36–44.
130. Douette P, Navet R, Gerkens P, de Pauw E, Leprince P, Sluse-Goffart C, Sluse FE. (2005) Steatosis-induced proteomic changes in liver mitochondria evidenced by two-dimensional differential in-gel electrophoresis. *J Proteome Res* 4:2024–2031.
131. Zeindl-Eberhart E, Haraida S, Liebmann S, Jungblut PR, Lamer S, Mayer D, Jager G, Chung S, Rabes HM. (2004) Detection and identification of tumor-associated protein variants in human hepatocellular carcinomas. *Hepatology* 39:540–549.
132. Younossi ZM, Baranova A, Ziegler K, Del Giacco L, Schlauch K, Born TL, Elariny H, Gorreta F, VanMeter A, Younoszai A, Ong JP, Goodman Z, Chandhoke V. (2005) A genomic and proteomic study of the spectrum of nonalcoholic fatty liver disease. *Hepatology* 42:665–674.
133. Sanyal AJ. (2002) AGA technical review on nonalcoholic fatty liver disease. *Gastroenterology* 123:1705–1725.
134. Charlton M. (2004) Nonalcoholic fatty liver disease: a review of current understanding and future impact. *Clin Gastroenterol Hepatol* 2:1048–1058.
135. Laurin J, Lindor KD, Crippin JS, Gossard A, Gores GJ, Ludwig J, Rakela J, McGill DB. (1996) Ursodeoxycholic acid or clofibrate in the treatment of non-alcohol-induced steatohepatitis: a pilot study. *Hepatology* 23: 1464–1467.
136. Poynard T, Ratziu V, Bedossa P. (2000) Appropriateness of liver biopsy. *Can J Gastroenterol* 14:543–548.
137. Bellentani S, Bedogni G, Miglioli L, Tiribelli C. (2004) The epidemiology of fatty liver. *Eur J Gastroenterol Hepatol* 16: 1087–1093.
138. Poynard T, Ratziu V, Charlotte F, Messous D, Munteanu M, Imbert-Bismut F, Massard J, Bonyhay L, Tahiri M, Thabut D, Cadranel JF, Le Bail B, de Ledinghen V. (2006) Diagnostic value of biochemical markers (NashTest) for the prediction of non alcohol steato hepatitis in patients with non-alcoholic fatty liver disease. *BMC Gastroenterol* 6:34.
139. Tarantino G, Conca P, Coppola A, Vecchione R, Di Minno G. (2007) Serum concentrations of the tissue polypeptide specific antigen in patients suffering from non-alcoholic steatohepatitis. *Eur J Clin Invest* 37:48–53.
140. Sheard MA, Vojtesek B, Simickova M, Valik D. (2002) Release of cytokeratin-18 and -19 fragments (TPS and CYFRA 21-1) into the extracellular space during apoptosis. *J Cell Biochem* 85:670–677.
141. Shimada M, Kawahara H, Ozaki K, Fukura M, Yano H, Tsuchishima M, Tsutsumi M, Takase S. (2007) Usefulness of a combined evaluation of the serum adiponectin level, HOMA-IR, and serum type IV collagen 7S level to predict the early stage of nonalcoholic steatohepatitis. *Am J Gastroenterol* 102:1931–1938.

Chapter 7

Biomarkers in Breast Cancer

María dM. Vivanco

Abstract

Breast cancer is one of the leading causes of death in women worldwide. During the last decade, great developments in our understanding of breast cancer at the molecular level have arisen from microarray data. Molecular profiling has supported the notion that breast cancer is not a simple disease with a single tumorigenic pathway but a rather heterogeneous one. Gene expression studies have identified and validated the existence of different breast cancer subtypes whose signatures correlate with the clinical outcome. Therefore, the identification of gene expression patterns has become a key issue in understanding the biological diversity of breast tumors, leading to new hope for diagnosis, prognosis, and future treatment. This chapter is a selection of some of the key results that have contributed to the advance toward this end.

Key words: Breast cancer, biomarkers, mammary gland, luminal cells, myoepithelial cells, stem cells.

1. Historical Introduction

Breast cancer has been affecting the lives of women for centuries, and it is the leading cause of cancer-related death in women. However, somehow the economical development and life choices of women in Western countries have increased its incidence well above the levels found in Asia, Africa, and Central and South America.

The history of breast cancer goes back to the Egyptians, who had described it earlier than 1,500 years B.C. in the Papyrus of Edward Smith that is conserved in the British Museum in London. For centuries breast cancer remained an untreatable disease. In the 17th century, the Italian doctor Ramazzini observed the

R. Matthiesen (ed.), *Bioinformatics Methods in Clinical Research*, Methods in Molecular Biology 593,
DOI 10.1007/978-1-60327-194-3_7, © Humana Press, a part of Springer Science+Business Media, LLC 2010

high incidence of breast cancer among convent nuns. Interestingly, there is a painting dating from the 13th century in which a surgeon is examining a nun, suggesting that breast cancer was recognized as a common disease among nuns 400 years earlier. The choice of celibacy was implicated in a higher risk of developing breast cancer. However, these revealing observations remained without consequence for a long time.

In the 19th century, the developments in surgical procedures led to the establishment of the radical mastectomy that was used for almost a century, until the 1970s, when conservative procedures started being explored together with the use of radiotherapy. In 1896, Beatson described the healing of locally recurrent cancer of the breast following ovariectomy. A rational explanation for this observation awaited the discovery of estrogens and the demonstration by Pearson and colleagues that the administration of estrogen could reverse the beneficial effect of ovariectomy in women with disease metastatic to bone, the progress of which was monitored by estimates of urinary calcium output (1). There is now no doubt that estrogenic hormones of ovarian origin promote the growth of human breast cancer.

This confirmation prompted the search for hormones that would antagonize the effects of estrogen, which led to the discovery of tamoxifen in 1966 (2). Tamoxifen was quickly validated, and it was also shown that the mechanism of action of tamoxifen is to block the activity of the estrogen receptor (ER) in tumor cells, therefore blocking their growth. In 1973, tamoxifen was approved in the UK for breast cancer treatment, followed in 1977 by the FDA. Interestingly, in 1998, the use of tamoxifen was further approved for the prevention of breast cancer in high-risk women, following the publication of the Breast Cancer Prevention Trial (BCPT) conducted by the National Surgical Adjuvant Breast and Bowel Project (NSABP) (3). The finding of a decrease in contralateral breast cancer incidence following tamoxifen administration for adjuvant therapy had led to the concept that the drug may play a role in breast cancer prevention.

Another critical step forward in our understanding of estrogen action was achieved from research in the 1960s and 1970s identifying high-affinity binding sites for estrogen and leading to the detection of this activity in excised human breast cancer (4). The great advances in molecular biology allowed the cloning of the receptor for estrogen by the group of Chambon (5). As a result, many studies were published that provided further insight into its molecular characterization and mode of action. It was only a decade later that a second receptor for estrogen was identified (6), referred to as ERbeta; as a consequence, the original receptor is known as ERalpha, or simply ER. The presence of two different ERs likely accounts for the complexity of estrogen and antiestrogen action, some of which is still not fully understood.

These molecular and cellular studies gave rise to the identification of ER, PR, p53, and HER-2 as valuable markers essential for the histopathological classification of tumors that facilitate the diagnosis and treatment of breast cancer and that are now routinely used in the clinic. In addition to the expression of these markers, treatment for individual patients is decided based on various criteria, such as patient's age, tumor size, the extent of tumor spread or staging (status of axillary lymph nodes), histological type of the tumor, and pathological grade. Although guidelines based on histopathological data are clearly established and standardized for current use in breast cancer management in Europe (7) and the United States (8), it has become apparent that patients with similar clinical and pathological features develop distinctly and show different response to therapy. With the growing collection of systemic therapy agents available, there is a perception that the existing prognostic factors are not sufficient to reflect the whole clinical and molecular heterogeneity of the disease. Although a multidisciplinary approach is common practice in cancer management, there is a clear need for additional prognostic factors to improve breast disease classification and prognosis.

2. Class Discovery in Breast Cancer (Classification by Subgroups)

During the last few years, the development of the technological advances that allowed the use of microarrays to study breast cancer cell lines and tumor tissues has opened the possibilities for identifying new biomarkers. The expectation in the short term was a more refined classification that would allow a better diagnosis of the disease, with the implications, in a longer term, to be able to provide a more personalized treatment that spares the aggressive therapies for women who would not obtain any benefit from them. The use of cDNA microarray technology to identify physiologically relevant gene expression patterns in simple biological samples has been widely used in recent years. However, the study of gene expression in primary breast tumors, as in most solid tumors, is more complicated for two reasons: First, breast tumors are heterogeneous; second, the breast carcinoma cells are diverse, both morphologically and genetically.

Despite these problems, the group from Stanford (9) proved the usefulness of this technology to study variations in gene expression in human cancers using human mammary epithelial cells growing in culture and primary breast tumors. Shortly afterwards, the same group published a classification of breast tumors into five molecular classes based upon their gene expression profiles and their similarity to normal cell counterparts (10). There are two main branches or clusters, reflecting the expected separation into ER+ and ER– disease. The ER+ group is characterized by a higher expression of a set of genes typically expressed by breast luminal epithelial cells (luminal cancer, representing the majority of tumor cases). The ER– branch comprises three subgroups: one overexpressing ERBB2 (HER2), one expressing genes characteristic of breast basal/myoepithelial cells (basal-like cancer, which may account for 3–15% of all breast tumors), and another with a profile resembling the normal breast tissue, which consistently clustered together with normal breast samples and fibroadenomas.

Ensuing studies confirmed this classification and added some groups to the luminal tumors, and the normal breast-like cancers appeared to be indistinguishable from the ER– cluster (11). Interestingly, this report showed significantly different outcomes for the patients belonging to the different groups, including a poor prognosis for the basal-like subtype and the finding that ER+ tumors may be subclassified into distinct subgroups with different outcomes. Therefore, relating gene expression patterns to clinical outcome was not only possible but has become a key issue in understanding the biological diversity of breast tumors.

This emerging classification of tumors was not based on single genes or a specific pathway; in fact, no single gene can identify these classes reliably. Instead, several genes are needed to define each class. This observation represented an interesting turn in the way cancer has been viewed and studied previously. Although they used cell lines rather than primary tissue, one group analyzed gene and protein expression profiling of 31 breast cell lines to identify as many as 1,233 genes that are differentially expressed between basal and luminal samples (12). This basal/luminal signature correctly reclassified the published series of tumor samples that originally served to identify the molecular subtypes, suggesting that the identified markers could be useful for tumor classification.

3. Prognosis Prediction

Another big step forward was taken with the work of a group from the Netherlands Cancer Institute with the publication of two relevant papers in 2002 (13, 14). These studies continued to focus on a very important problem, the fact that breast cancer patients with the same apparent stage of disease can have markedly different therapy responses and overall outcome. Furthermore, they showed that the outcome of gene expression profiling of breast tumors could be used to predict which patients will develop clinical metastases (the spread of the tumor to other sites in the body). In women with local disease, the treatment consists of removal of the tumor followed by radiotherapy and perhaps the antiestrogen tamoxifen if the tumor has been shown to be ER+. Unfortunately, some of these patients will later develop metastases. Chemotherapy as adjuvant therapy (after surgery) implies the use of cytotoxic drugs to reach cancer cells that might have spread to other parts of the body through the bloodstream. However, the secondary effects of these treatments can be rather toxic since they target all dividing cells. Offering these really unpleasant (and expensive) treatments to women who do not need them or will not benefit from them could be avoided if the truly responsive population could be properly identified. Using oligonucleotide microarrays on lymph-node-negative samples, the researchers applied supervised classification to identify a gene expression signature that strongly predicts metastasis and disease outcome. This suggested a strategy to select patients who would benefit from adjuvant therapy and significantly reduce the number of patients who receive redundant treatment (13).

The identification of these 70 marker genes ("poor prognosis" signature, predicts the appearance of clinical distant metastases within five years of surgery) suggested that molecular prediction of the outcome of cancer is possible and provided a significant advantage over existing prognostic methods. However, the sample was relatively small (97 sporadic cancers) and the results corresponded to two groups selected on the basis of outcome: the presence or absence of metastasis within five years. To provide a more accurate estimate of the risks of metastases associated with the expression signature defined, the same group studied the expression of 25,000 genes in a cohort of 295 young patients (less than 55 years of age) with breast cancer, including 151 patients with lymph-node-negative disease and 144 with lymph-node-positive disease (14). They evaluated the predictive power of the prognosis profile using univariate and multivariate statistical analyses and found that the profile performed best as a predictor of the appearance of distant metastases during the first five years after treatment, and also of the development of distant metastases in patients with lymph-node-positive disease.

Interestingly, a comparison of the probability that patients would remain free of distant metastasis among 151 patients with lymph-node-negative breast cancer with the use of the good/poor prognosis signature, the St. Gallen criteria or the NIH consensus showed that the prognosis profile was better at classifying the patients. For example, more patients with lymph-node-negative disease were assigned to the good prognosis signature, and these had a higher likelihood of metastasis survival than those classified according to the traditional criteria. Thus, both the St. Gallen and the NIH subgroups contained misclassified patients that would be either overtreated or undertreated in current clinical practice, while the prognosis signature was better at classifying the patients into high-risk and low-risk subgroups. This was a considerable finding with important implications for future cancer management. Furthermore, it was especially noteworthy that the prognosis profile was significantly associated with the histological grade of the tumor, the age, and the ER status; 97% of the tumors in the good prognosis category were positive for ER, thus confirming the key role of ER in predicting the outcome of breast cancer. In contrast, no association was found with the diameter of the tumor, the extent of vascular invasion, the number of positive lymph nodes, or treatment (14). Therefore, this work represents an excellent example of how gene expression profiling can provide very useful prognostic information, since it identifies molecular profiles that are linked to the response to treatment and thus raises the possibility that these can be used in the clinic to help select those patients who would really benefit from a particular type of therapy.

4. Class Prediction

Microarray technology has opened up many opportunities in breast cancer research with the discovery of classes of tumors based on their gene expression profiles. In addition, this approach has also been used to compare predefined classes or groups of tumors and to identify differentially expressed genes, and also for class prediction, which includes derivation of predictors of prognosis, response to therapy, or any other characteristic defined independently of the gene expression profile (15). This type of study has not only offered further detailed molecular characterization of tumors that were known to be phenotypically different, but has also provided the potential to identify new diagnostic and therapeutic targets. For example, a comparison between in situ and invasive disease, often combined with laser capture microdissection, showed that gene expression alterations conferring the potential for invasive growth are already present at the preinvasive stages and that tumors of similar histological grade cluster together (16–19). Furthermore, using class prediction algorithms and multivariate analysis, grade 2 tumors were clearly separated from grade 1 and grade 3 tumors, and their clinical heterogeneity further resolved into subtypes leading to a genetic grade signature that improved the detection of patients with less harmful tumors (20). All these findings support the view that low- and high-grade tumors are independent entities and follow distinct genetic pathways (21, 22). Similarly, the comparison of different histological types of breast cancer has confirmed the previously known altered expression of E-cadherin between ductal and lobular carcinoma (the most common histological types of breast cancer) and has identified new ones (23–25). In addition, microarray analysis has also been used to distinguish tumors of lymph-node-positive and -negative status (26–28).

The majority of breast tumors are sporadic; however, approximately 5–10% of breast cancers are related to an inherited gene mutation. Of these cases, 84% of hereditary breast cancer is caused by mutations in the *BRCA1* or *BRCA2* genes (29). These tumors are of great phenotypic and genotypic heterogeneity, given the association of hereditary breast cancer with a plethora of differing cancer syndromes. Increased knowledge about the genetics of breast cancer may contribute to the identification of high-risk patients who may benefit from early diagnosis. A signature was established that identifies tumors of *BRCA1* carriers (13). The histopathological changes in these cancers are often characteristic of the mutant gene; therefore, it was hypothesized that the genes expressed by these two types of tumors may also be distinctive, perhaps allowing the identification of hereditary breast cancer cases on the basis of gene expression profiles.

Permutation analysis of multivariate classification functions established that the gene expression profiles of tumors with BRCA1 mutations, tumors with BRCA2 mutations, and sporadic tumors differed significantly from one another. An analysis of variance between the levels of gene expression and the genotype of the samples identified 176 genes that were differentially expressed in tumors with BRCA1 mutations and tumors with BRCA2 mutations. These results suggest that a heritable mutation influences the gene expression profile of the cancer (30), although the *BRCA1* signature did not coincide with the one reported by Vant't Veer and colleagues, likely due to the different set of genes studied and the different type of analysis employed by both groups.

Furthermore, it has become clear that there must be additional breast cancer predisposition genes, although their identification has so far been unsuccessful, presumably because of genetic heterogeneity, low penetrance, or recessive/polygenic mechanisms. These non-BRCA1/2 breast cancer families comprise a histopathologically heterogeneous group that could be distinguished from BRCA1/2 mutation carriers by their global gene expression profile (31). These results suggest that the combination of large-scale gene expression profiling with conventional positional linkage/candidate gene analysis may be a more effective approach to identify novel breast cancer predisposition genes.

Since the separation by Perou and colleagues of breast tumors into various subtypes (namely luminal A/ER+, luminal B/ER+, normal breast-like, ERBB2+ and basal-like) (10), other studies have reported distinct molecular profiles in different patient populations. However, although the basal and the ERBB2 subtypes are repeatedly recognized, the identification of ER+ subtypes has been inconsistent. For example, the subclassification of luminal tumors has ranged from one to up to three groups; normal breastlike cancers seemed to be indistinguishable from the ER– cluster (10, 11), although in some studies these five molecular subtypes continued to be apparent (32). Furthermore, using a gene expression grade index (GCI), which defines the histological grade based on gene expression profiles (33), two ER+ molecular subgroups (high and low genomic grade) were defined and found to be associated with statistically distinct clinical outcome in both systemically untreated and tamoxifen-treated populations (34).

More refined molecular portraits were identified that distinguish ER+ and ER– tumors and demonstrate that these two subtypes display remarkably different gene expression phenotypes involving multiple critical events that were not only explained by differences in estrogen responsiveness (35, 13, 36). In addition, Wang and colleagues studied 286 cases of lymph-node-negative breast cancers from patients who had not received adjuvant therapy and identified a 76-gene signature consisting of 60 genes for ER+ and 16 genes for ER– patients. This signature, in comparison with the St. Gallen's and NIH consensus guidelines, was better at predicting those patients who should not receive adjuvant therapy, which should help clinicians avoid adjuvant systemic therapy or choose less aggressive therapeutic options (37).

Interestingly, this 76-gene expression-based prognostic signature (37) and the 70-gene signature described by van't Veer and colleagues (13) have now been further validated in large multicenter studies that confirm their prognostic improvement over the traditional guidelines for selecting patients for adjuvant systemic therapy (38–40). In fact, the 70-gene signature has been marketed for clinical use as a commercial assay. Despite the encouraging results of these microarray analyses and new hopes for improved prognosis, many researchers remain skeptical. Critical points are the potential bias in sample selection, in statistical analyses, and even in the analysis of data based on assumptions of outcome (41).

Another puzzling concern is the fact that the two major studies described above were designed to ask basically the same scientific question, identifying a gene signature to improve prognosis, but produced very limited overlap (only three genes) and did not even overlap in the signaling pathways implicated (37). Consequently, the Breast Cancer International Group has launched a European trial called MINDACT (Microarray in Node-negative Disease may Avoid ChemoTherapy) that aims to recruit 6,000 patients to assess whether an improved prognostic set will be revealed by analysis of a large cohort of tumor samples. Similarly, the U.S. Food and Drug Administration has recruited 137 participants from 51 academic and industrial institutions to address the reproducibility of the microarray measurements and their conclusions (42). No doubt this type of collaborative clinical trial will contribute to validate novel predictor assays before their widespread use in the clinic. Perhaps a balanced route would be the integration of this new technology within the traditional methods of histopathological assessment to complement current breast cancer management strategies and achieve improved treatment strategies for individual patients.

5. Redefining Gene Expression Profiles

With time, other prognostic signatures for breast cancer have appeared (43–47). Initial studies had focused on the genetic profiles of different types of tumors (classified by their histopathological properties) and their association with a particular survival rate (11) or in the identification of a poor-prognosis signature that could help to classify breast cancers according to the clinical outcome (14, 37). A recurrence signature has been defined based on the distant recurrence in patients that had node-negative breast cancer and who were treated with tamoxifen (43). More recent studies included strategies that take into account new concepts of cancer biology to predict clinical outcomes. For example, the observation that aggressive tumors may prosper independently of hypoxic conditions led to the identification of a hypoxia-response signature that is associated with poor prognosis in breast cancer (47, 48).

Similarly, advances in the knowledge of the relevance of fibroblasts in tumorigenesis led to the definition of the 512-gene wound-response signature that correlates with overall survival and is associated with poor prognosis. The analysis of the gene expression profiling of serum-stimulated fibroblasts in culture was based on the idea that it may be related to the transcriptional response elicited by fibroblasts in wounded tissues, such as the tissue wounded by an invasive tumor (44). Moreover, the suggestion that the tumor microenvironment can influence certain disease characteristics has guided the analysis of gene expression profiles of tumor stroma (49). Another study employed gene microarray analysis to compare expression levels in 53 patients with invasive breast cancer who were followed up for around four years, using lymph node status, tumor metastasis, responsiveness to therapy, and overall survival to define outcome. The

26-gene "stroma-derived prognostic predictor" stratifies disease outcome independently of the traditional prognostic factors and previously published expression-based signatures (50). Interestingly, its prognostic power increases substantially when the predictor is combined with existing outcome signatures, and this may be the way to reconcile the profusion of gene expression signatures that have been described to identify the best possible association with clinical outcome. These reports and the diverse signatures they identified suggest that the aggressiveness of the tumor may result from specific genetic changes that the tumors undergo in response to the adverse environments they encounter during tumorigenesis. Therefore, it may be less perplexing to consider that the different genetic signatures associated with clinical outcomes may reflect the varied adaptations of the tumors to outside pressures.

Gene expression profiling has shown that specific molecular subtypes in breast cancer are associated with particular clinical outcomes. In addition to prognosis [(11, 51), among others] a gene expression signature has also identified the response to a specific chemotherapy regimen (52). The 70-gene poor-prognosis signature identified was predictive of a short interval to distant metastasis in lymph-node-negative patients (13, 14, 38). Another important biological and clinical question is the distant site to which a tumor preferentially metastasizes. Microarray analyses have identified gene expression profiles for bone (53–55) and lung (45) metastasis in breast cancer. Smid and colleagues showed that the five major molecular subtypes in breast cancer are not only distinct with regard to primary tumor features, tumor aggressiveness, and response to certain types of chemotherapy; they also clearly differ in their ability to metastasize to distant organ(s) (56). They used the intrinsic gene list describing the breast cancer subtypes to classify 344 primary breast tumors of lymph-node-negative patients, Fisher exact tests to determine the association between a tumor subtype and a particular site of distant relapse, and Significance Analysis of Microarrays and Global Testing to identify the affected genes and pathways in the various groups. Taking all the types of analysis together, they showed consistent results when analyzing the subtypes and the site of relapse that relate to the biology of brain, lung, and bone relapse. For the lung relapse patients, the focal adhesion signaling cascade was an important modulator of organ-specific relapse, while patients with a relapse to the bone presented an association with ER status. Interestingly, the Wnt signaling pathway was associated with patients relapsing to the brain and bone, as well as with the basal and luminal B subtypes. The Wnt signaling pathway has been implicated in the development of normal brain tissue as well as in brain tumorigenesis, therefore supporting the view that tumor cells grow better in the microenvironment they resemble (57, 58).

6. The Cancer Stem Cell Hypothesis

In 2001, in an elegant and highly cited review, Weissman and colleagues presented accumulating evidence supporting the existence of stem cells in the hematopoietic system. This work has stimulated research into the prospective isolation of stem cells from other tissues, their initial characterization, and their potential use in regenerative medicine (59). These studies have also had a profound effect on the understanding of solid tumors, including breast cancer, and opened up new hopes for future therapeutic strategies.

Stem cells can be defined by their ability to self-renew and to differentiate into the different cell types of the tissue in which they reside. The property of self-renewal is particularly interesting for the striking parallels that can be drawn between stem cells and cancer cells: Tumors may originate from the transformation of normal stem cells, and similar signaling pathways may regulate self-renewal in stem cells and in cancer cells. As a consequence, the idea of the existence of "cancer stem cells" as a low-abundance population with unlimited potential for self-renewal that drive the growth of the tumor, while also giving rise to a large population of differentiated progeny that form the bulk of the tumor, appeared very intriguing and plausible. Cancer stem cells have been identified in a variety of human tumors, and specific signaling pathways are being identified that play a functional role in cancer stem cell self-renewal and/or differentiation. In addition, evidence is accumulating about the relevance of the microenvironmental niche to influence the properties of the stem cells.

In the case of breast tumors, cancer stem cells were first isolated as $CD44^{+}CD24^{-/low}Lin^{-}ESA^{+}$ and shown to be able to form new tumors in immunocompromised mice with low numbers, while cells lacking these cell surface markers failed to form tumors even at a much higher concentration (60). At the same time, normal stem cells were identified in the normal mammary gland (61, 62), and an in vitro cultivation system was developed that allows for propagation of human mammary epithelial cells in an undifferentiated state while growing in suspension as nonadherent mammospheres (63). This method of culture was based on the culture of brain tumor stem cells as neurospheres, which has been used to study their stem cell properties (64).

7. Stem Cell Signature

Two studies published in parallel compared the gene expression profiles of mouse embryonic, neural, and hematopoietic stem cells and found an overlapping set of genes that could be defined as a molecular signature for stem cells (65, 66). However, as has been observed with the analysis of the gene profiling of breast tumor samples, the genetic programs defined presented only a limited overlap. When the transcriptional profiles of cells isolated from nonadherent mammospheres were compared with the stem cell signature revealed by Ramalho-Santos and colleagues, overlapping genetic programs and specific genes (the global overlap between genes expressed in all three stem cells and mammospheres includes 10 upregulated genes), as well as new genes, were identified. Although this analysis was not performed with stem cells isolated based on specific markers (but instead using a cell population enriched for progenitor cells), the interesting point is that gene expression profiling was, once again, used as a tool in parallel with other technical advances to elucidate the molecular pathways that regulate normal mammary development and tumorigenesis.

Other researchers have continued to employ this approach to compare the gene profiles of breast stem cells and assess whether there is a correlation with the overall survival in patients with breast cancer, despite the technical difficulty in addressing this type of question. Breast stem cells, whether normal or cancer, are a rare population, and it is not trivial to collect sufficient tissue material to perform this type of assay in a reproducible and significant manner with a sufficiently large number of samples.

A comparison of the transcriptional profiles of the population of cells previously identified as breast cancer stem cells, $CD44^{+}CD24^{-/low}$, and normal breast epithelium generated what the authors defined as a 186-gene-invasiveness signature (67). This genetic program was associated with metastasis-free survival and overall survival not only for patients with breast cancer, but also among patients with lung cancer, medulloblastoma, or prostate cancer. This gene-invasiveness signature differs from previously defined gene profiles in breast cancer (11, 13) and shows little overlap (only six genes) with those of the wound-response signature reported previously (44). This finding is, once more, disconcerting and raises questions about their biological relevance and clinical implications. If the genetic adaptations to the adverse environments by the tumor cells are so extraordinarily varied, it may be difficult to define a simple signature of practical use in the clinic. This makes it even more important to take into account the results from the large multicenter clinical trials (41).

One interesting aspect of the study by Liu and colleagues is that they showed that the combination of two independent signatures improved the reliability of the association with clinical outcome. The invasiveness signature and the wound-response signature are representations of different biological phenomena and are based on nonoverlapping lists of genes. However, when combined, they perform better than either alone. Using the full data on 295 patients and both signatures, it was found that after 10 years of follow-up, metastatic disease had developed in 20, 31, and 53% of patients with tumors that were negative for both signatures, positive for one signature, or positive for both signatures, respectively (67). This observation is consistent with a model in which self-renewing cancer stem cells give rise to the cancer while the tumor microenvironment promotes their growth (58). Most importantly, it pointed to the relevance of combining the varied tools that have been developed in the cancer research field. Similarly, the stroma-derived prognostic predictor was also a better predictor when combined with other signatures, suggesting functional interactions between the biological processes underlying the different gene expression signatures (50).

Nevertheless, another combination of signatures has not resulted in improved performance. Several gene expression signatures – the intrinsic-subtype, the 70-gene poor-prognosis, wound-response, two-gene ratio (for patients treated with tamoxifen), and recurrence-score signatures – showed a significant agreement in the clinical outcome predicted for the same patients (68). Even though different gene sets were used for the prognosis in patients with breast cancer, four of the five tested showed significant agreement in the outcome predictions for individual patients and are probably tracking a common set of biologic phenotypes. However, the authors of the invasiveness signature agree that it was based on the isolation of breast cancer stem cells from only a few samples and that analysis of a much larger group of patients could result in an improved signature (67). Furthermore, the power of combined signatures may be improved with the use of decision-tree analysis methods (44) or the integration of new signatures (69). It is becoming apparent that it is necessary to integrate all aspects of tumor biology to achieve increased prognostic power and improved clinical outcomes.

8. Disease Recurrence

One intriguing aspect of the cancer stem cell hypothesis is its implications to a very relevant clinical problem, the development of resistance to therapy. Normal stem cells are known to be relatively quiescent, to be resistant to drugs and toxins through the expression (or overexpression) of drug efflux pumps, to possess an active DNA repair capacity, and to have increased resistance to apoptosis (70). The hypothesis implies that conventional therapies, which target rapidly cycling cells, will kill the majority of the bulk of the tumor, resulting in the apparent disappearance of the tumor. However, the cancer stem cells would remain unaffected and capable, due to their intrinsic properties, of repopulating the tumor and giving rise to cancer recurrence (71). Using the human breast cancer cell line MCF-7 as a model, it has been shown that radiation induces the enrichment of a population of stem cell progenitors, suggesting that progenitor cells have different cell survival properties that may give rise to recurrent disease (72). Understanding the molecular mechanisms underlying the resistance of cancer stem cells to conventional therapies might contribute to the development of anticancer strategies that specifically target the cancer stem cells or to finding ways to manipulate them to become resistant to these therapies.

Breast tumor cells with the cancer stem cell phenotype ($CD44^{+}CD24^{-/low}$) were analyzed for their global gene expression profile (SAGE, or serial analysis of gene expression) compared to the nontumorigenic cells ($CD44^{+/-}CD24^{+}$) from the same tumors (73). The cells with the cancer stem cell phenotype expressed genes associated with stem/progenitor properties, while the nontumorigenic cells expressed differentiation-associated genes. The authors identified the TGFbeta signaling pathway as specifically active in the cancer stem cell population (CD44+) and showed that inhibition of this pathway leads to differentiation. They also found an association of the gene expression signature characteristic of $CD44^{+}CD24^{-/low}$ with shorter distant metastasis-free and overall survival times. These findings strongly suggest that the presence and frequency of breast cancer stem cells in tumors have prognostic relevance (42).

9. Embryonic Stem Cell Signature and Breast Cancer

A recent meta-analysis provides a broad view of the presence of molecular imprints of stemness in cancer by examining the activity of gene sets associated with human embryonic stem cell identity in human tumors (74). This group analyzed the enrichment patterns of gene sets associated with embryonic stem (ES) identity in the expression profiles of various human tumor types and found that expression of the ES signature is associated with poor prognosis. Furthermore, activation targets of Nanog, Oct4, Sox2, and c-Myc are more frequently overexpressed in poorly differentiated tumors, high-grade ER– tumors, often of the basal-like subtype, and with aggressive tumor behavior. In general, an inverse relationship between the presence of the ES signature and the degree of tumor differentiation was found. Intriguingly, the authors admit that it is not possible to distinguish whether the ES signature is inherited from a stem cell-of-origin or is reactivated during the process of tumorigenesis. These results reveal a previously unknown link between genes associated with ES cell identity and the histopathological traits of tumors and support the possibility that these genes contribute to the stem cell–like phenotypes shown by many tumors.

10. Conclusions

Cancer gene expression signatures are beginning to be tested in the clinic. Predictions for a good prognosis could be part of the deciding criteria for the use of appropriately targeted adjuvant therapy. However, it is likely that, at least for the time being, the gene expression signatures will complement, rather than substitute, the current use of histopathological parameters for breast cancer management. Furthermore, it is possible that the identification of better-defined signatures may help to identify the specific molecular pathways affected in the more aggressive tumors and, as a consequence, highlight the potential benefits of drug combinations for the treatment of such a complex disease as breast cancer. Finally, although many questions remain to be answered before the role of cancer stem cells in tumor initiation and progression is fully understood, it appears that incorporating cancer stem cells in our view of tumorigenesis may contribute to our biological knowledge and opens up new hopes for improved breast cancer prognosis and treatment.

11. Notes

Some useful web pages and information for microarray analysis:

http://www.mged.org/ MIAME describes the Minimum Information About a Microarray Experiment that is needed to enable the interpretation of the results of the experiment unambiguously and potentially to reproduce the experiment.

http://gepas.bioinfo.cipf.es/ The Gene Expression Profile Analysis Suite (GEPAS) is one of the most complete integrated packages of tools for microarray data analysis available over the web.

http://genome-www.stanford.edu/ Hyperlinks to systematic analysis projects, resources, laboratories, and departments at Stanford University.

http://microarrays.nki.nl/ Microarray facility and links from the Netherlands Cancer Institute (NKI).

http://jura.wi.mit.edu/bio/microarrays/ Page from the Whitehead Institute for Biomedical Research, Cambridge, MA, on Bioinformatics and Research Computing.

Acknowledgments

The author would like to thank Robert Kypta for critically reading the manuscript. This work was funded, in part, by the Department of Industry, Tourism and Trade of the Government of the Autonomous Community of the Basque Country (Etortek Research Programs 2005/2006/2007) and the Innovation Technology Department of the Basque Country.

References

1. Pearson OH, West CD, Hollander VP, Treves NE. (1954) Evaluation of endocrine therapy for advanced breast cancer. *J Am Med Assoc* 154:234–239.
2. Harper MJ, Walpole AL. (1966) Contrasting endocrine activities of cis and trans isomers in a series of substituted triphenylethylenes. *Nature* 212:87.
3. Fisher B, Costantino JP, Wickerham DL, Redmond CK, Kavanah M, Cronin WM, Vogel V, Robidoux A, Dimitrov N, Atkins J, et al. (1998) Tamoxifen for prevention of breast cancer: report of the national surgical adjuvant breast and bowel project P-1 Study. *J Natl Cancer Inst* 90:1371–1388.
4. Jensen EV, Block GE, Smith S, Kyser K, DeSombre ER. (1971) Estrogen receptors and breast cancer response to adrenalectomy. *Natl Cancer Inst Monogr* 34:55–70.
5. Green S, Walter P, Kumar V, Krust A, Bornert JM, Argos P, Chambon P. (1986) Human oestrogen receptor cDNA: sequence, expression and homology to v-erb-A. *Nature* 320:134–139.
6. Kuiper GG, Enmark E, Pelto-Huikko M, Nilsson S, Gustafsson JA. (1996) Cloning of a novel receptor expressed in rat prostate and ovary. *Proc Natl Acad Sci USA* 93: 5925–5930.
7. Goldhirsch A, Wood WC, Gelber RD, Coates AS, Thurlimann B, Senn HJ. (2007) Progress and promise: highlights of the international expert consensus on the primary therapy of early breast cancer 2007. *Ann Oncol* 18:1133–1144.
8. Eifel P, Axelson JA, Costa J, Crowley J, Curran WJ, Jr., Deshler A, Fulton S, Hendricks CB, Kemeny M, Kornblith AB, et al. (2001) National Institutes of Health Consensus Development Conference Statement: adjuvant therapy for breast cancer, November 1–3, 2000. *J Natl Cancer Inst* 93:979–989.
9. Perou CM, Jeffrey SS, van de Rijn M, Rees CA, Eisen MB, Ross DT, Pergamenschikov A, Williams CF, Zhu SX, Lee JC, et al. (1999) Distinctive gene expression patterns in human mammary epithelial cells and breast cancers. *Proc Natl Acad Sci USA* 96: 9212–9217.
10. Perou CM, Sorlie T, Eisen MB, van de Rijn M, Jeffrey SS, Rees CA, Pollack JR, Ross DT, Johnsen H, Akslen LA, et al. (2000) Molecular portraits of human breast tumours. *Nature* 406:747–752.

11. Sorlie T, Perou CM, Tibshirani R, Aas T, Geisler S, Johnsen H, Hastie T, Eisen MB, van de Rijn M, Jeffrey SS, et al. (2001) Gene expression patterns of breast carcinomas distinguish tumor subclasses with clinical implications. *Proc Natl Acad Sci USA* 98; 10869–10874.
12. Charafe-Jauffret E, Ginestier C, Monville F, Finetti P, Adelaide J, Cervera N, Fekairi S, Xerri L, Jacquemier J, Birnbaum D, Bertucci F. (2006) Gene expression profiling of breast cell lines identifies potential new basal markers. *Oncogene* 25:2273–2284.
13. van't Veer LJ, Dai H, van de Vijver MJ, He YD, Hart AA, Mao M, Peterse HL, van der Kooy K, Marton MJ, Witteveen AT, et al. (2002) Gene expression profiling predicts clinical outcome of breast cancer. *Nature* 415:530–536.
14. van de Vijver MJ, He YD, van't Veer LJ, Dai H, Hart AA, Voskuil DW, Schreiber GJ, Peterse JL, Roberts C, Marton MJ, et al. (2002) A gene-expression signature as a predictor of survival in breast cancer. *N Engl J Med* 347:1999–2009.
15. Simon R, Radmacher MD, Dobbin K, McShane LM. (2003) Pitfalls in the use of DNA microarray data for diagnostic and prognostic classification. *J Natl Cancer Inst* 95:14–18.
16. Seth A, Kitching R, Landberg G, Xu J, Zubovits J, Burger AM. (2003) Gene expression profiling of ductal carcinomas in situ and invasive breast tumors. *Anticancer Res* 23:2043–2051.
17. Porter D, Lahti-Domenici J, Keshaviah A, Bae YK, Argani P, Marks J, Richardson A, Cooper A, Strausberg R, Riggins GJ, et al. (2003) Molecular markers in ductal carcinoma in situ of the breast. *Mol Cancer Res* 1:362–375.
18. Ma XJ, Salunga R, Tuggle JT, Gaudet J, Enright E, McQuary P, Payette T, Pistone M, Stecker K, Zhang BM, et al. (2003) Gene expression profiles of human breast cancer progression. *Proc Natl Acad Sci USA* 100:5974–5979.
19. Schuetz CS, Bonin M, Clare SE, Nieselt K, Sotlar K, Walter M, Fehm T, Solomayer E, Riess O, Wallwiener D, et al. (2006) Progression-specific genes identified by expression profiling of matched ductal carcinomas in situ and invasive breast tumors, combining laser capture microdissection and oligonucleotide microarray analysis. *Cancer Res* 66:5278–5286.
20. Ivshina AV, George J, Senko O, Mow B, Putti TC, Smeds J, Lindahl T, Pawitan Y, Hall P, Nordgren H, et al. (2006) Genetic reclassification of histologic grade delineates new clinical subtypes of breast cancer. *Cancer Res* 66:10292–10301.
21. Buerger H, Mommers EC, Littmann R, Simon R, Diallo R, Poremba C, Dockhorn-Dworniczak B, van Diest PJ, Boecker W. (2001) Ductal invasive G2 and G3 carcinomas of the breast are the end stages of at least two different lines of genetic evolution. *J Pathol* 194:165–170.
22. Roylance R, Gorman P, Hanby A, Tomlinson I. (2002) Allelic imbalance analysis of chromosome 16q shows that grade I and grade III invasive ductal breast cancers follow different genetic pathways. *J Pathol* 196:32–36.
23. Korkola JE, DeVries S, Fridlyand J, Hwang ES, Estep AL, Chen YY, Chew KL, Dairkee SH, Jensen RM, Waldman FM. (2003) Differentiation of lobular versus ductal breast carcinomas by expression microarray analysis. *Cancer Res* 63:7167–7175.
24. Zhao H, Langerod A, Ji Y, Nowels KW, Nesland JM, Tibshirani R, Bukholm IK, Karesen R, Botstein D, Borresen-Dale AL, Jeffrey SS. (2004) Different gene expression patterns in invasive lobular and ductal carcinomas of the breast. *Mol Biol Cell* 15:2523–2536.
25. Turashvili G, Bouchal J, Baumforth K, Wei W, Dziechciarkova M, Ehrmann J, Klein J, Fridman E, Skarda J, Srovnal J, et al. (2007) Novel markers for differentiation of lobular and ductal invasive breast carcinomas by laser microdissection and microarray analysis. *BMC Cancer* 7:55.
26. West M, Blanchette C, Dressman H, Huang E, Ishida S, Spang R, Zuzan H, Olson JA, Jr., Marks JR, Nevins JR. (2001) Predicting the clinical status of human breast cancer by using gene expression profiles. *Proc Natl Acad Sci USA* 98:11462–11467.
27. Ahr A, Karn T, Solbach C, Seiter T, Strebhardt K, Holtrich U, Kaufmann M. (2002) Identification of high risk breast-cancer patients by gene expression profiling. *Lancet* 359:131–132.
28. Huang E, Cheng SH, Dressman H, Pittman J, Tsou MH, Horng CF, Bild A, Iversen ES, Liao M, Chen CM, et al. (2003) Gene expression predictors of breast cancer outcomes. *Lancet* 361:1590–1596.
29. Marshall M, Solomon S. (2007) Hereditary breast-ovarian cancer: clinical findings and medical management. *Plast Surg Nurs* 27:124–127.
30. Hedenfalk I, Duggan D, Chen Y, Radmacher M, Bittner M, Simon R, Meltzer P, Gusterson B, Esteller M, Kallioniemi OP,

et al. (2001) Gene-expression profiles in hereditary breast cancer. *N Engl J Med* 344: 539–548.
31. Hedenfalk I, Ringner M, Ben-Dor A, Yakhini Z, Chen Y, Chebil G, Ach R, Loman N, Olsson H, Meltzer P, et al. (2003) Molecular classification of familial non-BRCA1/BRCA2 breast cancer. *Proc Natl Acad Sci USA* 100:2532–2537.
32. Calza S, Hall P, Auer G, Bjohle J, Klaar S, Kronenwett U, Liu ET, Miller L, Ploner A, Smeds J, et al. (2006) Intrinsic molecular signature of breast cancer in a population-based cohort of 412 patients. *Breast Cancer Res* 8:R34.
33. Sotiriou C, Wirapati P, Loi S, Harris A, Fox S, Smeds J, Nordgren H, Farmer P, Praz V, Haibe-Kains B, et al. (2006) Gene expression profiling in breast cancer: understanding the molecular basis of histologic grade to improve prognosis. *J Natl Cancer Inst* 98:262–272.
34. Loi S, Haibe-Kains B, Desmedt C, Lallemand F, Tutt AM, Gillet C, Ellis P, Harris A, Bergh J, Foekens JA, et al. (2007) Definition of clinically distinct molecular subtypes in estrogen receptor-positive breast carcinomas through genomic grade. *J Clin Oncol* 25:1239–1246.
35. Gruvberger S, Ringner M, Chen Y, Panavally S, Saal LH, Borg A, Ferno M, Peterson C, Meltzer PS. (2001) Estrogen receptor status in breast cancer is associated with remarkably distinct gene expression patterns. *Cancer Res* 61: 5979–5984.
36. Abba MC, Hu Y, Sun H, Drake JA, Gaddis S, Baggerly K, Sahin A, Aldaz CM. (2005) Gene expression signature of estrogen receptor alpha status in breast cancer. *BMC Genomics* 6:37.
37. Wang Y, Klijn JG, Zhang Y, Sieuwerts AM, Look MP, Yang F, Talantov D, Timmermans M, Meijer-van Gelder ME, Yu J, et al. (2005) Gene-expression profiles to predict distant metastasis of lymph-node-negative primary breast cancer. *Lancet* 365: 671–679.
38. Buyse M, Loi S, van't Veer L, Viale G, Delorenzi M, Glas AM, d'Assignies MS, Bergh J, Lidereau R, Ellis P, et al. (2006) Validation and clinical utility of a 70-gene prognostic signature for women with node-negative breast cancer. *J Natl Cancer Inst* 98: 1183–1192.
39. Foekens JA, Atkins D, Zhang Y, Sweep FC, Harbeck N, Paradiso A, Cufer T, Sieuwerts AM, Talantov D, Span PN, et al. (2006) Multicenter validation of a gene expression-based prognostic signature in lymph node-negative primary breast cancer. *J Clin Oncol* 24:1665–1671.
40. Desmedt C, Piette F, Loi S, Wang Y, Lallemand F, Haibe-Kains B, Viale G, Delorenzi M, Zhang Y, d'Assignies MS, et al. (2007) Strong time dependence of the 76-gene prognostic signature for node-negative breast cancer patients in the TRANSBIG multicenter independent validation series. *Clin Cancer Res* 13:3207–3214.
41. Rakha EA, El-Sayed ME, Reis-Filho JS, Ellis IO. (2008) Expression profiling technology: its contribution to our understanding of breast cancer. *Histopathology* 52: 67–81.
42. Shipitsin M, Polyak K. (2008) The cancer stem cell hypothesis: in search of definitions, markers, and relevance. *Lab Invest* 88: 459–463.
43. Paik S, Shak S, Tang G, Kim C, Baker J, Cronin M, Baehner FL, Walker MG, Watson D, Park T, et al. (2004) A multigene assay to predict recurrence of tamoxifen-treated, node-negative breast cancer. *N Engl J Med* 351:2817–2826.
44. Chang HY, Nuyten DS, Sneddon JB, Hastie T, Tibshirani R, Sorlie T, Dai H, He YD, van't Veer LJ, Bartelink H, et al. (2005) Robustness, scalability, and integration of a wound-response gene expression signature in predicting breast cancer survival. *Proc Natl Acad Sci USA* 102: 3738–3743.
45. Minn AJ, Gupta GP, Siegel PM, Bos PD, Shu W, Giri DD, Viale A, Olshen AB, Gerald WL, Massague J. (2005) Genes that mediate breast cancer metastasis to lung. *Nature* 436:518–524.
46. Bild AH, Yao G, Chang JT, Wang Q, Potti A, Chasse D, Joshi MB, Harpole D, Lancaster JM, Berchuck A, et al. (2006) Oncogenic pathway signatures in human cancers as a guide to targeted therapies. *Nature* 439:353–357.
47. Chi JT, Wang Z, Nuyten DS, Rodriguez EH, Schaner ME, Salim A, Wang Y, Kristensen GB, Helland A, Borresen-Dale AL, et al. (2006) Gene expression programs in response to hypoxia: cell type specificity and prognostic significance in human cancers. *PLoS Med* 3:e47.
48. Winter SC, Buffa FM, Silva P, Miller C, Valentine HR, Turley H, Shah KA, Cox GJ, Corbridge RJ, Homer JJ, et al. (2007) Relation of a hypoxia metagene derived from head and neck cancer to prognosis of multiple cancers. *Cancer Res* 67: 3441–3449.

49. West RB, Nuyten DS, Subramanian S, Nielsen TO, Corless CL, Rubin BP, Montgomery K, Zhu S, Patel R, Hernandez-Boussard T, et al. (2005) Determination of stromal signatures in breast carcinoma. *PLoS Biol* 3:e187.
50. Finak G, Bertos N, Pepin F, Sadekova S, Souleimanova M, Zhao H, Chen H, Omeroglu G, Meterissian S, Omeroglu A, et al. (2008) Stromal gene expression predicts clinical outcome in breast cancer. *Nat Med* 14:518–527.
51. Sorlie T, Tibshirani R, Parker J, Hastie T, Marron JS, Nobel A, Deng S, Johnsen H, Pesich R, Geisler S, et al. (2003) Repeated observation of breast tumor subtypes in independent gene expression data sets. *Proc Natl Acad Sci USA* 100: 8418–8423.
52. Rouzier R, Perou CM, Symmans WF, Ibrahim N, Cristofanilli M, Anderson K, Hess KR, Stec J, Ayers M, Wagner P, et al. (2005) Breast cancer molecular subtypes respond differently to preoperative chemotherapy. *Clin Cancer Res* 11: 5678–5685.
53. Kang Y, Siegel PM, Shu W, Drobnjak M, Kakonen SM, Cordon-Cardo C, Guise TA, Massague J. (2003) A multigenic program mediating breast cancer metastasis to bone. *Cancer Cell* 3:537–549.
54. Deckers M, van Dinther M, Buijs J, Que I, Lowik C, van der Pluijm G, ten Dijke P. (2006) The tumor suppressor Smad4 is required for transforming growth factor beta-induced epithelial to mesenchymal transition and bone metastasis of breast cancer cells. *Cancer Res* 66: 2202–2209.
55. Smid M, Wang Y, Klijn JG, Sieuwerts AM, Zhang Y, Atkins D, Martens JW, Foekens JA. (2006) Genes associated with breast cancer metastatic to bone. *J Clin Oncol* 24: 2261–2267.
56. Smid M, Wang Y, Zhang Y, Sieuwerts AM, Yu J, Klijn JG, Foekens JA, Martens JW. (2008) Subtypes of breast cancer show preferential site of relapse. *Cancer Res* 68: 3108–3114.
57. Paget S. (1989) The distribution of secondary growths in cancer of the breast. 1889. *Cancer Metastasis Rev* 8:98–101.
58. Fidler IJ. (2003) The pathogenesis of cancer metastasis: the "seed and soil" hypothesis revisited. *Nat Rev Cancer* 3:453–458.
59. Reya T, Morrison SJ, Clarke MF, Weissman IL. (2001) Stem cells, cancer, and cancer stem cells. *Nature* 414: 105–111.
60. Al-Hajj M, Wicha MS, Benito-Hernandez A, Morrison SJ, Clarke MF. (2003) Prospective identification of tumorigenic breast cancer cells. *Proc Natl Acad Sci USA* 100: 3983–3988.
61. Alvi AJ, Clayton H, Joshi C, Enver T, Ashworth A, Vivanco MM, Dale TC, Smalley MJ. (2003) Functional and molecular characterisation of mammary side population cells. *Breast Cancer Res* 5: R1–8.
62. Clayton H, Titley I, Vivanco M. (2004) Growth and differentiation of progenitor/stem cells derived from the human mammary gland. *Exp Cell Res* 297:444–460.
63. Dontu G, Abdallah WM, Foley JM, Jackson KW, Clarke MF, Kawamura MJ, Wicha MS. (2003) In vitro propagation and transcriptional profiling of human mammary stem/progenitor cells. *Genes Dev* 17: 1253–1270.
64. Singh SK, Clarke ID, Terasaki M, Bonn VE, Hawkins C, Squire J, Dirks PB. (2003) Identification of a cancer stem cell in human brain tumors. *Cancer Res* 63:5821–5828.
65. Ramalho-Santos M, Yoon S, Matsuzaki Y, Mulligan RC, Melton DA. (2002) "Stemness": transcriptional profiling of embryonic and adult stem cells. *Science* 298:597–600.
66. Ivanova NB, Dimos JT, Schaniel C, Hackney JA, Moore KA, Lemischka IR. (2002) A stem cell molecular signature. *Science* 298: 601–604.
67. Liu R, Wang X, Chen GY, Dalerba P, Gurney A, Hoey T, Sherlock G, Lewicki J, Shedden K, Clarke MF. (2007) The prognostic role of a gene signature from tumorigenic breast-cancer cells. *N Engl J Med* 356: 217–226.
68. Fan C, Oh DS, Wessels L, Weigelt B, Nuyten DS, Nobel AB, van't Veer LJ, Perou CM. (2006) Concordance among gene-expression-based predictors for breast cancer. *N Engl J Med* 355:560–569.
69. Massague J. (2007) Sorting out breast-cancer gene signatures. *N Engl J Med* 356:294–297.
70. Dean M, Fojo T, Bates S. (2005) Tumour stem cells and drug resistance. *Nat Rev Cancer* 5:275–284.
71. Ailles LE, Weissman IL. (2007) Cancer stem cells in solid tumors. *Curr Opin Biotechnol* 18:460–466.
72. Woodward WA, Chen MS, Behbod F, Alfaro MP, Buchholz TA, Rosen JM. (2007) WNT/beta-catenin mediates radiation resistance of mouse mammary progenitor cells. *Proc Natl Acad Sci USA* 104: 618–623.

73. Shipitsin M, Campbell LL, Argani P, Weremowicz S, Bloushtain-Qimron N, Yao J, Nikolskaya T, Serebryiskaya T, Beroukhim R, Hu M, et al. (2007) Molecular definition of breast tumor heterogeneity. *Cancer Cell* 11:259–273.

74. Ben-Porath I, Thomson MW, Carey VJ, Ge R, Bell GW, Regev A, Weinberg RA. (2008) An embryonic stem cell-like gene expression signature in poorly differentiated aggressive human tumors. *Nat Genet* 40:499–507.

Chapter 8

Genome-Wide Proximal Promoter Analysis and Interpretation

Elizabeth Guruceaga, Victor Segura, Fernando J. Corrales, and Angel Rubio

Abstract

High-throughput gene expression technologies based on DNA microarrays allow the examination of biological systems. However, the interpretation of the complex molecular descriptions generated by these approaches is still challenging. The development of new methodologies to identify common regulatory mechanisms involved in the control of the expression of a set of co-expressed genes might enhance our capacity to extract functional information from genomic data sets.

In this chapter, we describe a method that integrates different sources of information: gene expression data, genome sequence information, described transcription factor binding sites (TFBSs), functional information, and bibliographic data. The starting point of the analysis is the extraction of promoter sequences from a whole genome and the detection of TFBSs in each gene promoter. This information allows the identification of enriched TFBSs in the proximal promoter of differentially expressed genes. The functional and bibliographic interpretation of the results improves our biological insight into the regulatory mechanisms involved in a microarray experiment.

Key words: Functional genomics, DNA microarrays, transcriptional regulation, promoter analysis, data integration.

1. Introduction

Large-scale gene expression studies are one of the most recent breakthroughs in experimental molecular biology (57). Microarray technology allows the detection of time-dependent changes of the transcriptome or changes in gene expression between normal and diseased tissue samples (39). Bioinformatics tools manage to deal with the massive amount of data generated in microarray analysis. However, data interpretation continues to be the main bottleneck (**Fig. 8.1**) (26).

R. Matthiesen (ed.), *Bioinformatics Methods in Clinical Research*, Methods in Molecular Biology 593,
DOI 10.1007/978-1-60327-194-3_8, © Humana Press, a part of Springer Science+Business Media, LLC 2010

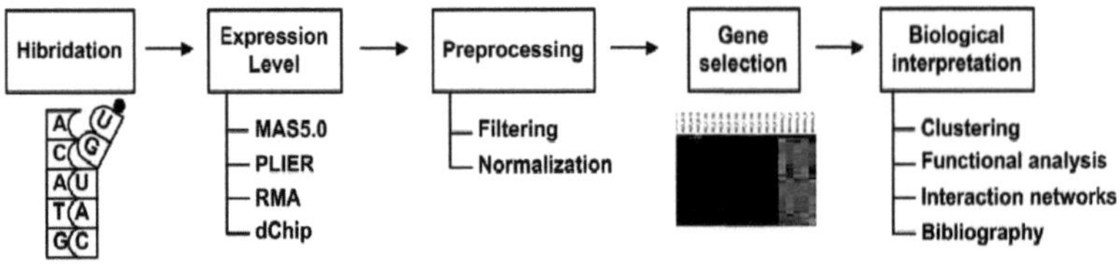

Fig. 8.1. Process of a microarray analysis: After the hybridization of the samples probe-set expression levels are calculated, then data are filtered and normalized, the selection of differentially expressed genes is performed, and the analysis finishes with the interpretation of the results.

An issue that is becoming increasingly important for the interpretation of genome and transcriptome data is the understanding of gene expression regulation (47, 45, 1). Gene expression data obtained from a microarray analysis and used to find genes with similar expression profiles can be the input data of a genome-wide promoter analysis (*see* **Note 1**). It is assumed that the obtained expression profiles are a manifestation of underlying common regulatory mechanisms (63). These regulatory mechanisms constitute the first level in the control of protein synthesis that is further modulated in subsequent steps, including mRNA splicing and translation. Transcriptional regulation determines a gene to be transcribed into RNA molecules in response to molecular signals. Both RNA polymerase and transcription factors (TFs) are needed for the initiation of transcription (51).

TFs bind to specific DNA sites among a vast number of structurally similar nonspecific sites. These TF binding sites (TFBSs) were initially represented in the form of consensus sequences, and later position weight matrices (PWM) were developed for a more precise description (60). One of the approaches to analyze and elicit the control mechanisms that explain the similar expression profile of a set of genes is to search for TFBSs located within the promoters of a significant number of co-expressed genes using known PWMs. Another option is to perform a multiple alignment on the promoter sequences, especially tailored to align small and variable sequences, obtaining putative regulatory sequences. Afterwards, these sequences need to be related with particular TFs (**Fig. 8.2**). For the interested reader, Tompa et al. (62) perform a thorough and complete comparison of different alignment methods to detect new TFBSs. This chapter focuses on the first procedure.

It is well known that enhancer and suppressor control elements can be located at sites tens of thousands of bases upstream or even downstream of the transcription start site (TSS) (8). In many cases, however, the essential control elements are present within the proximal promoter, a few hundred to several thousand bases upstream of TSS (46). In fact, many conserved TFBSs tend to concentrate in the approximately 1 kb region around the TSS (72).

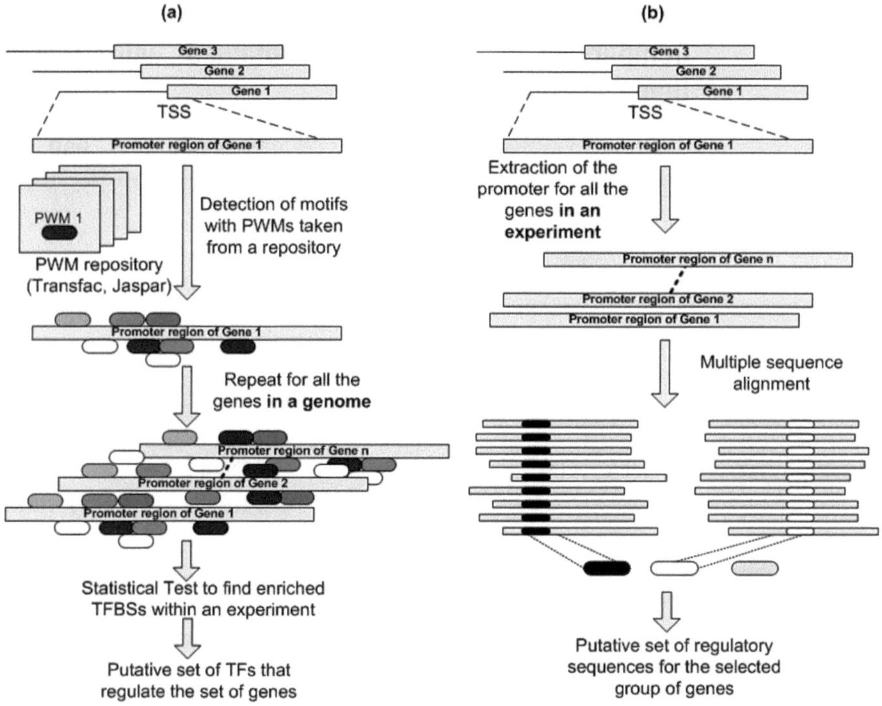

Fig. 8.2. There are two ways to discover the underlying common regulatory mechanism of a set of genes that are corexpressed. Both methods start with the extraction of the promoter sequences. (**a**) This procedure searches for TFBSs in the extracted sequences using known PWMs of databases such as Jaspar (67) or Transfac (43) and then applies an enrichment analysis. (**b**) In this case, a multiple-sequence alignment algorithm is performed to obtain a set of putative regulatory sequences.

A broad variety of algorithms and tools have been developed to tackle the promoter analysis. Some of them have been designed to analyze prokaryotic or yeast sequences (e.g., RSAT) (65). Other tools [e.g., rVISTA (41) and TraFaC (30)] work on sequences of higher eukaryotes. However, these applications are limited to the analysis of a single gene at a time, so that studying a large number of genes becomes impractical. Tools such as Toucan (4), Opossum (24), or PAP (10) have been developed to analyze multiple genomic regulatory regions. PAP and Opossum assume that essential TFBSs are conserved during evolution and detect the enriched TFBSs that are conserved in humans and mice. This assumption is being revised because not all functionally important TFBS are conserved even between closely related species (16, 13, 18), and not every conserved pattern is necessarily functional (11). Toucan performs TFBS statistical enrichment analysis of the entire gene upstream region in addition to the evolutionary conserved sequences, but the user cannot select the complete genome or a particular microarray as reference.

Some of these tools have been compared with two sets of genes described in the literature: muscle- (69) and liver-specific

Table 8.1
Comparison of the results obtained with Toucan, Opossum, and PAP

	Toucan	Opossum	PAP
	SRF	SRF	SRF
	Myogenin	TEF-1	MEF
Muscle-specific	MAZR	MEF2	Myf
	MAZ	Myf	SP1
	LBP1		
	SP1		
	MZF		
Liver-specific	HNF1	HNF1	HNF-1
	FOX	FREAC-2	HNF-3
			C/EBP
			HNF-4

In the analysis, we have used the recommended parameters for each bioinformatics tool.

genes (34). Regulatory elements known to be important in the analyzed set of genes can be detected with these tools, showing that the TFBS enrichment analysis provides biological insights into the regulatory mechanisms (**Table 8.1**).

Some of the motifs found in the muscle-specific genes are known to be muscle-specific factors (SRF, Myogenin, MYF, MEF2, and TEF1), and the detection of SP1 sites is not surprising, since this is a general TF. Similar results are obtained in the analysis of liver-specific genes, although fewer TFs have overrepresented TFBSs (HNF and C/EBP are known to be liver-specific).

While existing applications may allow genome-wide TFBS enrichment analysis, the interpretation of the results is a challenging task due to the obtained amount of data and the number of false positives. We present an analysis pipeline that integrates the promoter analysis results with gene expression data, and functional and bibliographic information, improving our capacity to extract biological conclusions from genomic data sets.

2. Materials

2.1. Required Bioinformatics Resources

The required resources for DNA microarray analysis, genome-wide proximal promoter analysis, and interpretation of the results are listed with their corresponding web pages.

1. *Ensembl* is a joint project between EMBL-EBI and the Sanger Institute to develop a software system that produces and maintains automatic annotation on selected eukaryotic genomes (http://www.ensembl.org).
2. *JASPAR* is a collection of TFBSs, modeled as matrices. These can be converted into PWMs, used for scanning genomic sequences (http://jaspar.genereg.net).
3. *TRANSFAC 6.0* contains data on TFs, their experimentally proven binding sites, and regulated genes. Its broad compilation of binding sites allows the derivation of PWMs (http://www.gene-regulation.com/pub/databases.html).
4. *MotifScanner* can be used to screen DNA sequences with precompiled motif models. The algorithm is based on a probabilistic sequence model in which motifs are assumed to be hidden in a noisy background sequence (http://homes.esat.kuleuven.be/~thijs/download.html).
5. *Cluster 3.0* and *Java TreeView* provide a computational and graphical environment for analyzing data from DNA microarray experiments or other genomic data sets (http://www.geo.vu.nl/huik/cluster.htm and http://jtreeview.sourceforge.net).
6. The *Gene Ontology* project is developing three structured, controlled vocabularies (ontologies) that describe gene products in terms of their associated biological processes, cellular components, and molecular functions in a species-independent manner (http://www.geneontology.org).
7. *Ingenuity Systems* enables researchers to model, analyze, and understand complex biological systems integral to human health and disease (http://www.ingenuity.com).
8. *PubGene* can retrieve information on genes and proteins. Gene and protein names are cross-referenced to each other and to terms that are relevant to understand their biological function, importance in disease, and relationship to chemical substances (http://www.pubgene.org).
9. *iHOP* provides a network of concurring genes and proteins through the scientific literature, touching on phenotypes, pathologies, and gene functions as a natural way of accessing millions of PubMed abstracts (http://www.pdg.cnb.uam.es/UniPub/iHOP).
10. *Bioconductor* is an open source and open development software project for the analysis and comprehension of genomic data (http://www.bioconductor.org).

3. Methods

The proposed methodology consists of the identification of enriched TFBSs in a set of differentially expressed genes. The enrichment analysis has been limited to the proximal promoter, and the information about the promoters of human genome required for the analysis is stored in a database. Bioinformatics tools used for the interpretation of the results are described.

3.1. Design of the Database and Data Collection

The authors have designed a database to store all the needed information for the proposed proximal promoter analysis. In this way, the enrichment analysis execution time is reduced and it is possible to choose a whole genome or microarray as statistical reference. This analysis takes longer in tools that do not create their own database, such as Toucan (4), because each time an analysis is performed, the promoter sequences have to be extracted from a public database and the TFBSs have to be detected before the statistical enrichment is calculated. The disadvantage of precomputing a database is the required process of actualization.

The study of regulatory DNA is more difficult than that of coding sequences because there are no well-known properties in regulatory DNA analogous to open reading frames and nonuniform codon usage in coding sequences. This makes it difficult to define the location of the gene promoter (1). There are several databases that contain the experimentally verified position of TSSs, such as EPD (56) or DBTSS (61). However, for a genome-wide promoter analysis, the Ensembl database is more appropriate due to its automatic annotation system (27). The annotation of the gene start in the Ensembl database is accurate enough to be used as TSS (3). Therefore, promoter sequences have been retrieved from the version of the EnsMart database (32) that corresponds to Ensembl release 42 (14), assuming the most 5' upstream position of the annotated transcripts to be the TSS.

We estimate that every kilobase of genomic DNA contains many dozens of potential TFBSs on the basis of random similarity (71). Consequently, it is important to limit the analyzed sequence to reduce this false positive rate (*see* **Note 2**). One option to remove false positives is the selection of sequences conserved in evolution by phylogenetic footprinting (24, 10). As stated in the introduction, this strategy is being revised.

Therefore, our database contains the proximal promoter sequences extracted from the human genome, taken as 1,000 bp upstream to TSS. We therefore assume that essential control elements are present within the proximal promoter (46, 72).

Publicly available information distributed by the Jaspar (67) and Transfac 6.0 (43) databases about known TFBSs is also stored in the database. There are more recent versions of the Transfac database; however, these are not public versions. Starting from the promoter sequences and mentioned information about PWMs, different algorithms can be applied for TFBS detection (1). These algorithms are classified according to the following criteria:

- *TFBS model.* Consensus sequences (66) or PWM of the motif can be used to represent the TFBSs (4).
- *Search method.* The algorithm can use different methods to find the TFBSs in the promoter sequences: greedy search (23), iterative methods (39), and Gibbs sampling (36).
- *Score function.* Several scores are used to discriminate each motif in a TFBS from background noise: expectation-maximization (37), information content (23), maximum a posteriori probability (40), group specificity (28), positional bias (28).

The prediction of new TFBSs is based on a score function that has to be evaluated against a certain threshold (6, 60). One approach for the calculation of this threshold compares the number of TFBS hits in the analyzed sequence with the number of hits in a randomly generated DNA sequence [i.e., MatInspector (52), Match (33)]. This detection of TFBSs has either low sensitivity or low specificity (19), resulting in a number of false positives. Other algorithms, such as MotifScanner (2), estimate the expected number of motifs for each TF in each sequence trying to minimize the false positive rate.

We decided to run MotifScanner (2) against the extracted promoter sequences and the stored PWMs of known TFBSs. This algorithm finds known TFBSs in DNA sequences based on a probabilistic sequence model. It is assumed that TFBSs are hidden in a noisy background sequence whose statistical properties are estimated offline. Afterwards, the existence of a TFBS is predicted with a hidden Markov Model (HMM). In the analysis, a background model of the vertebrate sequences of EPD (56) has been used in combination with MotifScanner default parameters.

The process of extracting promoter sequences and finding TFBS instances in each sequence can be automated to generate SQL scripts that create the database, as shown in **Fig. 8.3**. In this way, the task of periodical actualization of the database is resolved.

Functional information of the genomes has been added to the database (21) to be used in the interpretation of the results. This functional annotation is necessary to determine GO category enrichment in the selection of co-expressed genes that present a particular TFBS in their promoters.

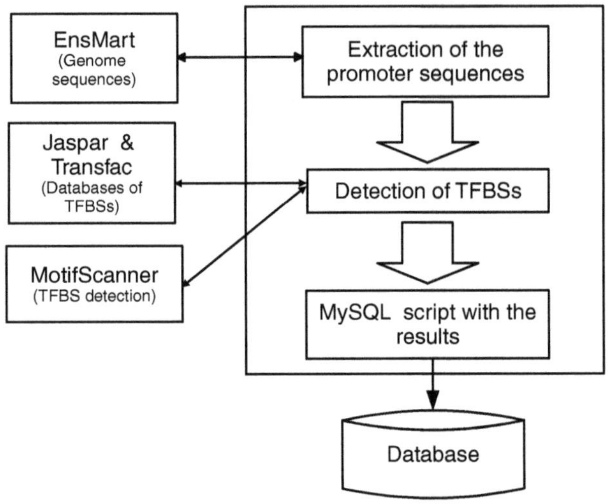

Fig. 8.3. The process for database construction is shown: Promoter sequences are extracted from EnsMart and sent to MotifScanner to detect Jaspar and Transfac TFBSs. This information is stored in a database.

3.2. Bioinformatics Analysis

The input data for the TFBS enrichment analysis can be Ensembl gene identifiers (14) or probesets from a DNA microarray that we translate to Ensembl identifiers (we use the Ensembl MySQL database to perform the translation). The observed distribution of a TFBS in the gene selection is then compared with its distribution in a reference gene group such as the corresponding whole genome or Ensembl genes represented in a particular microarray. Having a reference group with N genes, K genes will present the TFBS under study in their proximal promoter. If n genes from the total genes of the reference group are randomly selected, x genes will present the TFBS. In this kind of sampling without replacement, the probability of finding x genes with the TFBS in their proximal promoter is

$$P(X=x)=P(X)=\frac{\binom{K}{x}\binom{N-K}{n-x}}{\binom{N}{n}} \quad [1]$$

This expression is the density function of a hypergeometric distribution. The probability is calculated with the hypergeometric distribution, as shown in Equation [2], and can be interpreted as a p-value that indicates that a particular TFBS is especially enriched within the selected genes in the analysis if compared with the expected number of instances for each TFBS.

$$P(X > x) = 1 - P(X \leq x) = 1 - \sum_{X=0}^{x} P(X) \quad [2]$$

The computational cost of the hypergeometric distribution makes it necessary the use of an approximation. We propose one of the algorithms described by Ling and Pratt (38) that is 10 times faster and whose maximum absolute error is below 0.0001. The selected algorithm (Zpp) is a normal approximation of the hypergeometric distribution that performs some log-transformations on the four parameters of the hypergeometric distribution. Since all the TFBSs are tested simultaneously, it is important to deal with multiple-hypothesis testing, performing, for example, FDR correction (49) (*see* **Note 3**).

Once we have a list of enriched TFBSs, the Jaspar (67), Transfac (43), iHOP (25), Ensembl (14), and PubMed (50) databases provide access to the available information about each TFBS and its corresponding TF. These results can be used to discover the regulatory mechanism that is behind the observed gene co-expression (**Fig. 8.4**).

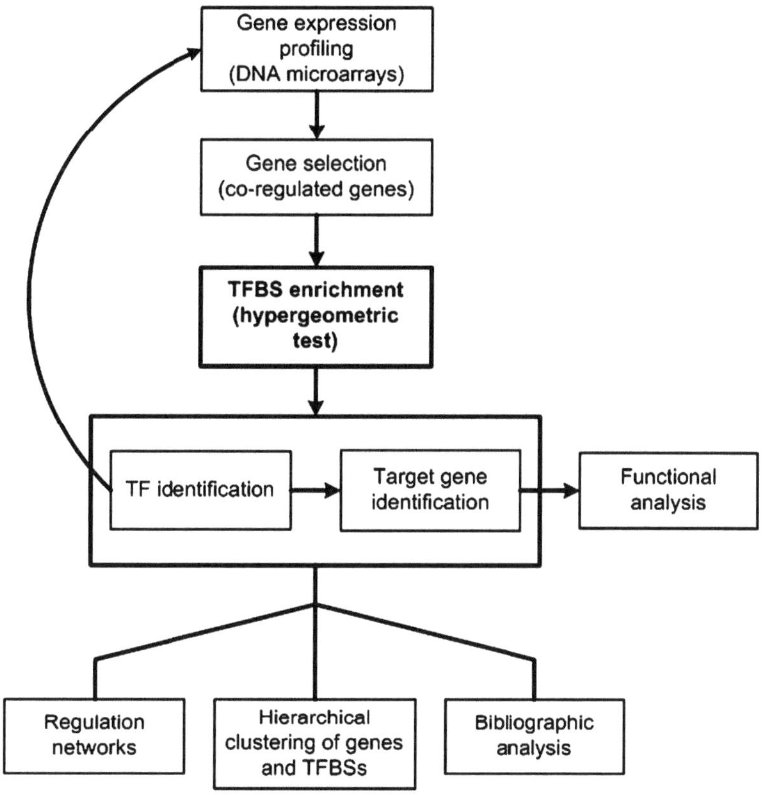

Fig. 8.4. The usage flowchart shows the inputs (probes of the microarray or Ensembl identifiers) and outputs of the analysis. We can perform a TFBS enrichment analysis to predict the TFs that are most likely to regulate the selected set of genes. In addition, a hierarchical clustering of genes and TFBSs, and functional or bibliographic analysis of the results, allow the biological interpretation of the experiment.

The GO enrichment analysis (73) of genes that present a TFBS in their proximal promoter determines which biological processes involved in our analysis can be regulated by the TF corresponding to the selected TBFS. In addition, other bioinformatics resources of functional information, such as Ingenuity (29), can be integrated to the functional analysis of the studied set of genes.

Cluster 3.0 (12, 53) allows the genes and TFBSs to be clustered according to their regulation profiles using a hierarchical clustering method (17, 53). This clustering can be used to find groups of genes with the same TFBSs in their proximal promoter. The observed groups could be due to the sequence similarity of TFBSs of the same family, or they could be evidence of a common regulatory mechanism.

Finally, PubMed (50) helps the researcher to look for papers where the genes and TFBSs relevant for the analysis are related. There are several co-citation and text mining tools, such as PubGene (48), that facilitate understanding the relationships between TFs and genes.

Integrating the results obtained from these sources of information can provide a global picture of the regulation mechanism responsible for the analyzed expression profile. We have developed a web application, FactorY (http://garban.tecnun.es/Factory) (23), which facilitates the complete bioinformatics analysis described in this section. In this way, the identification of common regulatory mechanisms involved in the transcriptional control of coexpressed genes will be carried out using one bioinformatics tool. In addition to the freely accessible website, the entire contents of the database required for the analysis could be downloaded from the main page of FactorY as a MySQL database dump that could be used to reconstitute a local copy of the database.

3.3. TFBS Enrichment Analysis

Several data sets from the literature have been analyzed to evaluate the reliability of the described TFBS enrichment analysis: muscle-specific (69) and liver-specific (34) genes. In both cases, we detect the experimentally verified TFBSs, validating our methodology (*see* **Note 4**). It is also well established that NFκB induces transcription of target genes in response to signal transduction pathways activated by TNF-α (64). A set of 21 genes has been analyzed that were twofold upregulated within 1 h of TNF-α treatment (31). As expected, the most significant TFBSs correspond to NFκB sites (p50, p65, NFkB, and c-REL). The results are summarized in **Table 8.2**.

Table 8.2
Enriched TFBSs in test cases

	TFBS	Name	*n*	*N*	*p*-value	FDR
Muscle-specific (27 genes)	M00184	MYOD	16	8,561	0.000124	0.053088
	M00215	SRF	7	2,378	0.000680	0.096872
	MA0055	MYF	9	3,970	0.001201	0.085492
	MA0052	MEF2	8	3,696	0.002932	0.104363
	M00008	SP1	17	12,210	0.003588	0.117861
	MA0090	TEF1	8	4,106	0.005989	0.134609
Liver-specific (14 genes)	MA0046	HNF1	8	4,679	0.000037	0.016150
	M00134	HNF4	7	5,529	0.001065	0.113739
	MA0047	HNF3	9	10,173	0.003421	0.182645
NFκB targets (21 genes)	M00051	p50	13	3,095	$5.3\ 10^{-10}$	$2.2\ 10^{-7}$
	MA0107	p65	10	3,609	0.000005	0.000825
	MA0061	NFκB	9	3,538	0.000038	0.002037
	MA0101	c-REL	8	5,186	0.004219	0.069304

Test cases were collected from the literature: 27 muscle-specific, 14 liver-specific, and 21 NFκB target genes. We calculated enrichment *p*-values for all TFBSs in Jaspar and Transfac by comparing the number of TFBSs detected in the promoter of collected gene sets (*n*) with the number of TFBSs detected in the whole genome (*N*).

3.4. A Genome-Wide Promoter Analysis: Effects of IFN-αcon1 and IFN-γ1b on Gene Expression in A549 Cells

The microarray experiment of Sanda et al. (55) has been used to validate the utility of our methodology in the interpretation of the obtained enriched TFBSs using publicly available microarray data. The authors examined the effects of a type I interferon (IFN-αcon1) and a type II IFN (IFN-γ1b) on gene expression in A549 cells. It was demonstrated that there is a common set of genes modulated by both IFNs as well as a set of genes specifically regulated by each of them separately, reflecting the activation of different signaling pathways. This experiment consisted of eight nonstimulated samples, eight samples of cells treated with IFN-αcon1, eight samples of cells treated with IFN-γ1b, and eight samples of cells treated with both IFNs. Affymetrix U133A GeneChip was used to interrogate gene expression, and the expression values were generated with MAS5.0 software. Data for all 32 arrays were obtained from Gene Expression Omnibus at NCBI, accession number GSE5542.

The common effect of IFNs and the differences between IFN-αcon1 and IFN-γ1b have been analyzed, excluding the samples treated with both IFNs simultaneously. The selection of those genes that were differentially expressed with respect to the control condition was performed using Limma, a statistical package of Bioconductor (70, 7). For the analysis of the common effect, 139 genes were selected that were differentially expressed in cells stimulated with either IFN-αcon1 or IFN-γ1b (p-value < 0.01, for which $FDR_{IFN-\alpha con1} = 0.019$ and $FDR_{IFN\gamma 1b} = 0.12$). In the analysis of the IFN-αcon1 effect, 23 genes were included, which are up- or downregulated upon IFN-αcon1 stimulation (p-value < 0.01) but were not altered by IFN-γ1b (p-value > 0.5). The effect of IFN-γb1 was analyzed in the same manner, resulting in 200 affected genes (**Fig. 8.5**). The promoter analysis of each set of genes would allow the identification of TFs involved in the cellular response triggered by the different treatments, which are likely responsible for the observed change in the expression of their target genes (**Table 8.3**).

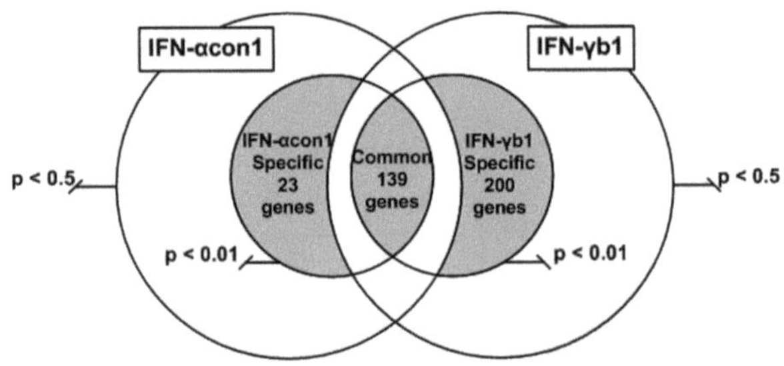

Fig. 8.5. Gene selection criteria based on the p-value are shown in a Venn diagram. The common effect of IFN-αcon1 (p-value < 0.01) and IFN-γ1b (p-value < 0.01) in A549 cells has been analyzed. In the study of the IFN-αcon1-specific effect, genes with a significant expression change in cells treated with IFN-αcon1 (p-value < 0.01) and with no change in cells treated with IFN-γ1b (p-value > 0.5) have been selected. On the other hand, in the study of the IFN-γ1b-specific effect, genes with a significant expression change in cells treated with IFN-γ1b (p-value < 0.01) and with no change in cells treated with IFN-αcon1(p-value > 0.5) have been selected.

In the proposed analysis workflow, first, the enriched TFBSs are detected (**Table 8.3**). Then the Jaspar and Transfac databases are consulted to know which TFs bind the enriched TFBSs. Next, a bibliographic analysis is performed to help understand the relationships between these TFs and the effect of IFNs. Finally, the expression of the TFs and the biological functions in which the differentially expressed genes are involved can be verified with GO enrichment analysis (73) and Ingenuity (29). This analysis provides a global representation of IFN signaling in A549 cells at different levels: TF regulation, regulation of the expression of target genes, and then the functional consequences of these alterations (*see* **Note 5**).

The stimulation of A549 cells with either IFN-αcon1 or IFN-γ1b induced the overexpression of STAT1, STAT2, STAT3, and ISGF3G. The interaction of these TFs with ISRE TFBS in the promoter region regulates the expression of target genes, including those involved in NFκB cascade, JAK-STAT cascade, protein

Table 8.3
Enriched TFBSs in the IFN case study

	TFBS	Name	*n*	*N*	*p*-value	FDR
Common effect	M00258	ISRE	44	1,752	$2.39{\cdot}10^{-9}$	$5.10{\cdot}10^{-7}$
(139 genes)	MA0050	Irf-1	47	2,322	$6.05{\cdot}10^{-7}$	$8.61{\cdot}10^{-5}$
	M00208	NFkB	24	1,285	0.001401	0.099775
	MA0107	p65	30	1,802	0.002591	0.138302
IFN-αcon1 effect	MA0051	IRF2	6	1,008	0.001090	0.155272
(23 genes)	M00415	AREB6	13	3,806	0.001095	0.116921
	M00258	ISRE	8	1,752	0.001540	0.109644
IFN-γb1 effect	M00187	USF	67	3,047	0.000144	0.061762
(200 genes)	MA0101	c-REL	57	2,549	0.000339	0.048351
	M00196	SP1	137	7,817	0.000875	0.062304

In A549 cells, 139 genes were differentially expressed upon stimulation with IFN-αcon1 or IFN-γb1; 23 genes were differentially expressed upon stimulation with IFN-αcon1 but remain unchanged with IFN-γb1; and 200 genes were differentially expressed with IFN-γb1 but remain unchanged with IFN-αcon1. We calculated enrichment *p*-values for all TFBSs in the Jaspar and Transfac databases by comparing the number of TFBSs detected in the promoter of selected genes (*n*) with the number of TFBSs detected in the genes of the HG-U133A microarray (*N*).

ubiquitination, and immune response, specifically antigen processing and presentation (**Fig. 8.6**). These data greatly agree with other studies describing IFN signaling (59, 9). This parallelism suggests that the bioinformatics strategy based on the proposed

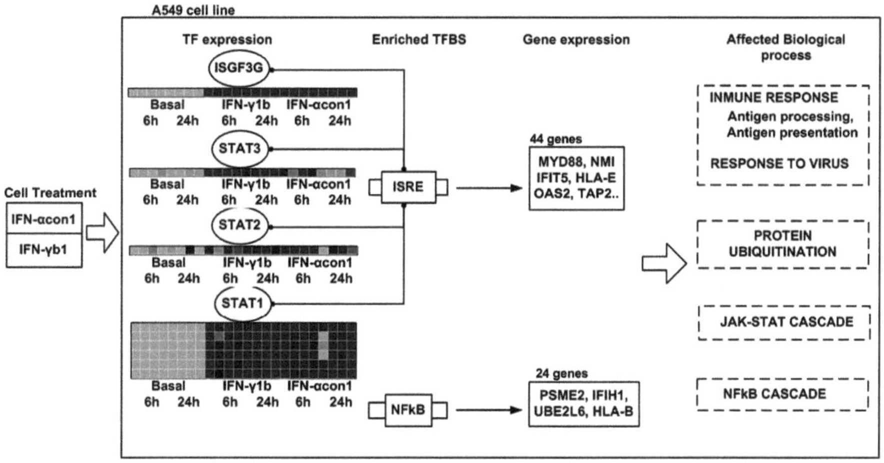

Fig. 8.6. The genome-wide promoter analysis of the probes differentially expressed in A549 cells treated with IFN-αcon1 or IFN-γ1b detects four enriched TFBSs: ISRE, Irf-1, NFκB, and p65. The induced expression of TFs such as STAT1, STAT2, and ISGF3G that bind ISRE is also shown. The results obtained with functional and bibliographic analysis confirm, as expected, that the common effect of IFN-αcon1 and IFN-γ1b agree with the IFN signaling described in the literature.

methodology allows the reliable interpretation of genomic data at the functional level.

Few genes were differentially expressed in response to IFN-αcon1 only, suggesting a weak specific effect in A549 cells. In fact, none of the TFBSs has a *p*-value less than 0.05 after FDR *p*-value correction (**Table 8.3**).

Three regulatory motifs were found to be statistically enriched upon FactorY analysis of genes differentially expressed after A549 cell incubation with IFN-γ1b only: USF, c-REL, and SP1 (**Table 8.3**). USF is a family of ubiquitous transcription factors implicated in the control of cellular proliferation (44) that interacts with STAT1 in the IFN-γ activation of MHC II expression (42). While c-REL is part of the NFκB complex (35), SP1 has been involved in the regulation of several TNF family members, such as TNF-α and TNFSF10 (68), and IFN-γ can modulate SP1 activity by phosphorylation (5, 54). In this case, TNFAIP1 – a member of the TNF-α family – is upregulated. Although SP1 binding to the promoters of HLA-B, HLA-C, and HLA-G has been described, there is no evidence for an important role of SP1 in constitutive or IFN-γ-induced MHC class I transactivation (22). The functional analysis revealed that the treatment with IFN-γ1b activated the immune response by additional pathways to those described under the common effects of IFNs (complement activation, inflammatory response, and antimicrobial humoral response). Programmed cell death signaling is also specifically modulated by IFN-γ1b.

Recently, efforts have been made to predict cis-regulatory modules (CRMs), consisting of more than one TF, by statistical evaluations of the co-occurrence of binding sites within the regulatory regions of a set of genes (58). We should identify the component elements of CRMs, because if two TFs cooperate to influence the expression of a set of genes, the TFBSs of both TFs will be identified as enriched and will be clustered together in the hierarchical clustering. It is simple then to determine which of the TFBSs map to nearby regions of those same genes. In the TFBS clustering of IFN-γ1b specific effect, there is not enough evidence for a CRM between USF and SP1, although physical and functional interactions between them have previously been described (20, 15).

4. Notes

1. The genome-wide promoter analysis complements the information obtained in a microarray experiment, improving the biological conclusions extracted after the interpretation of the results.

2. The prediction of new TFBSs results in a high false positive rate, so it is necessary to restrict the studied sequence. One option is to analyze only the proximal promoter, or it is possible to study the promoter sequence conserved in evolution. If the researcher does not want to lose information, both approaches can be combined.
3. One of the contributions of our methodology with respect to the existing tools for genome-wide promoter analysis (4, 10, 24, 30, 41, 65) is to consider the problem of multiple hypothesis testing. The enrichment of all the Jaspar and Transfac TFBSs is tested simultaneously in this type of analysis, increasing the probability of false positives. In order to reduce the number of false positives, FDR correction of enrichment *p*-value has been used.
4. The proposed methodology detects enriched TFBSs in the proximal promoter of a set of genes. The objective of the analysis is to find the common regulatory mechanism that explains the similar expression profile of a gene set. Our approach has been evaluated with three sets of genes described in the literature: muscle- and liver-specific genes, and NFκB target genes. Regulatory elements known to be important in the analyzed set of genes have been detected, demonstrating that our strategy provides biological insights into the regulatory mechanisms.
5. While existing applications may allow TFBS enrichment analysis, they do not assist the user in the interpretation of the results of TFBS enrichment analysis, which is still a challenging task. The greatest potential of our approach has been demonstrated by the analysis of gene expression after IFN treatment in A549 cells. Publicly available microarray data allowed the selection of genes differentially expressed after IFN-αcon1 treatment alone, genes affected only by the IFN-γ1b, and genes altered with both IFN treatments. Our methodology correctly identified TFBSs expected to be important in the common effect of IFNs and discovered TFs that could explain the additional cellular responses triggered by IFN-γ1b. In addition, combining all the information obtained with our methodology allowed us to summarize the global representation of IFN signaling in A549.

Acknowledgments

This work was funded by the University of Navarra, Fundación para la Investigación Médica Aplicada (FIMA), and the Torres Quevedo program of the Ministerio de Educación y Ciencia in Spain, through the Social European Fund.

References

1. Abnizova I, Gilks WR. (2006) Studying statistical properties of regulatory DNA sequences, and their use in predicting regulatory regions in the eukaryotic genomes. *Brief Bioinform* 7(1):48–54.
2. Aerts S, Thijs G, Coessens B, Staes M, Moreau Y, De Moor B. (2003) Toucan: deciphering the cis-regulatory logic of coregulated genes. *Nucleic Acids Res* 31(6):1753–1764.
3. Aerts S, Thijs G, Dabrowski M, Moreau Y, De Moor B. (2004) Comprehensive analysis of the base composition around the transcription start site in Metazoa. *BMC Genomics* 5(1):34.
4. Aerts S, Van Loo P, Thijs G, Mayer H, de Martin R, Moreau Y, De Moor B. (2005) TOUCAN 2: the all-inclusive open source workbench for regulatory sequence analysis. *Nucleic Acids Res* 33(Web Server issue): 393–396.
5. Amin MR, Malakooti J, Sandoval R, Dudeja PK, Ramaswamy K. (2006) IFN-gamma and TNF-alpha regulate human NHE3 gene expression by modulating the Sp family transcription factors in human intestinal epithelial cell line C2BBe1. *Am J Physiol Cell Physiol* 291(5):887–896.
6. Berg OG, von Hippel PH. (1987) Selection of DNA binding sites by regulatory proteins. Statistical-mechanical theory and application to operators and promoters. *J Mol Biol* 193(4):723–750.
7. Bioconductor. http://www.bioconductor.org.
8. Blackwood EM, Kadonaga JT. (1998) Going the distance: a current view of enhancer action. *Science* 281(5373):60–63.
9. Brierley MM, Fish EN. (2002) Review: IFN-alpha/beta receptor interactions to biologic outcomes: understanding the circuitry. *J Interferon Cytokine Res* 22(8):835–845.
10. Chang LW, Nagarajan R, Magee JA, Milbrandt J, and Stormo GD. (2006) A systematic model to predict transcriptional regulatory mechanisms based on overrepresentation of transcription factor binding profiles. *Genome Res* 16(3):405–413.
11. Cheng J, Kapranov P, Drenkow J, Dike S, Brubaker S, Patel S, Long J, Stern D, Tammana H, Helt G, Sementchenko V, Piccolboni A, Bekiranov S, Bailey DK, Ganesh M, Ghosh S, Bell I, Gerhard DS, Gingeras TR. (2005) Transcriptional maps of 10 human chromosomes at 5-nucleotide resolution. *Science* 308(5725):1149–1154.
12. Cluster 3.0. http://www.geo.vu.nl/huik/cluster.htm.
13. Costas J, Casares F, Vieira J. (2003) Turnover of binding sites for transcription factors involved in early *Drosophila* development. *Gene* 310(May):215–220.
14. Cuff JA, Coates GM, Cutts TJ, Rae M. (2004) The Ensembl computing architecture. *Genome Res* 14(5):971–975.
15. deGraffenried LA, Hopp TA, Valente AJ, Clark RA, Fuqua SA. (2004) Regulation of the estrogen receptor alpha minimal promoter by Sp1, USF-1 and ERalpha. *Breast Cancer Res Treat* 85(May):111–120.
16. Dermitzakis ET, Clark AG. (2002) Evolution of transcription factor binding sites in mammalian gene regulatory regions: conservation and turnover. *Mol Biol Evol* 19(7): 1114–1121.
17. Eisen MB, Spellman PT, Brown PO, Botstein D. (1998) Cluster analysis and display of genome-wide expression patterns. *Proc Natl Acad Sci USA* 95(25):14863–14868.
18. Emberly E, Rajewsky N, Siggia ED. (2003) Conservation of regulatory elements between two species of *Drosophila*. *BMC Bioinformatics* 4(Nov):57.
19. Frech K, Quandt K, Werner T. (1997) Finding protein-binding sites in DNA sequences: the next generation. *Trends Biochem Sci* 22(3):103–104.
20. Ge Y, Jensen TL, Matherly LH, Taub JW. (2003) Physical and functional interactions between USF and Sp1 proteins regulate human deoxycytidine kinase promoter activity. *J Biol Chem* 278(50):49901–49910.
21. GO Consortium. (2006) The Gene Ontology (GO) project in 2006. *Nucleic Acids Res* 34(Database issue):322–326.
22. Gobin SJ, van Zutphen M, Woltman AM, van den Elsen PJ. 1999. Transactivation of classical and nonclassical HLA class I genes through the IFN-stimulated response element. *J Immunol* 163(3):1428–1434.
23. Hertz GZ, Stormo GD. (1999) Identifying DNA and protein patterns with statistically significant alignments of multiple sequences. *Bioinformatics* 15(7–8):563–577.
24. Ho Sui SJ, Mortimer JR, Arenillas DJ, Brumm J, Walsh CJ, Kennedy BP, Wasserman WW. (2005) oPOSSUM: identification of over-represented transcription factor binding sites in co-expressed genes. *Nucleic Acids Res* 33(10):3154–3164.
25. Hoffmann R, Valencia A. (2004) A gene network for navigating the literature. *Nat Genet* 36(7):664–664.
26. Hoheisel JD. (2006). Microarray technology: beyond transcript profiling and

genotype analysis. *Nat Rev Genet* 7(3): 200–210.
27. Hubbard TJ, Aken BL, Beal K, Ballester B, Caccamo M, Chen Y, Clarke L, Coates G, Cunningham F, Cutts T, Down T, Dyer SC, Fitzgerald S, Fernandez-Banet J, Graf S, Haider S, Hammond M, Herrero J, Holland R, Howe K, Howe K, Johnson N, Kahari A, Keefe D, Kokocinski F, Kulesha E, Lawson D, Longden I, Melsopp C, Megy K, Meidl P, Ouverdin B, Parker A, Prlic A, Rice S, Rios D, Schuster M, Sealy I, Severin J, Slater G, Smedley D, Spudich G, Trevanion S, Vilella A, Vogel J, White S, Wood M, Cox T, Curwen V, Durbin R, Fernandez-Suarez XM, Flicek P, Kasprzyk A, Proctor G, Searle S, Smith J, Ureta-Vidal A, Birney E. (2007) Ensembl 2007. *Nucleic Acids Res* 35(Database issue):610–617.
28. Hughes JD, Estep PW, Tavazoie S, Church GM. (2000) Computational Identification of cis-regulatory elements associated with groups of functionally related genes in *Saccharomyces cerevisiae*. *J Mol Biol* 296(5):1205–1214.
29. Ingenuity ®Systems. http://www.ingenuity.com.
30. Jegga AG, Sherwood SP, Carman JW, Pinski AT, Phillips JL, Pestian JP, Aronow BJ. (2002) Detection and visualization of compositionally similar cis-regulatory element clusters in orthologous and coordinately controlled genes. *Genome Res* 12(9):1408–1417.
31. Karanam S, Moreno CS. (2004) CONFAC: automated application of comparative genomic promoter analysis to DNA microarray datasets. *Nucleic Acids Res* 32(Web server issue):475–484.
32. Kasprzyk A, Keefe D, Smedley D, London D, Spooner W, Melsopp C, Hammond M, Rocca-Serra P, Cox T, Birney E. (2004) EnsMart: a generic system for fast and flexible access to biological data. *Genome Res* 14(1):160–169.
33. Kel AE, Gössling E, Reuter I, Cheremushkin E, Kel-Margoulis OV, Wingender E. (2003) MATCH: a tool for searching transcription factor binding sites in DNA sequences. *Nucleic Acids Res* 31(13):3576–3579.
34. Krivan W, Wasserman WW. (2001) A predictive model for regulatory sequences directing liver-specific transcription. *Genome Res* 11(9):1559–1566.
35. Kunsch C, Ruben SM, Rosen CA. (1992) Selection of optimal kappa B/Rel DNA-binding motifs: interaction of both subunits of NF-kappa B with DNA is required for transcriptional activation. *Mol Cell Biol* 12(10):4412–4421.
36. Lawrence CE, Altschul SF, Boguski MS, Liu JS, Neuwald AF, and Wootton JC. (1993) Detecting subtle sequence signals: a Gibbs sampling strategy for multiple alignment. *Science*, 262(5131):208–214.
37. Lawrence CE, Reilly AA. (1990) An expectation maximization (em) algorithm for the identification and characterization of common sites in unaligned biopolymer sequences. *Proteins* 7(1):41–51.
38. Ling RF, Pratt JW. (1984) The accuracy of Peizer approximations to the hypergeometric distribution, with comparisons to some other approximations. *J Am Stat Assoc* 79(385):49–60.
39. Liu ET. (2005) Gene array technologies in biological investigations. *Proc IEEE* 93(4):737–749.
40. Liu XS, Brutlang DL, Liu JS. (2002) An algorithm for finding protein-DNA binding sites with applications to chromatin-immunoprecipitation microarray experiments. *Nat Biotechnol* 20(8):835–839.
41. Loots GG, Ovcharenko I, Pachter L, Dubchak I, Rubin EM. (2002) rVista for comparative sequence-based discovery of functional transcription factor binding sites. *Genome Res* 12(5):832–839.
42. Luo X, Sawadogo M. (1996) Antiproliferative properties of the USF family of helix-loop-helix transcription factors. *Proc Natl Acad Sci USA* 93(Feb):1308–1313.
43. Matys V, Kel-Margoulis OV, Fricke E, Liebich I, Land S, Barre-Dirrie A, Reuter I, Chekmenev D, Krull M, Hornischer K, Voss N, Stegmaier P, Lewicki-Potapov B, Saxel H, Kel AE, Wingender E. (2006) TRANSFAC and its module TRANSCompel: transcriptional gene regulation in eukaryotes. *Nucleic Acids Res* 34(Database issue):108–110.
44. Muhlethaler-Mottet A, Di Berardino W, Otten LA, Mach B. (1998) Activation of the MHC class II transactivator CIITA by interferon-gamma requires cooperative interaction between Stat1 and USF-1. *Immunity* 8(2):157–166.
45. Nardone J, Lee DU, Ansel KM, Rao A. (2004) Bioinformatics for the "bench biologist": how to find regulatory regions in genomic DNA. *Nat Immunol* 5(8):768–774.
46. Novina CD, Roy AL. (1996) Core promoters and transcriptional control. *Trends Genet* 12(9):351–355.
47. Ohler U, Niemann H. (2001) Identification and analysis of eukaryotic promoters: recent computational approaches. *Trends Genet* 17(2):56–60.
48. Pearson H. (2001) Biology's name game. *Nature* 411(June):631–632.

49. Pounds SB. (2006) Estimation and control of multiple testing error rates for microarray studies. *Brief Bioinform* 7(1):25–36.
50. PubMed. http://www.pubmed.gov.
51. Qiu P. (2003) Recent advances in computational promoter analysis in understanding the transcriptional regulatory network. *Biochem Biophys Res Commun* 309(3):495–501.
52. Quandt K, Frech K, Karas H, Wingender E, Werner T. (1995) MatInd and MatInspector: new fast and versatile tools for detection of consensus matches in nucleotide sequence data. *Nucleic Acids Res* 23(23):4878–4884.
53. Saldanha AJ. (2004) Java TreeView – extensible visualization of microarray data. *Bioinformatics* 20(17):3246–3248.
54. Sancéau J, Kaisho T, Hirano T, Wietzerbin J. (1995) Triggering of the human interleukin-6 gene by interferon-gamma and tumor necrosis factor-alpha in monocytic cells involves cooperation between interferon regulatory factor-1, NF kappa B, and Sp1 transcription factors. *J Biol Chem* 270(46):27920–27931.
55. Sanda C, Weitzel P, Tsukahara T, Schaley J, Edenberg HJ, Stephens MA, McClintick JN, Blatt LM, Li L, Brodsky L, Taylor MW. (2006) Differential gene induction by type I and type II interferons and their combination. *J Interferon Cytokine Res* 26(7): 462–472.
56. Schmid CD, Périer R, Praz V, Bucher P. (2006) EPD in its twentieth year: towards complete promoter coverage of selected model organisms. *Nucleic Acids Res* 34(Database issue):82–85.
57. Schulze A, Downward J. (2001) Navigating gene expression using microarrays – a technology review. *Nat Cell Biol* 3(8): 190–195.
58. Sharan R, Ben-Hur A, Loots GG, Ovcharenko I. (2004) CREME: Cis-Regulatory Module Explorer for the human genome. *Nucleic Acids Res* 32(Web server issue):253–256.
59. Stark GR, Kerr IM, Williams BR, Silverman RH, Schreiber RD. (1998) How cells respond to interferons. *Annu Rev Biochem* 67:227–264.
60. Stormo GD. (2000) DNA binding sites: representation and discovery. *Bioinformatics* 16(1):16–23.
61. Suzuki Y, Yamashita R, Sugano S, Nakai K. (2004) DBTSS, DataBase of Transcriptional Start Sites: progress report 2004. *Nucleic Acids Res* 32(Database issue):78–81.
62. Tompa M, Li N, Bailey TL, Church GM, De Moor BD, Eskin E, Favorov AV, Frith MC, Fu Y, Kent JJ, Makeev VJ, Mironov AA, Noble WS, Pavesi G, Pesole G, Régnier M, Simonis N, Sinha S, Thijs G, van Helden J, Vandenbogaert M, Weng Z, Workman C, Ye C, Zhu Z. (2005) Assessing computational tools for the discovery of transcription factor binding sites. *Nat Biotechnol*, 23(1): 137–144.
63. Ueda HR, Chen W, Adachi A, Wakamatsu H, Hayashi S, Takasugi T, Nagano M, Nakahama K, Suzuki Y, Sugano S, Iino M, Shigeyoshi Y, Hashimoto S. (2002) A transcription factor response element for gene expression during circadian night. *Nature* 418(6897):534–539.
64. van Antwerp DJ, Martin SJ, Verma IM, Green DR. (1998) Inhibition of TNF-induced apoptosis by NF-kappa B. *Trends Cell Biol* 8(3):107–111.
65. van Helden J, André B, Collado-Vides J. (2000). A web site for the computational analysis of yeast regulatory sequences. *Yeast* 16(2):177–187.
66. van Helden J, Rios AF, Collado-Vides J. (2000) Discovering regulatory elements in non-coding sequences by analysis of spaced dyads. *Nucleic Acids Res* 28(8):1808–1818.
67. Vlieghe D, Sandelin A, De Bleser PJ, Vleminckx K, Wasserman WW, van Roy F, Lenhard B. (2006) A new generation of JASPAR, the open-access repository for transcription factor binding site profiles. *Nucleic Acids Res* 34(Database issue):95–97.
68. Wang Q, Ji Y, Wang X, Evers BM. (2000) Isolation and molecular characterization of the 5'-upstream region of the human TRAIL gene. *Biochem Biophys Res Commun* 276(2):466–471.
69. Wasserman WW, Palumbo M, Thompson W, Fickett JW, Lawrence CE. (2000) Human-mouse genome comparisons to locate regulatory sites. *Nat Genet* 26(2):225–228.
70. Wettenhall JM, Smyth GK. (2004) limmaGUI: a graphical user interface for linear modeling of microarray data. *Bioinformatics* 20(18):3705–3706.
71. Wray GA, Hahn MW, Abouheif E, Balhoff JP, Pizer M, Rockman MV, Romano LA. (2003) The evolution of transcriptional regulation in eukaryotes. *Mol Biol Evol* 20(9):1377–1419.
72. Xuan Z, Zhao F, Wang J, Chen G, Zhang MQ. (2005) Genome-wide promoter extraction and analysis in human, mouse, and rat. *Genome Biol* 6(8):R72.
73. Zhang B, Kirov S, Snoddy J. (2005) WebGestalt: an integrated system for exploring gene sets in various biological contexts. *Nucleic Acids Res* 33(Web server issue): 741–748.

Chapter 9

Proteomics Facing the Combinatorial Problem

Rune Matthiesen and António Amorim

Abstract

A large number of scoring functions for ranking peptide matches to observed MS/MS spectra have been discussed in the literature. In contrast to scoring functions, search strategies have received less attention, and an accurate description of search algorithms is limited. Proteomics is becoming more and more commonly used in potential clinical applications; for such approaches to be successful, the combinatorial problems from amino acid modifications and somatic and heredity SAPs (single amino acid substitutions) need to be seriously considered. The modifications and SAPs are problematic since MS and MS/MS search algorithms are optimization processes, which means that if the correct match is not iterated through during the search, then the data will be matched incorrectly, resulting in serious downstream flaws. This chapter discusses several search algorithm strategies in more detail.

Key words: MS/MS, algorithms, search engine, database-dependent search, de novo sequencing.

1. Introduction

The details on how to obtain MS and MS/MS data from proteomics samples have been reviewed elsewhere (1–3). Interpreting MS and MS/MS data obtained from protein samples is a challenging pattern-matching problem, and several papers have discussed scoring functions that use hidden Markov models and other probabilistic models that recognize mass and peak intensity patterns of typically a-, b-, and y-ions for collision induced dissociation (CID) (4–7) (**Fig. 9.1**). Others have focused on accurate calibration of MS or MS/MS against all potential matches in a database (8, 9). The focus of this chapter is to review and discuss search algorithms for MS/MS and MS data. This includes database-dependent searches (10–12),

R. Matthiesen (ed.), *Bioinformatics Methods in Clinical Research*, Methods in Molecular Biology 593,
DOI 10.1007/978-1-60327-194-3_9, © Humana Press, a part of Springer Science+Business Media, LLC 2010

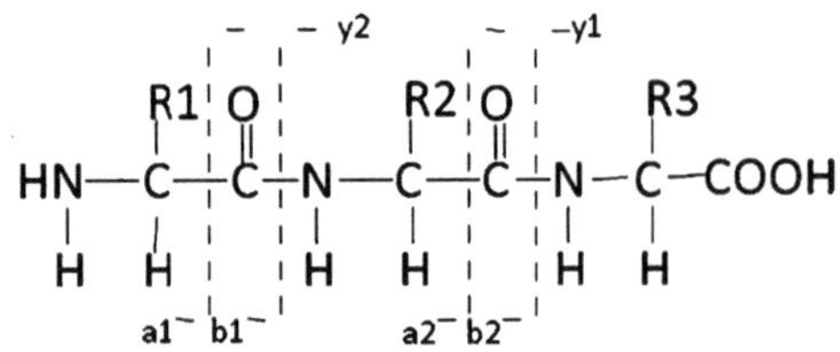

Fig. 9.1. Typical ion fragments observed for low-energy, collision-induced ion dissociation (CID).

de novo algorithms (13), tagging algorithms (14), library searches, and genome searches (15, 16).

2. Algorithms

2.1. Database-Dependent Searches

The database-dependent search algorithms iterate through a database to find optimal solutions that match input MS/MS spectra. If the database is indexed, then the search will be faster, but the indexed databases often assume a certain protein cleavage method and will require that a new index be built for each supported cleavage method. The theory provided here is therefore given for an algorithm that iterates through the raw FASTA sequence files for every search. The outline here is similar to Matthiesen et al. (8), but more details are provided.

Assuming *C*-specific cleavage sites (most likely trypsin cleavage) and *U* as the upper limit of missed cleavage sites ($U \leq C$), then the number of *in silico* peptide generated is given by

$$t = (U + 1)(C + 1) - \frac{u^2 + u}{2}. \qquad [1]$$

It is an advantage to design the algorithm to use regular expression for the recognition of cleavage sites so that the algorithm is as flexible as possible. The total number *N* of potential modified peptides from a specific protein given a set of variable modifications and a number of maximum missed cleavages *C* is given by the following equation:

$$N = \sum_{j=1}^{t} \prod_{i=1}^{n} (V_{i,j} + 1), \qquad [2]$$

where *j* is iterated over all *t* peptides given by Equation [**1**]. V_i is the number of possible variable modifications at residue *i* in peptide *j*. The problem with Equation [**2**] is that it will iterate through many solutions for which we have no parent ion, resulting in unnecessary time-consuming loops. A more clever iteration scheme is required. The trick is to split the iteration into two parts: (i) one that iterates over possible parent ion masses; if the parent ion mass exists, then (ii) the iteration over possible positions of the variable modifications will be searched. N_m, the number of possible combinations of variable modifications independent of position in the sequence, is given by

$$N_m = \prod_{i=1}^{N_v} \frac{\prod_{j=1}^{N_{V,aa}-1} (N_{aa,i} + j)}{(N_{V,aa})!}, \qquad [3]$$

where N_V is the number of variable modifications specified in the search, $N_{V,aa}$ is the number of potential modifications for a particular amino acid where the unmodified amino acid is not counted, and $N_{aa,i}$ is the number of amino acids for which the variable modification i is possible. Equation [**3**] is complicated to implement and is presented here only for theoretical reasons. An alternative and simpler presentation, which gives the same result but is easier to implement, is

$$N_m = \prod_{i=1}^{N_v} (N_{v,i} + 1), \qquad [4]$$

where $N_{v,i}$ is the total number of modification sites in the peptide for the variable modification i. The calculation of N_m possible parent ion masses can now be iterated through by using the same principle as counting. First, two vectors of length N_v are created:

$$UL = [N_{v,1}, N_{v,2}, \ldots, N_{v,n}],$$

$$C = [0, 0, \ldots, 0] \cdot$$

The first vector defines the UL upper limits for a specific variable modification and the second vector C the starting values. The zero values indicated the combination where none of the variable modifications is selected, and the parent ion mass calculated will correspond to the unmodified peptide. The iteration is exemplified below for $UL = [1, 1, 1]$:

$$[0,0,0]_0 => [0,0,1]_1 => [0,1,0]_2 => [0,1,1]_3 =>$$

$$[1,0,0]_4 => [1,0,1]_5 => [1,1,0]_6 => [1,1,1]_7 \cdot$$

Equation [**4**] gives 2 * 2 * 2 = 8 combinations. The following pseudo-algorithm can be used to generate the combinations:

```
NextComposition(C, UL)
For I = length(C) to 1
    If Ci< ULi then
       Ci=Ci+1
       Return Ci
      Ci= 0


AllParentMasses(Mp, UL, C, Nm)
C = [0, ,0]
For i=1 to Nm
    M=SumMassModifications(C) + Mp
    If M exist in mass list then
       ScorePeptidesWithCurrentModificationComposition
  NextComposition(C, UL)
```

The number of combinations due to different positions of the variable modifications for a certain composition in Equation [**5**] is given by

$$N_C = \prod_{i}^{N_V} \binom{N_{Caa,i}}{N_{aa,i}}, \qquad [5]$$

where $N_{\mathrm{Caa},i}$ is the number of variable modifications *i* existing in a certain composition. M in the function AllParentMasses are the calculated parent ion masses of the modified peptides. M_p is the parent ion of the unmodified peptide. The iteration through all possible positions of the modification can be avoided by using spectral convolution by using the Fast Fourier Transform. However, this approach will not resolve the position of the modifications. The generation of the N_C peptide starts with the creation of the matrix below. This example is given for the composition where the peptide has one oxidated methionine and two phosphorylations. For clarity, matrices are shown, but in reality sparse matrices are computationally more efficient. "1" indicates that the modification at this position is selected. Bold residues indicate amino acids that can be modified by phosphorylation, and underlined residues can be modified by oxidation.

```
                       THTLTFTLMLK
Phosphorylation        00001010000
Methionine oxidation   00000000100
```

The N_C peptide combinations can now be created by the matrix iteration below:

```
Iteration 0:
                       THTLTFTLMLK
Phosphorylation        00001010000
Methionine oxidation   00000000100
Iteration 1:
                       THTLTFTLMLK
Phosphorylation        00100010000
Methionine oxidation   00000000100
```

```
Iteration 2:
                        THTLTFTLMLK
Phosphorylation         00101000000
Methionine oxidation    00000000100

Iteration 3:
                        THTLTFTLMLK
Phosphorylation         10000010000
Methionine oxidation    00000000100

Iteration 4:
                        THTLTFTLMLK
Phosphorylation         10001000000
Methionine oxidation    00000000100

Iteration 5:
                        THTLTFTLMLK
Phosphorylation         10100000000
Methionine oxidation    00000000100
```

The algorithm works by starting at the first "1" on the right and iterates to the left until a "1" that can be moved to the next possible position is encountered (it cannot be moved if the next position has a "1" and the amino acid at the new position should be able to accommodate the modification). For example, for iteration 0 to 1 for phosphorylation, the "1" on the right cannot be moved to the left since F cannot accommodate a phosphorylation and the next possible position is already occupied by a "1." For each iteration, all the ones ("1"s) right of the moved "1" are moved back to the start position on the right (this happens in iteration 3 above).

The total number of peptides for a protein is given by

$$N = \sum_{j=1}^{t} \sum_{k=1}^{N_m} \prod_{i=1}^{N_V} \binom{N_{Caa,i,k,j}}{N_{aa,i,k,j}} \qquad [6]$$

The advantage with the above full combinatorial approach is that it is a "brute-force algorithm," which means that it considers all possible candidates in the database that fulfill the search criteria (*see* **Note 1**). If the used scoring function is 100% correct, the sequence database is complete, and if all relevant modifications, unspecific cleavages, and SAPs are considered, then it is guaranteed to find the optimal solution. The disadvantage is that it is computationally prohibitive to search all possible modifications and SAPs.

2.2. De novo Algorithms

De novo sequencing would be easy if, first, the peptide fragmentation in the MS/MS spectra are ideal in the sense that only one type of cleavage between the amino acids occurs and the intensity of each cleavage fragment is uniformly distributed and, second, only one of the two fragments generated for each cleavage is detected (*see* **Note 2**). However, the reality is that we have multiple cleavage products generating typically a-, b-, and y-ions for CID data. In addition, one often observes neutral losses of H_2O (−18 *m/z*) and NH_3 (−17 *m/z*) and combinations of H_2O and

NH_3 losses. If the peptide contains modifications, then additional neutral losses can be observed. In the low-mass region, diagnostic ions from amino acids and modifications can aid in the interpretation of the MS/MS spectrum. Another complication is that some MS/MS spectra contain a fragmentation spectrum from several peptides with similar parent ion mass and reverse-phase retention time; *see* Bunkenborg et al. for a review on MS/MS spectra fragmentation (**3**). Early approaches for de novo sequencing were based on the generation of all amino acid sequences and corresponding *theoretical spectra*. Calculating all possible sequences will generate large databases, which is not very practical. The number of sequences is given by 20^l, where l is the length of the peptide sequence (*see* **Note 3**).

To alleviate this problem, de novo sequencing using graph theory was introduced in SHERENGA (**13**). SHERENGA constructs optimal path scoring in the graphics representations of MSMS spectra (*see* **Fig. 9.2**).

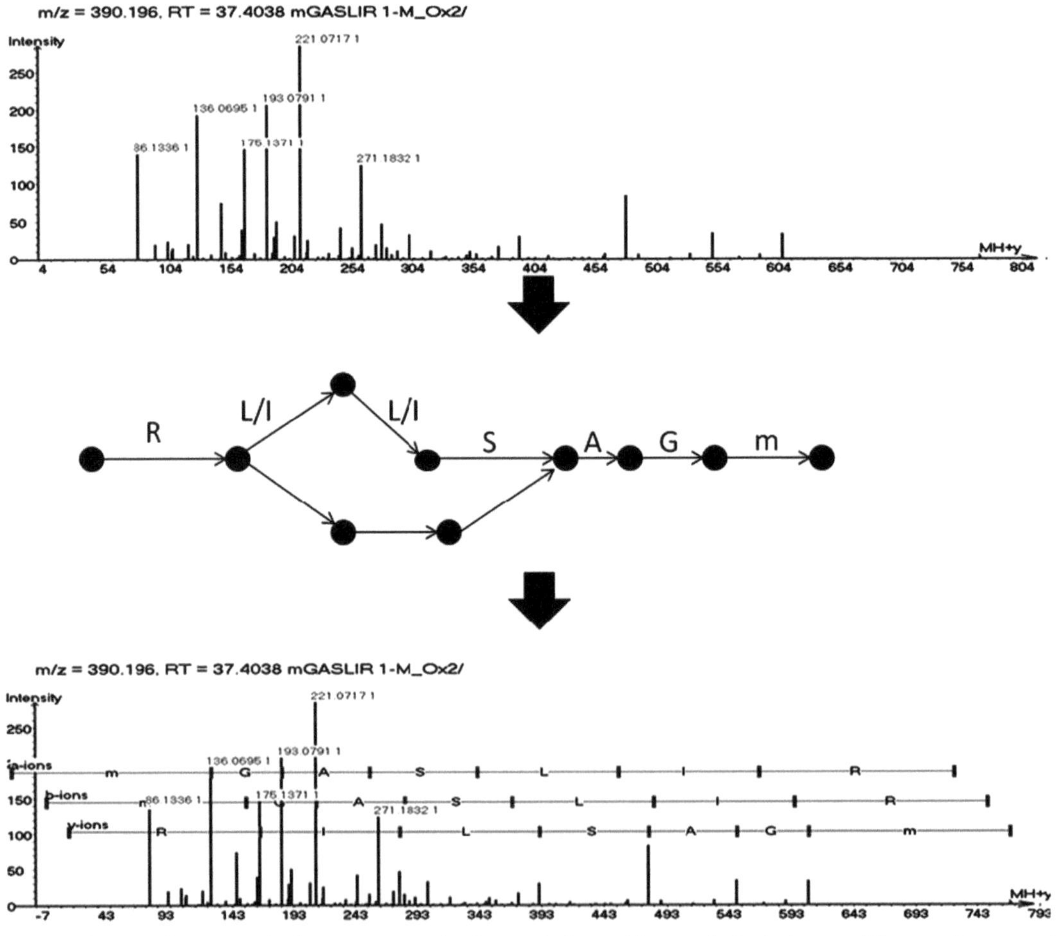

Fig. 9.2. Outline of the steps in de novo sequencing by graph theory. The graph illustrates possible ways (sequences) along the spectrum. Graphs obtained from real spectrum data are often much more complex than outlined here due to noise peaks, missing peaks, and the different ion types.

The first step in the graph-theoretic approach is to find all paths that start from 1.007276 Da (proton mass/start point for calculating the b-ion series) or 19.01 Da (H_2O + proton/start point for calculating the y-ion series) to the parent ion mass. It is generally recommended to consider and exclude complement masses ($m_{p2+}-m_{y+}$ or $m_{p2+}-m_{b+}$) and neutral losses ($-NH_3$, $-H_2O$, and loss of both $-NH_3$ and $-H_2O$) during the construction of the graph so that the number of possible graphs to be considered is lowered (*see* **Note 4**). It is not always possible to start at 1.007276 Da or 19.0178411381 Da and end at the parent ion mass if there are missing fragment ions. In this case, one has to consider all reasonable peaks as a start and end node and report the result as m1-sequence-m2 (a sequence tag). The outline of the graph theory provided here is conceptual since several publications give a full account of graph algorithms for de novo sequencing (13, 17–20).

2.3. Tag Algorithms

In the sequence tag approach, it is not necessary to specify enzyme cleavage. The aim is to define a subsequence with a high probability of being correct (5, 14). Note that this aim is slightly different from the aim in de novo sequencing, where one attempts to define the full peptide sequence. A tag algorithm can use a de novo algorithm as a starting point and then use the results from the de novo algorithm to define a subsequence that explains a certain percentage of total ion intensity of the spectrum. The scoring algorithm should also consider if the relative ion intensity of a-, b-, y-ions and neutral loss ions is reasonable. This can, for example, be done by a hidden Markov model (7) or by estimating the frequency offset function (13). The neutral loss ions frequently have a lower intensity than the corresponding a-, b-, or y-ion, although exceptions to this rule of thumb exist. For example, the neutral loss of phosphoric acid from the parent ion of a phosphopeptide is often the most intense fragment ion in the MS/MS spectra.

The defined subsequence or sequence tag can now be used to search a text-indexed sequence database. This can be done, for example, by building a suffix tree or suffix array of the FATSA database (21). Since the sequence tag is often small, many possible candidate peptides are extracted. The masses m_1, m_2, and m_p are therefore useful for further filtering the candidate sequences (**Fig. 9.3**).

An altered tag algorithm can also be used in a two-stage approach to identify more modifications and amino acid substitutions. In the first step, either a database-dependent search or the tag algorithm is used to identify confident proteins. In the second step, an altered version of the tag algorithm is used to search the confident-identified proteins from the first step. In the second step, any subset of—one or two modifications from UniMod or amino acid substitutions is allowed for each peptide, where a protein subsequence is matched to a sequence tag, but where m_1, m_2, and m_p do not fit.

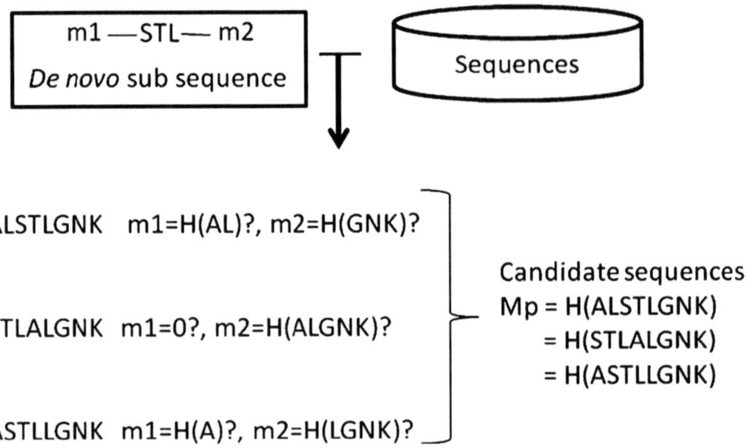

Fig. 9.3. Outline of the sequence tag approach. H is a function that calculates the mass of a peptide sequence. The list of candidate sequences on the right can be further filtered by considering m_1 and m_2.

2.4. Library Search

Recently, a new search strategy was introduced that is based on a library of all observed tryptic peptides in a large set of experiments (22). These observed peptides are referred to as *proteotypic* peptides. The proteotypic peptides can conveniently be stored in a relational database using the accurate parent ion mass of the peptide as an index. Given the high-mass accuracy of the observed parent ion mass, obtained by MS instruments nowadays, only a few *in silico* peptide candidates' theoretical spectra need to be compared with the observed spectra. The library search is useful for well-characterized samples that have been analyzed multiple times. Furthermore, the library can be used to design a subset of interesting peptides to follow quantitatively. The PeptideAtlas project provides a database from which proteotypic peptides can be extracted based on their *empirical observability score* (*EOS*), which acts as an approximate likelihood (23). For example, in QconCAT, one clones concatamers of proteotypic peptides and expresses the concatamers in *E. coli* using ^{13}C Arg- and Lys-enriched media (24). The trypsin-digested concatamers can then be spiked into biological samples to obtain absolute quantitation values. In this case, EOS can be used to define which peptides to clone.

Currently, the relative abundances of hundreds of peptides can be measured by selected reaction monitoring [SRM; also called multiple reaction monitoring (MRM)] methods. In SRM, the mass spectrometer only follows preselected ions and thereby achieves less redundancy, high sensitivity, and high throughput (25). The extraction and choice of proteotypic peptides should be carefully selected since not only is observation ability important, but the uniqueness of the peptide sequence also needs to be considered. For proteins that have not yet been sampled, predictors based on the information in PeptideAtlas (26) can be applied.

2.5. Genome Searches

It has been shown that many transcribed genes have alternative translation start sites that currently are not correctly annotated in NCBI and Ensembl (27). It is furthermore anticipated that proteomics data together with transcriptome data can aid in improving the annotation of transcription start sites and define exon–intron boundaries. The PeptideAtlas project maps MS/MS-identified peptide sequences to the genome sequences (28). Although useful in other contexts, it is of little use in terms of identifying new genome regions that are transcribed or translated. Searching the genome directly with all combinations of modifications and exon–intron boundary is a time-intensive computation. A sensible approach would be to reuse gene predictors to predict a larger set of genes, many of which have predicted scores below the normal acceptable threshold scores. The sequence output from such a prediction could be used as input in a database-dependent search approach.

3. Discussion

The introduction of mass spectrometry in clinical studies has to be done with care. One major issue is the high number of somatic and germinal mutations that are not fully covered in current sequence databases. Although many human inherited SNPs are cataloged in dbSNP (29), we currently do not know all the possible combinations of SNPs that can occur in real sequences. Take, for instance, the human genomic region containing the major histology complexes (MHC), which has more than 7,500 common SNPs associated (30). It is possible that careful linkage disequilibrium analysis can lower the number of combinations to be considered. However, the SNPs together with the high number of modifications continue to be a challenge for database-dependent search algorithms since current search algorithms work by optimization. The search engines find the best hit based on a sequence database and a specified set of modifications. However, if the correct match is not present, the search engine will present the next-best hit. The next-best match can obtain a score that is very close to the correct match.

A bioinformatics solution to alleviate this problem is to design search engines that consider modifications and SNPs that are already annotated in, for example, Swiss Prot and the Human Proteome Reference Database (HPRD) (31). This strategy is currently being used by X!tandem (32), Phenyx (4), and VEMS (33). VEMS allows a full combinatorial approach of the annotated modifications and SNPs with the limitation of a maximum of 10 SNPs per peptide. The full combinatorial search for SNP variants requires distribution of the search to several computers in order to accomplish the search within a day.

A major objective, from the experimental point of view, is to be able to obtain better fragmentation coverage in the MS/MS spectra. If a more uniform fragmentation is obtainable, then all problems with SAPs and modifications will be manageable and de novo algorithms will become the superior interpretation option. Furthermore, specialized fragmentation methods for peptides with posttranslational modifications are likely to be further improved. For example, ETD (electron transfer dissociation) and ECD (electron capture dissociation) have already proved useful for the analysis of phospho- and glyco-peptides (34). It is therefore likely that the next generation of MS proteomics search algorithms needs to handle the combined information from spectra obtained by multiple techniques on the same peptide.

4. Notes

1. A brute-force algorithm finds the optimal solution by trying all possible solutions and are therefore computationally expensive. In contrast, heuristic algorithms are based on applying rules to guide the search for the optimal solution. Heuristic algorithms are computationally less expensive but often cannot guarantee that the optimal solution will be found.
2. An ideal fragmentation method creates one fragment ion for each peptide bond cleavage, and the intensity of the fragment ions should be uniform. Such spectra would be straightforward to interpret by de novo algorithms and would give maximum sensitivity. The reality is that the most frequent fragmentation method (which is collision-induced dissociation, or CID) gives the most uniform fragmentation of doubly charged peptides. These doubly charged peptides fragment mostly into two single-charged peptides that are recorded by the detector, giving rise to the y- and b-ion series (remember that the fragmentation occurs at multiple peptide bonds, which gives rise to series of ions for which the mass difference between peaks in a series correspond to amino acid residue masses). The single-charged y- and b-ions can fragment further, giving rise to ions such as

internal fragment ions, a-ions, and ions for which the neutral loss of H_2O, NH_3, or both has occurred.

3. The number of combinations to consider is even worse if posttranslational modifications, missed cleavage patterns (a cleavage site that in theory should be cleaved but is observed not to be), and miss cleavage patterns (an unspecific cleavage site) are considered as well.
4. It is generally recommended to consider and exclude complement masses (e.g., for doubly charged peptides, the complementary ions can be calculated as $m_{p2+}-m_{y+}$ and $m_{p2+}-m_{b+}$) and neutral losses ($-NH_3$, $-H_2O$, and loss of both $-NH_3$ and $-H_2O$) during the construction of the graph so that the number of possible graphs to be considered is minimized.

Acknowledgments

Fundação para a Ciência e a Tecnologia (Ciência2007), Portugal (C2007-IPATIMUP/AA2). IPATIMUP is partially supported by "Programa Operacional Ciência e Inovação 2010" (POCI 2010), VI Programa Quadro (2002–2006).

References

1. Matthiesen R, Mutenda KE. (2006) Introduction to proteomics. *Methods Mol Biol* 367:1–36.
2. Hjerno K. (2006) Protein identification by peptide mass fingerprinting. *Methods Mol Biol* 367:61–76.
3. Bunkenborg J, Matthiesen R. (2006) Interpretation of collision-induced fragmentation tandem mass spectra of posttranslationally modified peptides. *Methods Mol Biol* 367:169–194.
4. Colinge J, Masselot A, Giron M, Dessingy T, Magnin J. (2003) OLAV: towards high-throughput tandem mass spectrometry data identification. *Proteomics* 3: 1454–1463.
5. Frank A, Tanner S, Bafna V, Pevzner P. (2005) Peptide sequence tags for fast database search in mass-spectrometry. *J Proteome Res* 4:1287–1295.
6. Tabb DL, Saraf A, Yates JR, 3rd. (2003) GutenTag: high-throughput sequence tagging via an empirically derived fragmentation model. *Anal Chem* 75:6415–6421.
7. Wan Y, Yang A, Chen T. (2006) PepHMM: a hidden Markov model based scoring function for mass spectrometry database search. *Anal Chem* 78:432–437.
8. Matthiesen R, Trelle MB, Hojrup P, Bunkenborg J, Jensen ON. (2005) VEMS 3.0: algorithms and computational tools for tandem mass spectrometry based identification of post-translational modifications in proteins. *J Proteome Res* 4:2338–2347.
9. Zhang W, Chait BT. (2000) ProFound: an expert system for protein identification using mass spectrometric peptide mapping information. *Anal Chem* 72:2482–2489.
10. Henzel WJ, Billeci TM, Stults JT, Wong SC, Grimley C, Watanabe C. (1993) Identifying proteins from two-dimensional gels by molecular mass searching of peptide fragments in protein sequence databases. *Proc Natl Acad Sci USA* 90:5011–5015.
11. Mann M, Hojrup P, Roepstorff P. (1993) Use of mass spectrometric molecular weight information to identify proteins in sequence databases. *Biol Mass Spectrom* 22:338–345.

12. Pappin D, Hojrup P, Bleasby A. (1993) Rapid identification of proteins by peptide-mass finger printing. *Curr Biol* 3:327–332.
13. Dancik V, Addona TA, Clauser KR, Vath JE, Pevzner PA. (1999) De novo peptide sequencing via tandem mass spectrometry. *J Comput Biol* 6:327–342.
14. Mann M, Wilm M. (1994) Error-tolerant identification of peptides in sequence databases by peptide sequence tags. *Anal Chem* 66:4390–4399.
15. Choudhary JS, Blackstock WP, Creasy DM, Cottrell JS. (2001) Interrogating the human genome using uninterpreted mass spectrometry data. *Proteomics* 1:651–667.
16. Colinge J, Cusin I, Reffas S, Mahe E, Niknejad A, Rey PA, Mattou H, Moniatte M, Bougueleret L. (2005) Experiments in searching small proteins in unannotated large eukaryotic genomes. *J Proteome Res* 4: 167–174.
17. Frank AM, Savitski MM, Nielsen ML, Zubarev RA, Pevzner PA. (2007) De novo peptide sequencing and identification with precision mass spectrometry. *J Proteome Res* 6:114–123.
18. Hernandez P, Gras R, Frey J, Appel RD. (2003) Popitam: towards new heuristic strategies to improve protein identification from tandem mass spectrometry data. *Proteomics* 3:870–878.
19. Ma B, Zhang K, Hendrie C, Liang C, Li M, Doherty-Kirby A, Lajoie G. (2003) PEAKS: powerful software for peptide de novo sequencing by tandem mass spectrometry. *Rapid Commun Mass Spectrom* 17: 2337–2342.
20. Yan B, Pan C, Olman VN, Hettich RL, Xu Y. (2005) A graph-theoretic approach for the separation of b and y ions in tandem mass spectra. *Bioinformatics* 21:563–574.
21. Lu B, Chen T. (2003) A suffix tree approach to the interpretation of tandem mass spectra: applications to peptides of non-specific digestion and post-translational modifications. *Bioinformatics* 19(Suppl 2):II113–II121.
22. Kuster B, Schirle M, Mallick P, Aebersold R. (2005) Scoring proteomes with proteotypic peptide probes. *Nat Rev Mol Cell Biol* 6: 577–583.
23. Deutsch EW, Lam H, Aebersold R. (2008) PeptideAtlas: a resource for target selection for emerging targeted proteomics workflows. *EMBO Rep* 9:429–434.
24. Pratt JM, Simpson DM, Doherty MK, Rivers J, Gaskell SJ, Beynon RJ. (2006) Multiplexed absolute quantification for proteomics using concatenated signature peptides encoded by QconCAT genes. *Nat Protoc* 1: 1029–1043.
25. Lange V, Picotti P, Domon B, Aebersold R. (2008) Selected reaction monitoring for quantitative proteomics: a tutorial. *Mol Syst Biol* 4:222.
26. Mallick P, Schirle M, Chen SS, Flory MR, Lee H, Martin D, Ranish J, Raught B, Schmitt R, Werner T, et al. (2007) Computational prediction of proteotypic peptides for quantitative proteomics. *Nat Biotechnol* 25:125–131.
27. Oyama M, Kozuka-Hata H, Suzuki Y, Semba K, Yamamoto T, Sugano S. (2007) Diversity of translation start sites may define increased complexity of the human short ORFeome. *Mol Cell Proteomics* 6:1000–1006.
28. Desiere F, Deutsch EW, King NL, Nesvizhskii AI, Mallick P, Eng J, Chen S, Eddes J, Loevenich SN, Aebersold R. (2006) The PeptideAtlas project. *Nucleic Acids Res* 34:D655–D658.
29. Sherry ST, Ward MH, Kholodov M, Baker J, Phan L, Smigielski EM, Sirotkin K. (2001) dbSNP: the NCBI database of genetic variation. *Nucleic Acids Res* 29:308–311.
30. de Bakker PI, McVean G, Sabeti PC, Miretti MM, Green T, Marchini J, Ke X, Monsuur AJ, Whittaker P, Delgado M, et al. (2006) A high-resolution HLA and SNP haplotype map for disease association studies in the extended human MHC. *Nat Genet* 38: 1166–1172.
31. Peri S, Navarro JD, Amanchy R, Kristiansen TZ, Jonnalagadda CK, Surendranath V, Niranjan V, Muthusamy B, Gandhi TK, Gronborg M, et al. (2003) Development of human protein reference database as an initial platform for approaching systems biology in humans. *Genome Res* 13: 2363–2371.
32. Craig R, Beavis RC. (2004) TANDEM: matching proteins with tandem mass spectra. *Bioinformatics* 20:1466–1467.
33. Matthiesen R. (2007) Methods, algorithms and tools in computational proteomics: a practical point of view. *Proteomics* 7: 2815–2832.
34. Wiesner J, Premsler T, Sickmann A. (2008) Application of electron transfer dissociation (ETD) for the analysis of post-translational modifications. *Proteomics* 8: 4466–4483.

Chapter 10

Methods and Algorithms for Relative Quantitative Proteomics by Mass Spectrometry

Rune Matthiesen and Ana Sofia Carvalho

Abstract

Protein quantitation by mass spectrometry (MS) is attractive since it is possible to obtain both the identification and quantitative values of novel proteins and their posttranslational modifications in one experiment. In contrast, protein arrays only provide quantitative values of targeted proteins and their modifications. There are an overwhelming number of quantitative mass spectrometry (MS) methods for protein and peptide quantitation. The aim here is to provide an overview of the most common MS-based quantitative methods used in the proteomics field and discuss the computational algorithms needed for the robust quantitation of proteins, peptides, and their posttranslational modifications.

Key words: Protein quantitation, stable isotope labeling, LC-MS, label-free quantitation

1. Introduction

Although quantitative proteomics is considered an advanced and costly mass spectrometry technique, the importance of quantitative proteomics cannot be questioned. The advantage is that quantitative proteomics techniques are becoming more user-friendly, sensitive, and robust, and some techniques, such as label-free quantitation by Liquid Chromatography- Mass Spectrometry (LC-MS), are inexpensive. In fact, the quantitative information is often available in the LC-MS/MS raw data even though the experiment was not run with the intention to quantify the sample constituents. Nevertheless, label-free quantitation does require some experimental planning, particularly due to the need for replicate runs to obtain accurate statistical measures (1, 2).

There are a vast number of MS-based protein quantitative methods; they can be divided into two main categories: (i) stable

R. Matthiesen (ed.), *Bioinformatics Methods in Clinical Research*, Methods in Molecular Biology 593,
DOI 10.1007/978-1-60327-194-3_10, © Humana Press, a part of Springer Science+Business Media, LLC 2010

isotope labeling strategies and (ii) label-free methods. The stable isotope labeling strategies can be further divided into metabolic labeling and chemical labeling (3, 4). Each of the quantitative methods has its own characteristics, meaning that the same algorithms cannot be used for all quantitative methods. The quantitative algorithms need to be corrected for different artifacts created by the different methods (4). Lau et al. (3) recently reviewed a number of programs for quantitative MS-based proteomics, most of which are commercial. The aim here is to review the underlying algorithms of the most common MS-based relative protein quantitation methods (**Fig. 10.1**). This division is not 100% justified since the relative methods based on chemical modifications can in principle be used for absolute quantification. By spiking, into the sample, a known quantity of stable isotope labeled peptides or proteins.

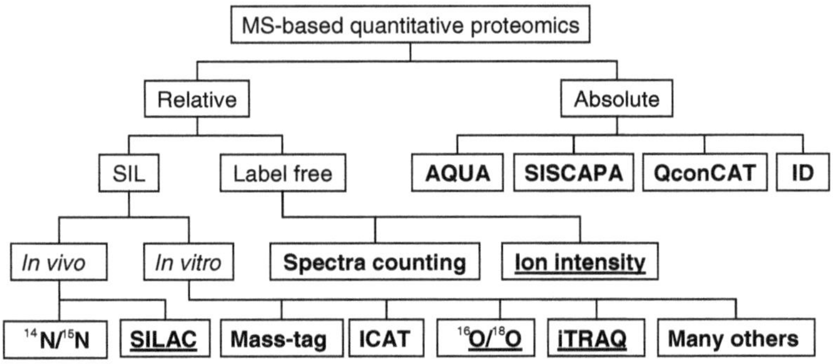

Fig. 10.1. Schema of MS-based quantitative methods. Methods are highlighted in bold and the underlined methods are reviewed in this chapter. A number of in vitro chemical stable labeling methods that are not mentioned here are reviewed by Julkar and Regnier (5). ID [isotope dilution (6)], SILAC [stable isotope labeling with amino acids in cell culture (7)], QconCAT [absolute quantitation by spiking in of stable isotope-labeled artificial express proteins "QconCATs," which are comprised of concatenated proteotypic peptides (8–10)], SISCAPA [stable isotope standards with capture by anti-peptide antibody (11, 12)], AQUA [absolute quantification (13)], ICAT [isotope-coded affinity tag (14)], iTRAQ (a primary amine-specific stable isotope label method for relative protein quantitation using mass spectrometry (15, 16)]. The figure is adapted from Lau et al. (3).

SISCAPA and AQUA are powerful methods for absolute quantitation, but they are costly and not easily applied to modified peptides. QconCAT is more cost-effective once the plasmids for the overexpression of concatenated proteotypic peptides have been established. However, QconCAT does not adequately consider posttranslational modifications. The methods described in the next section do not discuss QconCAT, SISCAPA, and AQUA; however, the final data output from these methods is quite similar to the data produced by the SILAC method. The computational methods for SILAC, if properly implemented, can therefore directly be applied to QconCAT, SISCAPA, and AQUA.

2. Algorithms for Quantitation

This section describes algorithms for specific proteomics quantitative methods but does not provide an extensive overview of common computational and statistical issues for all quantitative techniques, such as background subtraction, noise filtering, mass calibration, transformation, normalization, scaling, peak detection, missing values, classification, and power estimation, which are covered in many other reviews (4).

2.1. SILAC

From a computational point of view, SILAC is a very simple quantitative method. The unlabeled and labeled peptides are well separated in the LC-MS intensity profile (**Fig. 10.2**) and quantitative ratios can simply be calculated by dividing the integrated intensities over retention time of the unlabeled and labeled peptides.

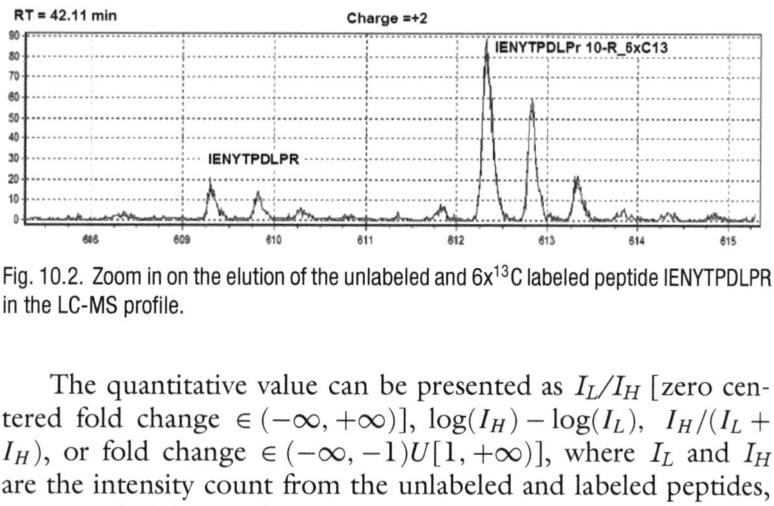

Fig. 10.2. Zoom in on the elution of the unlabeled and 6x^{13}C labeled peptide IENYTPDLPR in the LC-MS profile.

The quantitative value can be presented as I_L/I_H [zero centered fold change $\in (-\infty, +\infty)$], $\log(I_H) - \log(I_L)$, $I_H/(I_L + I_H)$, or fold change $\in (-\infty, -1)U[1, +\infty)$], where I_L and I_H are the intensity count from the unlabeled and labeled peptides, respectively. The two last quantitative value representations have the advantage that they can be calculated even if the labeled peptide intensity is zero, which is not possible for I_L/I_H. The quantitative values are most accurate if the intensity over several MS scan numbers is integrated (integration over the retention time dimension). It is also worth mentioning that one can use intensity in the LC-MS profile from other charge states of the peptide to obtain more accurate quantitation even though the peptide is only identified from an MS/MS spectrum corresponding to a specific charge state of the peptide.

The above described simplicity is not always a reality. In practice complications occur when complex protein samples are analyzed, which often means that there is a high probability of having overlapping peaks in the LC-MS profile. Such overlaps give rise to two different types of problems, which require two separate types of algorithms. The first problem is shown in **Fig. 10.3**.

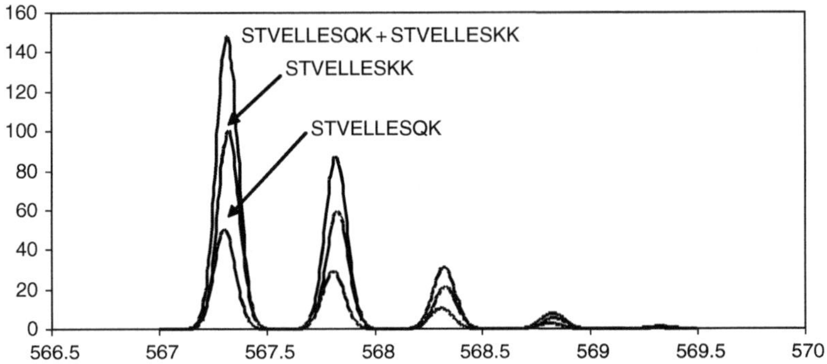

Fig. 10.3. Overlap between two isotopic distributions from the peptides STVELLESQK and STVELLESKK. The peak overlap shown here should be dealt with during the processing of the continuous MS data to mass and intensity peak lists.

In this case, the isotopic distribution from the peptides STVELLESQK and STVELLESKK, which have a mass difference of ~0.03638 Da, overlaps. Depending on the mass accuracy and resolution of the instrument, such peaks will reveal one (Orbitrap/FT-ICR) or two apparent isotopic distributions (TOF and Ion-traps). An overlap of isotopic peak distributions of the kind shown in **Fig. 10.3** can only be accurately resolved by modeling the peak width and, for example, fitting a multi-Gaussian or a multi-Gaussian/Lorentzian mixture peak model (17). An aid for this type of fitting can be obtained from the corresponding MS/MS spectrum, which in some cases can reveal the sequence of several tryptic peptides. Current quantitative proteomics software does not handle the above-mentioned issue in an optimal way.

Another type of overlap of isotopic distributions can occur when the double-charged peptides have approximately the same observed parent ion mass as a peptide with charge state +4 (**Fig. 10.4**).

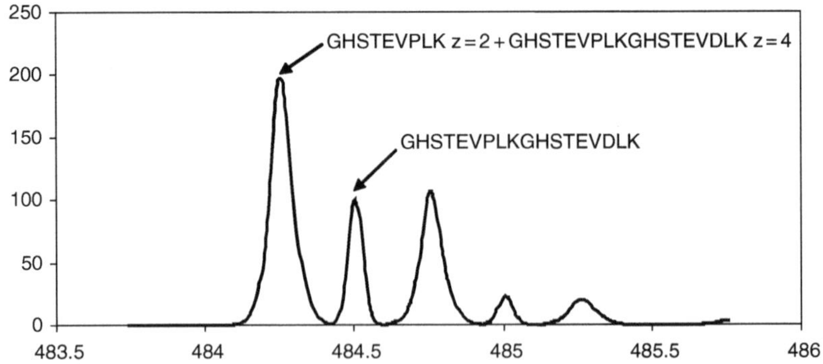

Fig. 10.4. Peak overlap that can be resolved by fitting multiple theoretical isotopic distributions of charge state $z = +1–5$.

In this case, the overlap can be resolved by fitting multiple theoretical isotopic distributions to the observed peak pattern. This can be done by using linear regression (18) or a nonlinear, Newton–Gauss unweighted least-squares method, for example. A third possibility is that two peptides are isobaric, in which case it will not be possible to resolve the peak overlap and the quantitation is doomed to fail.

An additional number of experimental artifacts can occur depending on the quality of the sample labeling. In **Fig. 10.2**, a small peak at ~0.5 *m/z* (–1 Da) below the monoisotopic mass for the parent ion of the stable isotope labeled peptide can be observed. This peak can be caused by incomplete labeling of lysine and arginine with ^{13}C and ^{15}N. The peak can also be caused by transaminases in vivo, which can exchange the amino group next to the α-carbon, which is therefore mainly observed if one uses $2x^{15}N$, $6x^{13}C$ lysine or $4x^{15}N$, $6x^{13}C$ arginine. The intensity of the –1 Da peak can in some cases be more than 10% of the intensity of the monoisotopic peak and in general should be included in the integration of intensity over the labeled peptides′ isotopic distributions. Another in vivo generated artifact is the conversion of $6x^{13}C$ arginine to $5x^{13}C$ proline, which gives rise to a small peak +5 Da above the monoisotopic peak of the stable isotope labeled peptide that contains proline (**Fig. 10.5**). In principle, multiple peaks with +5 Da intervals can be observed depending on how many proline residues are contained in the peptide. This error can be corrected by including these peaks in the integration procedure. Alternatively, one can approximate the probability that proline is ^{13}C labeled by dividing the intensity at +5 Da with the total observed intensity for a stable isotope labeled peptide with one proline and then use a binomial distribution to calculate the total percentage of these satellite peaks for other proline-containing peptides.

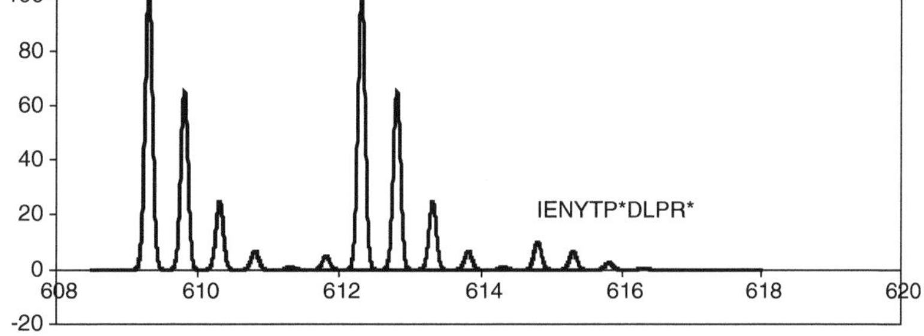

Fig. 10.5. Zoom in on the elution of the unlabeled and $6x^{13}C$ labeled Arg peptide IENYTPDLPR in the LC-MS profile. In this case, $6x^{13}C$ labeled Arg have been in vivo converted to $5x^{13}C$ Pro, giving rise to a small peak +2.5 *m/z* (+5 Da) compared to the $6x^{13}C$ labeled Arg peptide IENYTPDLPR. * indicates ^{13}C-labeled residues.

The latter approach has the advantage of lowering the probability for peak overlaps, which can introduce errors in the quantitation procedure. Software for SILAC quantitation includes programs such as MSquant (http://msquant.sourceforge.net/), RelEx (19), ASAPratio (20), and VEMS (21).

A third sample-related artifact is due to incomplete labeling of the peptide. This can be observed if the number of cell doublings is not sufficient or if the cells are able to in vivo synthesize arginine or lysine (**Fig. 10.6**). The dilution factor can be estimated as $2n$ if one assumes no in vivo synthesis of arginine and lysine (22). This means that five cell doublings leads to ~0.03% nonlabeled amino acids in the stable isotope labeled cell culture. Incomplete labeling of the peptide can only be realized when observing peptides with two or more labeled amino acids. The peptides with two or more stable isotope labeled amino acids can be used to estimate a correction factor. This can be done by fitting a binomial distribution to the observed intensity of the unlabeled, single-labeled, and double-labeled peptides. The estimated probability (incorporation efficiency) can then be used in the binomial probability function to correct all the observed intensities.

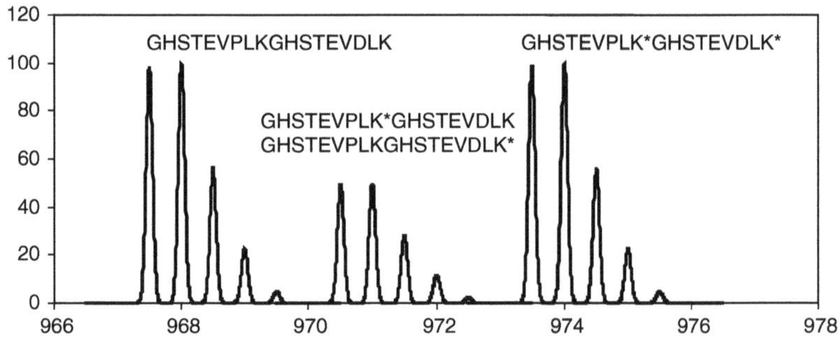

Fig. 10.6. Zoom in on the elution of the unlabeled, one and two 6x^{13}C-labeled Lys peptide GHSTEVPLKGHSTEVDLK in the LC-MS profile. * indicates ^{13}C-labeled residues.

2.2. ^{18}O Labeling

There are two protocols for ^{18}O labeling of peptides for MS-based quantitation: (i) In the original protocol for ^{18}O labeling, the proteins are digested in ^{18}O-enriched water (23), and (ii) in the alternative approach, the digestion is performed in normal water followed by a lyophilization and ^{18}O water labeling (24). The major complication in both ^{18}O labeling protocols is incomplete labeling (**Fig. 10.7**). The peptides from sample 2 will overlap with the isotopic distribution of the unlabeled peptides from sample 1. The peptides from sample 2 with no ^{18}O incorporated will perfectly overlap with the unlabeled peptides, whereas the peptides (sample 2) with one ^{18}O incorporated will partially overlap with the isotopic distribution from the peptides of sample 1 and the peptides from sample 2, which has two ^{18}O incorporated.

A number of algorithmic methods have been proposed for quantitation by ^{18}O labeling (23, 25–28). These methods

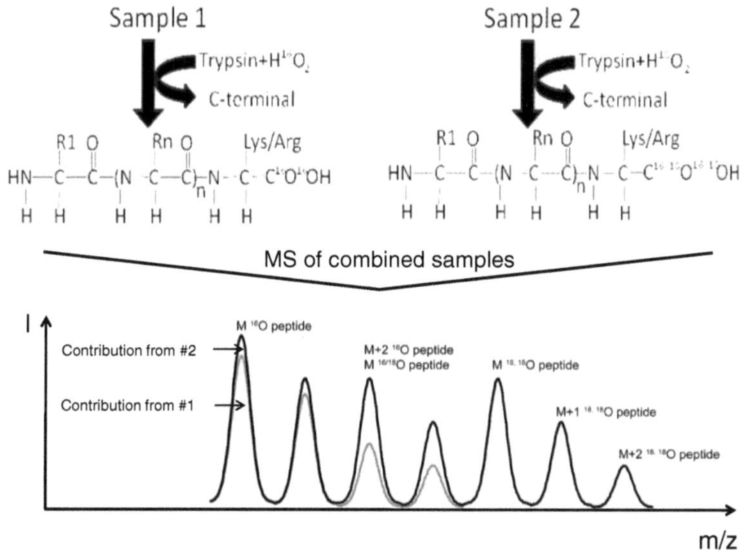

Fig. 10.7. Outline of the ^{18}O labeling technique. M is the monoisotopic mass. The monoisotopic peak (M 160 peptide) presents a contribution from sample 1, mainly, and partly from sample 2. The peak area difference between the dark (partially labeled) and grey (unlabeled) lines corresponds to the contribution of sample 2. This peak area difference is variable between peptides making quantitation by 180 labeling rather complex.

can be divided into two categories, which consider different assumptions. If one assumes that it is not possible to correlate or model a relationship between the incorporation of the first ^{18}O with the second, then the intensity of the light and heavy forms of the peptide can be corrected by the following equation (23):

$$\left(\frac{^{16}O}{^{18}O}\right) = \frac{I'_m}{I'_{m+4} - \frac{I_{m+4}}{I_m} I'_m + I'_{m+2}\left(1 - \frac{I_{m+2}}{I_m}\right) - I'_m \frac{I_{m+2}}{I_m}\left(1 - \frac{I_{m+2}}{I_m}\right)}, \quad [1]$$

where the I'_m are the observed apparent relative intensities of total intensity for the peptide ion and the I_m are the theoretical relative intensities of total intensity for the peptide ion. m indicates the monoisotopic mass and (m + 1, … ,m + 4) are the second to fifth peaks in the isotopic envelope. The theoretical isotopic distributions can be calculated by linear approximations (*see* **Note 1**). However, for quantitative purposes, it is better to use more accurate calculations and it is not recommended to use the linear approximations when the quantitated peptide sequence is known from MS/MS spectra. In order to calculate theoretical isotopic distributions, it is important to understand how the different atoms contribute to the isotopic distribution. Biological molecules are mainly composed of atoms of carbon (C), hydrogen (H), nitrogen (N), oxygen (O), and sulfur (S). Some biological molecules also bind metal ions, and

Table 10.1
Masses and relative isotopic abundance values of biologically relevant isotopes (29)

Isotope	A	%	Isotope	A+1	%
^{12}C	12	98.93(8)	^{13}C	13.0033548378(1)	1.07(8)
^{1}H	1.0078250321(4)	99.9885(7)	^{2}H	2.0141017780(4)	0.0115(7)
^{14}N	14.0030740052(9)	99.632(7)	^{15}N	15.0001088984(9)	0.368(7)
^{16}O	15.9949146221(15)	99.757(2)	^{17}O	16.99913150(2)	0.038(1)
^{32}S	31.97207069(12)	94.93(3)	^{33}S	32.97145850(1)	0.76(2)
Isotope	A+2	%	Isotope	A+4	%
^{14}C	14.003241988(4)	–	–	–	–
^{3}H	3.0160492675(11)	–	–	–	–
^{18}O	17.9991604(9)	0.205(1)	–	–	–
^{34}S	33.96786683(11)	4.29(3)	^{36}S	35.96708088(3)	0.02(1)

Uncertain digits are shown in parentheses.

these include proteins and DNA. Naturally occurring isotopes of biological compounds occur at an almost constant relative abundance (**Table 10.1**).

A more extensive list of biological relevant isotopes can be found at http://www.ionsource.com/Card/Mass/mass.htm.

The relative isotopic abundance values for isotopes given in **Table 10.1** can be used to calculate the relative isotopic abundance of different biological molecules composed of many isotopes. The relative isotopic abundance of the monoisotopic mass of a molecule with the composition CxHyNzOvSw can be calculated using the following expression (30, 31):

$$I_m = P_C^x \times P_H^y \times P_N^z \times P_O^v \times P_S^w, \quad [2]$$

where I_m is the relative abundance of the monoisotopic peak for the molecule; P_C, P_H, P_N, P_O, and P_S are the abundance of the monoisotopic masses of the C, H, N, O, and S elements; and x, y, z, v, and w are positive integer values indicating the number of occurrences of the corresponding atom in the biological compound. The expression is simply the probability that all the elements in the molecule have the monoisotopic mass. A similar expression can be made for the monoisotopic mass +1:

$$I_{m+1} = \binom{x}{1} P_C^{x-1} P_{C+1} P_H^y P_N^z P_O^v P_s^W + P_C^x \binom{y}{1} P_H^{y-1} P_{H+1} P_N^z P_O^v P_S^w + \cdots + P_c^x P_H^y P_N^z P_O^v \binom{w}{1} P_S^{w-1} P_{S+1}, \quad [3]$$

where P_{C+1}, P_{H+1}, P_{N+1}, P_{O+1}, and P_{S+1} are the abundance of the monoisotopic mass +~1 Da of the elements. Again, the expression is the probability that one atom in the molecule is the monoisotopic mass plus one. Equation [3] can be further expanded to calculate I_{m+2}, I_{m+3}, and I_{m+4} by using the same technique as used to expand Equation [2] to [3]. It is important to note that this calculation of the isotopic distribution is an approximation, which works well when comparing with the observed isotopic distribution from mass spectrometers that are unable to resolve the different elements' contribution to the m + 1 ion. However, the above expression can be expanded using the same concept considering all possible unique masses rather than only the isotopic abundance for the approximate masses m, $m + 1$, $m + 2$, $m + 3$, and $m + 4$. For most common mass spectrometers and for $^{16}O/^{18}O$ labeling, the above approximation is adequate.

$^{16}O/^{18}O$ ratios calculated by Equation [1] are a good approximation as long as the peptides in the heavy labeled sample are labeled with either one or two ^{18}O. If a peptide in the heavy sample has too high a percentage of completely unlabeled peptide, then Equation [1] will give an erroneous quantitation for that peptide. The incorporation efficiency of ^{18}O by trypsin is lower if charged residues are present in the −2, −1, +1, or +2 position relative to the trypsin cleavage site. However, most peptides label well, and since the protein quantitation is based on several peptide quantitative values, the low incorporation efficiency is mainly a problem for proteins that are only detected by one or two peptides. To resolve these shortcomings, Mirgorodskaya and colleagues suggested to indirectly estimate the incorporation rate by running the labeled sample separately and using the experimentally observed peak heights as the expected abundance distribution (32). They proposed a linear matrix equation to calculate the concentration of labeled and unlabeled peptides. Eckel-Passow et al. (26) elaborated on the method proposed by Mirgorodskaya and colleagues. They proposed a regression model that does not require running the labeled sample independently in order to obtain the expected distribution of the labeled peptide that accounts for peptide-specific incorporation rates of the ^{18}O label. In this model, the incorporation rate is estimated directly from the multivariable regression model:

$$\hat{\theta}_c = (W_c^T W_c)^{-1} W_c^T y_c, \qquad [4]$$

where θ is a 2×1 vector containing the concentrations θ_{c1} (unlabeled) and θ_{c2} (labeled) for peptide c. The hat on top of θ represents the predicted values from the model. The matrix W_c contains the intercept vector and the expected (theoretical) isotopic abundances for the cth peptide in, the unlabeled and labeled samples. Eckel-Passow et al. (26) elegantly split the matrix W_c into $W_c = X_c S_c$, where

$$X_c = \begin{bmatrix} 1 & I_m & 0 & 0 \\ 1 & I_{m+1} & 0 & 0 \\ 1 & I_{m+2} & I_m & 0 \\ 1 & I_{m+3} & I_{m+1} & 0 \\ 1 & I_{m+4} & I_{m+2} & I_m \\ \vdots & \vdots & \vdots & \vdots \\ 1 & I_{n-1} & I_{n-3} & I_{n-5} \end{bmatrix}, \quad S_c = \begin{bmatrix} 1 & 0 & 0 \\ 0 & 1 & (1-p)^2 \\ 0 & 0 & 2p(1-p) \\ 0 & 0 & p^2 \end{bmatrix}.$$

The first column in X_c is the intercept, and the remaining columns are the theoretical isotopic distributions for the peptide c with the labels $^{18}O_0$, $^{18}O_1$, or $^{18}O_2$. The first column in S_c is the intercept and columns two and three are the concentration parameters for the labeled and unlabeled samples. p is the purity of the ^{18}O water. This model assumes that it is possible to correlate or model a relationship between the incorporation of the first ^{18}O with the second, which means that the intensity of the light and heavy forms of the peptide can be corrected even if the heavy sample contributes to the monoisotopic peak of ^{16}O unlabeled peptide. Vázquez et al. (27) assumed a similar kinetic model for the correlation between the incorporation of the first ^{18}O with the second ^{18}O. However, they use a nonlinear, Newton–Gauss unweighted least-squares method iteratively to estimate the parameters of the model rather than the multivariate linear regression described by Eckel-Passow et al. (26).

2.3. Primary Amine Labeling

Primary amine labeling constitutes a broad range of chemical labeling strategies that are widely used. It exploits the nucleophilicity of amino groups to displace a leaving group from an activated acid. The N-terminal α-amino group is less nucleophilic than the ϵ-amino group on lysine. This means that the α-amino group has a lower degree of protonation and therefore is less derivatized than the ϵ-amino group at neutral pH. High or low pH gives an equal degree of ionization. The phenolic hydroxyl group on tyrosine can also be derivatized at high pH but is easily hydrolyzed again. Regnier and Julka provide an excellent overview of the chemistry behind primary amine labeling (33).

One big advantage with amine labeling is the possibility to label all peptides in the sample, which leads to more accurate protein quantitation values. Coding through derivatization of amines can be done on both a protein and a peptide level. The advantage of labeling at the protein level is that samples to be compared can be mixed and digested simultaneously, eliminating differential proteolysis of samples. The disadvantage is that some coding labels prevent trypsin hydrolysis at lysine residues, which means fewer and longer peptides, which affects the accuracy of the protein quantitation. Many of the amine-specific labeling strategies use deuterium to obtain a differential mass. It is known that deuterium-labeled peptides elute slightly earlier on a reverse-phase column than nonlabeled peptides. This problem can be minimized by having polar groups close to deuterium atoms to inhibit interaction with the column (34). Using ^{13}C- or ^{18}O-labeled coding is another possibility for solving the differential elution from reverse-phase columns; however, for many amine labels, only the deuterium form is commercially available (35). Deuterium-coded N-acetoxysuccinimide is a simple and low-cost acetylating agent. However, acetylation reduces the peptide charge and diminishes the ionization efficiency. Regnier and Julka (33) state that this problem is mainly observed for MALDI-MS rather than for ESI-MS.

The computational analysis of peptides labeled at amines is straightforward and gives few artifacts. However, one can get an overlap of the isotopic distribution from the unlabeled and labeled peptides if the number of labeled atoms is equal to or less than 4. This problem can be solved by the same computational methods presented in Section 2.2.

A special type of primary amine labeling reagents is the so-called tandem mass tags (TMT). The concept of this labeling strategy has been detailed by Thompson et al. (36) and is briefly summarized here. The TMT labels are isobaric, which means they cannot distinguish the samples on the MS level. This gives increased sensitivity and less sample complexity. The structure of TMT reagents is reporter ion-mass balancer-derivatizing agent specific for primary amines. These units are linked by labile bonds to produce intense reporter ions. TMT reagents come with —two to eight different reporter ions, meaning that up to eight samples can be analyzed simultaneously. The isobaric feature is achieved by the mass balancer, which compensates for the mass difference of the reporter ions.

2.3.1. iTRAQ Chemical Lbeling

The iTRAQ (isobaric tags for relative and absolute quantification) reagents supplied by the manufacturers are not 100% pure but come with a datasheet for each batch indicating the percentages of each reporter ion reagent that differ by –2, –1, +1, and +2 Da from the quoted mass. These percentages f (*see* **Note 2**) need to be considered and used to correct the apparent intensity count $I'r$ to obtain the corrected intensity for each reporter ion $I'r$, where r indicates a specific reporter ion. The true intensities can be obtained by solving the following set of linear equations:

$$\begin{pmatrix} I'_{114.1} = f_{m,114.1} \times I_{114.1} + f_{m-1,115.1} \times I_{115.1} + f_{m-2,116.1} \times I_{116.1} + f_{m-3,117.1} \times I_{117.1} \\ I'_{115.1} = f_{m+1,114.1} \times I_{114.1} + f_{m,115.1} \times I_{115.1} + f_{m-1,116.1} \times I_{116.1} + f_{m-2,117.1} \times I_{117.1} \\ I'_{116.1} = f_{m+2,114.1} \times I_{114.1} + f_{m+1,115.1} \times I_{115.1} + f_{m,116.1} \times I_{116.1} + f_{m-1,117.1} \times I_{117.1} \\ I'_{117.1} = f_{m+3,114.1} \times I_{114.1} + f_{m+2,115.1} \times I_{115.1} + f_{m+1,116.1} \times I_{116.1} + f_{m,117.1} \times I_{117.1} \end{pmatrix}, \quad [5]$$

where *m* indicates the monoisotopic mass. The number of linear equations can be adjusted depending on the number of reporter ion intensities that needs to be recovered. The above strategy is the same as that presented by Shadforth et al. (37) describing the i-Tracker tool (*see* **Note 3**). However, the above presentation and the way to solve the linear equations are more general. The above equations can be written by matrix notations as

$$y = X\beta, \quad [6]$$

where y is a vector containing the apparent intensities, X is a matrix with the percentages of each reporter ion, and β is a parameter vector containing the true intensities of reporter ions that need to be determined. Note that no error term can be included in the above model in Equations [5] and [6] (*see* **Note 4**). The parameter vector β can now be estimated by

$$\hat{\beta} = \left(X^T X\right)^{-1} X^T \vec{y}, \quad [7]$$

and the corrected intensities are now given by

$$\hat{\mathbf{y}} = \mathbf{X}\hat{\beta}. \quad [8]$$

A number of software tools are available for iTRAQ quantitation, such as ProQUANT (Applied Biosytems, Foster City, CA), i-TRACKER/TandTRACK (37, 38), Multi-Q (39, 40), and VEMS (4, 41).

2.4. Label-Free Quantitation

Label-free quantitation based on comparing LC-MS intensity profiles has started to gain acceptance in the field. This can mainly be due to more reproducible chromatography systems and more stable ion spray. Label-free quantitation of samples is preferably done by running the samples sequentially using exactly the same conditions. If the chromatography buffer compositions are different, then it is likely that one will observe different relative intensities between the charge states of the peptides. If the chromatography column is different, the elution gradient will be disturbed. If the mass spectrometer or ion spray needle is changed, then it is likely that one can observe a large effect on the intensity values, again making the data analysis more challenging.

There are a number of different experimental designs, and the algorithms needed for quantitation are heavily dependent on the design of the experiment. One experimental design is based on LC-MS/MS runs (42). In this design, the identification obtained from the LC-MS/MS run can be used as the anchor point for the alignment of the LC-MS profiles, giving very accurate alignment (**Fig. 10.8**, Exp. Design A). Accurate alignment is essential for proper quantitation. It is mainly the retention time dimension that needs to be aligned. The mass dimension is normally very accurate due to the frequent use of lock spray calibration or calibration using buffer contaminants. The LC-MS/MS runs contain both LC-MS (which may be used for quantitation) and LC-MS/MS (used for identification) data. This means that the intensity counts obtained in the LC-MS part are lower than what would be obtained in LC-MS runs since the sample is split between MS and MS/MS scans. This has led some authors to state that the experimental design in **Fig. 10.8** (Exp. Design A) gives undersampled quantitative values. However, the repeated runs in step 3 need not to be LC-MS/MS runs.

Experimental Design	
A	**B**
1. LC-MSMS run sample 1	
2. LC-MSMS run sample 2	
3. Repeat step 1 and 2 N number of times	
4. Identify peptides by database dependent or de novo sequencing	**4.** Align runs
5. Align runs by using identified peptides as anchor points	**5.** Integrate intensity of all peaks
6. Integrate intensity of all peaks	**6.** Compare intensity between samples
7. Compare intensity between samples	**7.** Extract mass retention time tags for significantly changed peak intensity between samples
	8. LC-MSMS runs using inclusion list obtained in step 7
	9. Identify the regulated peptides by database dependent search or de novo sequencing

Fig. 10.8. Two examples of experimental design for label-free quantitation. The number of technical replicates, *N*, can be obtained from statistical power estimations.

It is also possible to make an inclusion list of peptides that were not fragmented in step 1 and extend the method with additional LC-MS/MS runs since all quantitative values are extracted in step 7 (**Fig. 10.8**, Exp. Design A).

In another experimental design, the focus is in the first case on the LC-MS data. The aim is to identify peaks that have changed relative abundance between different sample types (**Fig. 10.8**, Exp. Design B). The peaks that show changes in relative abundance are then included in an inclusion list for an LC-MS/MS experiment with the purpose of identifying the peptides that show differential abundance. The computational algorithms needed for the two above approaches are different, but many of the substeps are common, such as background subtraction, noise filtering, mass calibration, transformation, normalization, scaling, peak detection, and replacement of missing values. Programs for label-free quantitation are MSquant (http://msquant.sourceforge.net/), Mzmine (43, 44), SpecArray (45), OpenMS (46), PEPPeR (47), MSinspect (48), SuperHirn (49), and VEMS (42).

3. Software

A number of software tools exist for quantitative proteomics. Most of these are commercial and have recently been summarized in a review by Lau et al. (3). The main problem with the commercial tools is that they often only work for specific vendors' mass spectrometers and only for a limited number of quantitative techniques. Another problem is that the flexibility of commercial software is often not good enough for scientific purposes. The freely available VEMS program supports all the quantitative techniques mentioned in this chapter (4). VEMS has recently been updated and tested with a protein reference set mixed in specific ratios. Our tests demonstrated that VEMS gives similar or more accurate quantitative results than commercial systems such as Mascot and PEAKS Q. VEMS is available from "http://www.portugene.com/software.html". MSquant is another application that supports a broad range of quantitative methods from different instruments.

4. Discussion

The ultimate aim is that protein quantitations obtained in various cell types in a given timeframe under specific conditions can support the growing field of system biology. The main bottleneck for a more detailed approach is the cost of the quantitative experiments. QconCAT is a cost-effective approach and could be used for more extensive studies; however, PTMs will be difficult to accurately quantify by such an approach. The computational methods for SILAC that are discussed in this chapter can be reused for AQUA and QconCAT since the final output of the data is very similar. An extra software layer may be needed for AQUA and QconCAT, however, to deal with the accurate calculation of protein concentration using a standard curve made from dilution series.

5. Notes

1. The theoretical isotopic distributions can be calculated by linear approximations as described by Wehofsky et al. (50). These linear approximations are not recommended for $^{16}O/^{18}O$ labeling when the quantitated peptide is known. The linear approximations can be used to analyze peaks in LC-MS run where the peptide sequence is unknown. In such cases, one can at best approximate the isotopic distribution based on the observed m/z and charge state of the detected peptide in the LC-MS run. The linear approximations are therefore appropriate and can provide faster relative isotopic abundance values. Wehofsky et al. (50) provide the following approximations:

$$I_{m+1} = -1.25446 + 0.05489 \times I_m \text{ and } I_{m+2} = 0.13977 + 0.00613 \times I_m + 1.49147E^{-5} \times M^2.$$

2. The percentage values $f_{m\pm i,x}$ are provided by the manufacturer of the iTRAQ reagent. The percentage for the monoisotopic peak can be calculated as

$$f_{m,x} = 100\% - \sum_{i}^{n} f_{m\pm i,x},$$

where i is an nonzero integer and x a specific iTRAQ reporter ion.

3. The above strategy is the same as that presented by Shadforth et al. (37) describing the i-Tracker tool. However, the equations provided by Shadforth et al. only work for iTRAQ with four reporter ions. The equations provided here are general and can be used for iTRAQ with any number of reporter ions.
4. It is not possible to model an error term in the above case since the number of parameters equals the number of linear equations (*see* Equation [5]), which means that the system will be underdetermined. There is no unique solution to an underdetermined system. However, in some cases, it is possible to set up more equations than presented here and it is then worthwhile to add a unit vector as a column to the matrix x in Equation [6], which will model an error term.

Acknowledgments

Support for RM was provided from Ramon y Cajal (RYC-2006-001446) and Fundação para a Ciência e a Tecnologia, Programa CIÊNCIA 2007 (C2007-IPATIMUP/AA2). Ana Sofia Carvalho gratefully thanks Fundação para a Ciência e a Tecnologia for a postdoctoral fellowship (SFRH/BPD/36912/2007).

References

1. Molloy MP, Brzezinski EE, Hang J, McDowell MT, VanBogelen RA. (2003) Overcoming technical variation and biological variation in quantitative proteomics. *Proteomics* 3: 1912–1919.
2. Karp NA, Lilley KS. (2007) Design and analysis issues in quantitative proteomics studies. *Proteomics* 7(Suppl 1):42–50.
3. Lau KW, Jones AR, Swainston N, Siepen JA, Hubbard SJ. (2007) Capture and analysis of quantitative proteomic data. *Proteomics* 7:2787–2799.
4. Matthiesen R. (2007) Methods, algorithms and tools in computational proteomics: a practical point of view. *Proteomics* 7: 2815–2832.
5. Julkar S, Regnier F. (2004) Quantification in proteomics through stable isotope coding: a review. *J Proteome Res* 3:350–363.
6. Bronstrup M. (2004) Absolute quantification strategies in proteomics based on mass spectrometry. *Expert Rev Proteomics* 1:503–512.
7. Ong SE, Blagoev B, Kratchmarova I, Kristensen DB, Steen H, Pandey A, Mann M. (2002) Stable isotope labeling by amino acids in cell culture, SILAC, as a simple and accurate approach to expression proteomics. *Mol Cell Proteomics* 1:376–386.
8. Mirzaei H, McBee J, Watts J, Aebersold R. (2007) Comparative evaluation of current peptide production platforms used in absolute quantification in proteomics. *Molecular & Cellular Proteomics* 7:813–823, 2008.
9. Pratt JM, Simpson DM, Doherty MK, Rivers J, Gaskell SJ, Beynon RJ. (2006) Multiplexed absolute quantification for proteomics using concatenated signature peptides encoded by QconCAT genes. *Nat Protoc* 1:1029–1043.
10. Rivers J, Simpson DM, Robertson DH, Gaskell SJ, Beynon RJ. (2007) Absolute multiplexed quantitative analysis of protein expression during muscle development using QconCAT. *Mol Cell Proteomics* 6:1416–1427.
11. Anderson L, Hunter CL. (2006) Quantitative mass spectrometric multiple reaction monitoring assays for major plasma proteins. *Mol Cell Proteomics* 5:573–588.
12. Anderson NL, Anderson NG, Haines LR, Hardie DB, Olafson RW, Pearson TW. (2004) Mass spectrometric quantitation of peptides and proteins using Stable Isotope Standards and Capture by Anti-Peptide Antibodies (SISCAPA). *J Proteome Res* 3: 235–244.
13. Kirkpatrick DS, Gerber SA, Gygi SP. (2005) The absolute quantification strategy: a general procedure for the quantification of proteins and post-translational modifications. *Methods* 35:265–273.
14. Han DK, Eng J, Zhou H, Aebersold R. (2001) Quantitative profiling of differentiation-induced microsomal proteins using isotope-coded affinity tags and mass spectrometry. *Nat Biotechnol* 19:946–951.
15. Aggarwal K, Choe LH, Lee KH. (2006) Shotgun proteomics using the iTRAQ isobaric tags. *Brief Funct Genomic Proteomic* 5:112–120.
16. Shadforth IP, Dunkley TP, Lilley KS, Bessant C. (2005) i-Tracker: for quantitative proteomics using iTRAQ. *BMC Genomics* 6:145.
17. Matthiesen R. (2006) Extracting monoisotopic single-charge peaks from liquid chromatography-electrospray ionization-mass spectrometry. *Methods Mol Biol* 367: 37–48.
18. Meija J, Caruso JA. (2004) Deconvolution of isobaric interferences in mass spectra. *J Am Soc Mass Spectrom* 15:654–658.
19. MacCoss MJ, Wu CC, Liu H, Sadygov R, Yates JR, 3rd. (2003) A correlation algorithm for the automated quantitative analysis of shotgun proteomics data. *Anal Chem* 75:6912–6921.
20. Li XJ, Zhang H, Ranish JA, Aebersold R. (2003) Automated statistical analysis of protein abundance ratios from data generated

by stable-isotope dilution and tandem mass spectrometry. *Anal Chem* 75:6648–6657.
21. Matthiesen R. (2006) Virtual expert mass spectrometrist v3.0: an integrated tool for proteome analysis. *Methods Mol Biol* 367:121–138.
22. Blagoev B, Mann M. (2006) Quantitative proteomics to study mitogen-activated protein kinases. *Methods* 40:243–250.
23. Yao X, Freas A, Ramirez J, Demirev PA, Fenselau C. (2001) Proteolytic ^{18}O labeling for comparative proteomics: model studies with two serotypes of adenovirus. *Anal Chem* 73:2836–2842.
24. Yao X, Afonso C, Fenselau C. (2003) Dissection of proteolytic ^{18}O labeling: endoprotease-catalyzed ^{16}O-to-^{18}O exchange of truncated peptide substrates. *J Proteome Res* 2:147–152.
25. Mason CJ, Therneau TM, Eckel-Passow JE, Johnson KL, Oberg AL, Olson JE, Nair KS, Muddiman DC, Bergen HR, 3rd. (2007) A method for automatically interpreting mass spectra of ^{18}O-labeled isotopic clusters. *Mol Cell Proteomics* 6:305–318.
26. Eckel-Passow JE, Oberg AL, Therneau TM, Mason CJ, Mahoney DW, Johnson KL, Olson JE, Bergen HR, 3rd. (2006) Regression analysis for comparing protein samples with $^{16}O/^{18}O$ stable-isotope labeled mass spectrometry. *Bioinformatics* 22: 2739–2745.
27. Ramos-Fernandez A, Lopez-Ferrer D, Vazquez J. (2007) Improved method for differential expression proteomics using trypsin-catalyzed ^{18}O labeling with a correction for labeling efficiency. *Mol Cell Proteomics* 6:1274–1286.
28. Halligan BD, Slyper RY, Twigger SN, Hicks W, Olivier M, Greene AS. (2005) ZoomQuant: an application for the quantitation of stable isotope labeled peptides. *J Am Soc Mass Spectrom* 16:302–306.
29. Coursey J, Schwab D, Dragoset R. (2001) Atomic weights and isotopic compositions (version 2.3.1). National Institute of Standards and Technology, Gaithersburg, MD. Available at http://physicsnistgov/Comp
30. Matthiesen R, Mutenda KE. (2006) Introduction to proteomics. *Methods Mol Biol* 367:1–36.
31. Snyder A (Ed.). (2001) *Interpreting Protein Mass Spectra, A Comprehensive Resource.* Oxford University Press, Oxford.
32. Mirgorodskaya O, Kozmin Y, Titov M, Körner R, Sönksen C, Roepstorff P (2000) Quantitation of peptides and proteins by matrix-assisted laser desorption/ionization mass spectrometry using ^{18}O-labeled internal standards. *Rapid Commun Mass Spectrom*, 14:1226–1232.
33. Regnier FE, Julka S. (2006) Primary amine coding as a path to comparative proteomics. *Proteomics* 6:3968–3979.
34. Zhang R, Sioma CS, Thompson RA, Xiong L, Regnier FE. (2002) Controlling deuterium isotope effects in comparative proteomics. *Anal Chem* 74:3662–3669.
35. Zhang R, Regnier FE. (2002) Minimizing resolution of isotopically coded peptides in comparative proteomics. *J Proteome Res* 1:139–147.
36. Thompson A, Schafer J, Kuhn K, Kienle S, Schwarz J, Schmidt G, Neumann T, Johnstone R, Mohammed AK, Hamon C. (2003) Tandem mass tags: a novel quantification strategy for comparative analysis of complex protein mixtures by MS/MS. *Anal Chem* 75:1895–1904.
37. Shadforth I, Crowther D, Bessant C. (2005) Protein and peptide identification algorithms using MS for use in high-throughput, automated pipelines. *Proteomics* 5:4082–4095.
38. Laderas T, Bystrom C, McMillen D, Fan G, McWeeney S. (2007) TandTRAQ: an open-source tool for integrated protein identification and quantitation. *Bioinformatics* 23:3394–3396.
39. Yu CY, Tsui YH, Yian YH, Sung TY, Hsu WL. (2007) The Multi-Q web server for multiplexed protein quantitation. *Nucleic Acids Res* 35:W707–W712.
40. Lin WT, Hung WN, Yian YH, Wu KP, Han CL, Chen YR, Chen YJ, Sung TY, Hsu WL. (2006) Multi-Q: a fully automated tool for multiplexed protein quantitation. *J Proteome Res* 5:2328–2338.
41. Matthiesen R, Trelle MB, Hojrup P, Bunkenborg J, Jensen ON. (2005) VEMS 3.0: algorithms and computational tools for tandem mass spectrometry based identification of post-translational modifications in proteins. *J Proteome Res* 4:2338–2347.
42. Beck HC, Nielsen EC, Matthiesen R, Jensen LH, Sehested M, Finn P, Grauslund M, Hansen AM, Jensen ON. (2006) Quantitative proteomic analysis of post-translational modifications of human histones. *Mol Cell Proteomics* 5:1314–1325.
43. Katajamaa M, Miettinen J, Oresic M. (2006) MZmine: toolbox for processing and visualization of mass spectrometry based molecular profile data. *Bioinformatics* 22:634–636.
44. Katajamaa M, Oresic M. (2005) Processing methods for differential analysis of LC/MS profile data. *BMC Bioinformatics* 6:179.
45. Li XJ, Yi EC, Kemp CJ, Zhang H, Aebersold R. (2005) A software suite for the gen-

eration and comparison of peptide arrays from sets of data collected by liquid chromatography-mass spectrometry. *Mol Cell Proteomics* 4:1328–1340.

46. Kohlbacher O, Reinert K, Gropl C, Lange E, Pfeifer N, Schulz-Trieglaff O, Sturm M. (2007) TOPP – the OpenMS proteomics pipeline. *Bioinformatics* 23:e191–e197.
47. Jaffe JD, Mani DR, Leptos KC, Church GM, Gillette MA, Carr SA. (2006) PEPPeR, a platform for experimental proteomic pattern recognition. *Mol Cell Proteomics* 5: 1927–1941.
48. Bellew M, Coram M, Fitzgibbon M, Igra M, Randolph T, Wang P, May D, Eng J, Fang R, Lin C, et al. (2006) A suite of algorithms for the comprehensive analysis of complex protein mixtures using high-resolution LC-MS. *Bioinformatics* 22:1902–1909.
49. Mueller LN, Rinner O, Schmidt A, Letarte S, Bodenmiller B, Brusniak MY, Vitek O, Aebersold R, Muller M. (2007) SuperHirn – a novel tool for high resolution LC-MS-based peptide/protein profiling. *Proteomics* 7:3470–3480.
50. Wehofsky M, Hoffmann R, Hubert M, Spengler B. (2001) Isotopic deconvolution of matrix-assisted laser desorption/ionization mass spectra for substances-class specific analysis of complex samples. *Eur J Mass Spectrom* 7:39–46.

Chapter 11

Feature Selection and Machine Learning with Mass Spectrometry Data

Susmita Datta and Vasyl Pihur

Abstract

Mass spectrometry has been used in biochemical research for a long time. However, its potential for discovering proteomic biomarkers using protein mass spectra has aroused tremendous interest in the last few years. In spite of its potential for biomarker discovery, it is recognized that the identification of meaningful proteomic features from mass spectra needs careful evaluation. Hence, extracting meaningful features and discriminating the samples based on these features are still open areas of research. Several research groups are actively involved in making the process as perfect as possible. In this chapter, we provide a review of major contributions toward feature selection and classification of proteomic mass spectra involving MALDI-TOF and SELDI-TOF technology.

Key words: MALDI-TOF, SELDI-TOF, isotopic, filter, wrapper, LDA, QDA, SVM, KNN, R, Poisson, logistic, random forest, ROC, classification, peak detection.

1. Introduction

Protein profiling by high-throughput, matrix-assisted laser desorption/ionization time-of-flight mass spectrometry (MALDI-TOF MS) and surface-enhanced laser desorption/ionization time-of-flight mass spectrometry (SELDI-TOF MS) is a powerful tool for biohazard and biomedical research. On the one hand, there are several analytical bottlenecks that make the results largely nonreproducible (1). On the other hand, Stühler et al. (2) revealed that mass spectral analysis of label-free samples provides high-throughput protein quantification with comparable sensitivity and specificity to other quantification technologies, for example, gel-based analysis and isotopically labeled mass spectral analysis, etc. Also, the label-free approach is faster and has

R. Matthiesen (ed.), *Bioinformatics Methods in Clinical Research*, Methods in Molecular Biology 593,
DOI 10.1007/978-1-60327-194-3_11, © Humana Press, a part of Springer Science+Business Media, LLC 2010

the potential of better automation. Sorace and Zhan (10) concluded, however, that proper experience is needed at the levels of data collection and data analysis. In mass spectrometry, a sample is co-crystallized with energy-absorbing molecules and analyzed by MALDI/SELDI-TOF MS, which generates a mass spectrum [mass-to-charge ratio *m/z* (on the *x*-axes) and intensity (on the *y*-axes)]. Each spectrum contains massive vectors of *m/z* and *y*. High noise levels, high dimensionality, and improper chemical justification of the features make the automatic analysis of proteomic mass spectra a very challenging task. In general, a meaningful identification of proteins or peptides from the differential identifiers between the studied groups of the mass spectra is difficult (3). The automatic analysis and discovery of biomarkers from proteomic mass spectra is an open research topic today. The following quote from a recent review article by Hilario et al. (4) summarizes the situation: "Despite intensive ongoing research on preprocessing and classification of protein mass spectra for biomarker discovery, the field is still very much in its infancy." Careful calibration of the mass spectrometric parameters and proper processing steps, such as (i) basic preprocessing to reduce noise, such as filtering and baseline subtraction, (ii) feature extraction (often the same as peak detection), and (iii) normalization and alignment of spectra, are necessary along with appropriate classification techniques. In Section **2**, we describe the basic preprocessing of mass spectrum data since it goes hand in hand with the feature selection method. However, we keep this section fairly short, as it is not the main focus of our chapter. Section **3** discusses some of the significant research in the area of feature selection. In Section **4**, we provide a comprehensive review of the classification techniques that are used to separate the mass spectra of the case and control samples. In Section **5**, we provide a list of free statistical software to analyze mass spectrometry data. Section **6** concludes with a discussion of existing challenges in the analysis of mass spectrometry data.

2. Basic Quality Control and Preprocessing

A typical mass spectrum from a low-resolution MALDI-TOF mass spectrometer may contain about 15,500 mass-to-charge ratio (*m/z*) values or features and their corresponding intensity values, *y*. These numbers are much higher on a high-resolution mass spectrometer. On the other hand, the sample sizes are much smaller than these features. Hence, dimension reduction and/or feature selection are among the major steps necessary to analyze the data in a meaningful way.

Before starting a discussion on the importance of feature selection for mass spectrometry data, researchers must be mindful of some of the following facts regarding mass spectrometry. Mass spectrometry has the potential to identify more sensitive biomarkers of a disease than existing ones. However, the process is extremely sensitive to changes in the protocol of sample and spectra collection. In other words, extreme caution has to be followed in order to maintain the same protocol throughout the study. Introducing any systematic bias into the data collection and sample handling will impact the study significantly even if very sophisticated feature selection tools and classification techniques are used to detect biomarkers. Hilario et al. (4) provide a comprehensive list of systematic sources of possible biases in mass spectrometry data. We are not going to discuss them in detail here. However, it is recommended that researchers be aware of them. Additionally, experimenters must follow the proper experimental design for getting reproducible results. In spite of all the best intentions of being careful to perform the above-mentioned steps, these experiments are still error-prone. Hence, the first and most important step of preprocessing the data is to draw heat maps of similar samples side by side in order to detect outliers, alignment issues, and nonuniform sample collection protocols of the mass spectra, etc. (5).

The data mining and bioinformatics work in the area began to grow after the seminal work by Petricoin et al. (6). So it is natural that this area of research is still in its infancy, and better quantitative work in the area will be forthcoming. Most quantitative work in the area of analyzing mass spectrometry data involves preprocessing of the data, including baseline correction, normalization, denoising, and then peak detection and peak alignment (7–9, among others). Proper preprocessing of the spectra is needed in order to get meaningful biological conclusions (10).

As a first step of preprocessing, the baseline signal usually has to be subtracted from the raw spectrum because the detector sometimes overestimates the number of ions arriving at its surface, especially in the low-molecular-weight regions. It is likely that the detector actually receives a lot of ions that are just chemical noise. **Figure 11.1** shows a raw spectrum and a baseline-corrected spectrum. Wu et al. (11) use a local linear regression technique to estimate the nonuniform background intensity. A semimonotonic baseline correction method was used by Baggerly et al. (12) for the analysis of SELDI data. Some researchers (13–15) use a nonlinear filter known as the “top-hat” operator (http://cmm.ensmp.fr/~serra/cours/index), which is used in the mathematical morphology literature. Breen et al. (14) subtract the computed convex hull from the raw spectrum to find the baseline-corrected spectrum. Satten et al. (17) use a local standardization technique of the original spectra that produced spectra with a uniform flat baseline, and standardized features and constant noise intensity across the spectrum. In their method, the standardized spectrum is a ratio of intensities, and so standardized spectra can be directly compared between two different samples. Shao et al. (18) use wavelet theory to estimate the baseline.

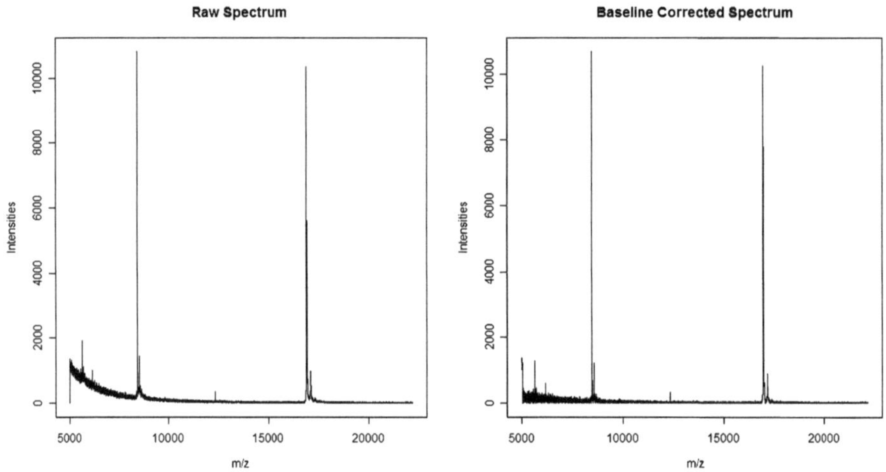

Fig. 11.1. Raw spectrum on the left and baseline-corrected spectrum on the right.

Baseline-corrected spectra consist of some features that are true signals and some that are random noise. The purpose of using mass spectrometry data for biomarker discovery involves classifying the case and control samples in terms of the differential features of the whole spectra. However, some of those features may be pure noise and not true signals. Hence, careful considerations should be given for estimating the noises in the data and removing them. Here we will consider the denoising effort as a part of the peak detection method as well. There are several methods to remove the noise. The features left after the removal of the random noise are the *selected features* and are often called *peaks*.

3. Feature Selection

We now discuss several feature selection techniques used with mass spectrometry data. All but the last section, Section **3.5**, are general data-analytic techniques where the knowledge of the underlying chemistry of a peptide is not taken into consideration. Saeys et al. (19) summarize the feature selection techniques in mass spectrometry data. There are three categories of basic feature selection techniques: (i) the filter method, (ii) the wrapper method, and (iii) the embedded method. In addition to those categories, we will also include (iv) the hybrid method and (v) the feature selection method, which considers the isotopic distribution of the peptides. We want to point out that we describe only the feature selection method in the following section, and not the feature reduction method. The basic difference between a feature reduction and a feature selection is that the feature selection method completely removes the unwanted features. However, a feature reduction method maps all the features into a lower-dimensional space. For example, in a feature reduction method for a given set of data points of p variables $\{x_1, x_2 \ldots, x_p\}$, one computes their representation in a lower dimension: $x \in \mathfrak{R}^p \rightarrow y \in \mathfrak{R}^d (p \gg d)$. One of the widely used dimension reduction techniques in the context of mass spectrometry data is principal components analysis (PCA) (20).

3.1. Filter Method

As the name suggests, in this method baseline-corrected features are filtered to get rid of the random noise. Only a subset of the original features is selected after the data have been filtered. There are several filters, including the linear filters of Savitzky and Golay (21), the penalized least-squares method of Eilers and Marx (22), Kast et al.'s Fourier transform filters (23), and the wavelets filter discussed by Morris et al. (24). Yasui et al. (8, 25) divide the range of mass spectra into several intervals and consider the local maximum within those specified regions that have a higher intensity than the average intensity in the specified regions to be the peaks or selected features, and the rest is considered noise. Breen et al. (14) consider local maxima to be the candidate peaks and filter those whose absolute intensities are smaller than a threshold. Coombes et al. (9) use a discrete wavelet with a hard thresholding method, which worked well with low-resolution SELDI spectra. Breen et al. (13, 14) estimate the background by top-hat filters (26) and then use a sequential alternating filter (26) of closings and openings to remove unwanted maxima and minima. Next, the watershed segmentation (mathematical morphology) technique is used and the centroid of each peak is determined at 70% of its maximum height. Satten et al. (17) estimate the noise or standard error from the negative standardized features and use multipliers of that as a filter.

There are several other papers where the important features are selected using *statistical cutoffs* based on Type I error or false discovery rate control of univariate and multivariate tests similar to microarray data. For example, Wu et al. (11) and Bhanot et al. (27) consider two-sample t-tests for every baseline-corrected feature and then perform, for every feature, a univariate t-test to find the significant difference between the case and control samples. Ideally, the features that pass the FDR cutoff (28) or overall Type I error rate (29–31), considering the multiple-hypothesis correction, are called *peaks*. However, the problem with these methods is that as the number of features gets larger and larger, it becomes harder to find useful features amid the large number of noisy features (11). Hence, Bhanot et al. (11) and Datta and DePadilla (32) rank the features with respect to their extreme *t*-statistic scores and then use the 15 and 25 top-ranked features, respectively, as important peaks for classification. Zhu et al. (33) use the *t*-test on normalized features and then perform the multiple-hypothesis correction based on random field theory. Wagner et al. (7) use the F-test to select important features. Izmirlian (34) uses the *t*-test on individual features and then uses multiple-hypothesis correction by controlling the false-discovery rate (28). Yu et al. (35) use the nonparametric Kolmogorov–Smirnov test on each feature and then select the features that are marginally significant at a *p*-value less than 0.05. Then they further restrict the number of features in terms of the restriction on the estimated coefficient of variation on the already-selected features.

Feature selection using ranking of the features is easy to implement. It is efficient on the order of $O(N)$ with dimensionality N. However, finding a suitable cutoff is a problem, and also correlations between the features are largely ignored. Filter methods result in selected features without the goal of optimizing the performance of any particular classification algorithm. Hence, the selected features can be used with any arbitrary classifier (Liu et al., (16)).

3.2. Wrapper Method

Wrapper methods wrap around a specific learning algorithm that assesses the selected feature subsets in terms of the estimated classification errors and then build the final classifier. Wrapper methods evaluate features in the context of a particular task.

In the wrapper method, the problem may become intractable, as the numbers of features are huge for mass spectrometry data. In order to avoid intractability issues, researchers use greedy or heuristic selection methods to find possible subsets of features. For example, one can train a classifier using a selected subset of features and then check their prediction accuracy. There are forward and backward selection algorithms (SFS – sequential forward selection, and SBS – sequential backward selection) for selecting the features sequentially. An SFS starts from an empty set of features and at each state adds a feature that produces the best performance accuracy. The backward selection, on the other hand, starts with the full set and then sequentially removes the features. Levner (36) discusses the possible intractability of the SBS procedure and suggests a modified SBS procedure that starts with all of the features and stops at the first feature whose removal does not affect the overall performance of the classification procedure. It is determined by the standard leave-one-out cross-validation (LOOCV) approach. At each loop of the SBS, after finding the first candidate feature, the features are reordered on the basis of the probability of each feature being irrelevant or redundant. This probability is based on the Kolmogorov–Smirnov (KS) test. Next, this modified SBS algorithm involves recording the stopping position of the ranked features. At every iteration, instead of checking all the features, SBS starts the feature search from the previous stopping position. On a similar note, the latest version of ClinProTools 2.1 (Brucker Daltonics) uses stochastic optimization techniques like Genetic Algorithm (GA) to pick the peaks that are most relevant to classify the samples. The features selected by the genetic algorithm are used as selected features for the classifiers like support vector machine (SVM) to classify the samples. However, it also uses SVM and then cluster analysis to rank the features. Ressom et al. (37) combine particle swarm optimization (PSO) to select the features for the SVM classifier. Ressom et al. (38) use ant colony optimization (ACO) and SVM to select a parsimonious set of peaks. Ant colony optimization was developed by simulating the behavior of real ant colonies (39). The ACO is used in this context to select a combination of features that are useful for classifying the samples with respect to the SVM classification algorithm. Real ants cooperate among themselves by depositing pheromone on the ground. This algorithm integrates prior information into the algorithm for better peak selection. An ant selects, for every iteration, a set of n features from a set of L total features each with a probability

$$P_i(t) = \frac{(\tau_i(t))^{\alpha}\eta_i^{\beta}}{\sum_{i=1}^{L}(\tau_i(t))^{\alpha}\eta_i^{\beta}},$$

where $\tau_i(t)$ can be thought of as the amount of pheromone trail deposited by feature i at iteration t and η_i is the prior information of feature i. This prior information can be the value of the t-statistic for that feature. α and β are the parameters involving the relative influence of the pheromone trail and the prior information. Initially, at $t = 0$, the $\tau_i(t)$'s are constants; then at every consecutive iteration, the $\tau_i(t)$'s are updated according to their classification accuracies. At the first iteration, each ant chooses n distinct features or a trail out of L features with probabilities proportional to the prior information. Let S_j be the jth ant with n distinct features. The performance of S_j is measured by its performance of classification accuracy using these n features determined by some cross-validation method. The amount of pheromone for each feature in S_j is updated each iteration by

$$\tau_i(t+1) = \rho \cdot \tau_i(t) + \Delta\tau_i(t),$$

where ρ is a constant between 0 and 1. $\Delta\tau_i$ is proportional to the classification accuracy of pheromen trail S_j. If feature i is absent in S_j then $\Delta\tau_i$ is zero. This updating is done for all the ants (i.e., all such S_j $j = 1, 2, \ldots, N$ ants). This updating scheme rewards the features with a larger amount of pheromone trails and strong prior information. This in turn influences the probability function to lead the ants toward them. This process increases the classification accuracy by increasing the pheromone trail.

Wrapper methods can be used with any classifier and can reduce the number of features. However, wrapper methods do not incorporate knowledge about the specific structure of the classification or regression function. They are extremely computationally expensive since they need to evaluate classification accuracies at every iteration following some cross-validation scheme. Overfitting could be a problem in this method. Also, the selected features are dependent on the particular classification algorithm used in the procedure.

3.3. Embedded Method

The embedded method of feature selection is different from any other feature selection method mentioned above in terms of the way feature selection and learning interact with each other. Filter methods do not incorporate learning. Wrapper methods use a learning algorithm to measure the accuracy of classification for the subsets of features without incorporating knowledge about the specific structure of the classification. On the other hand, in embedded methods, the learning part and the feature selection part cannot be separated. In this method, a classifier is trained by a feature subset in such a way that it optimizes an objective function that rewards the accuracy of a classifier and penalizes the usage of redundant features. Lal et al. (40) provide the details of the mathematical formulation of the embedded method. Weston et al. (41) measure the importance of a feature using a bound specifically for SVM. These operations are embedded in the classifier itself. For example, in the random forest classifier, many single-decision trees are embedded in such a way to make it possible to calculate the importance of each feature, and the features are ranked in terms of their decreasing importance or decreasing Gini Index. The features with relatively little importance can be removed from the classifier, and the remaining features can be treated as peaks. Levner (36) introduced the boosted feature extraction method (boostedFE), which is also known as one of the embedded feature selection procedures. In this algorithm, it searches throughout the features during each round of boosting and selects a single best feature upon which it builds the weighted nearest-centroid classifier.

One of the very well-known embedded methods for feature selection is *support vector machine recursive feature elimination* (*SVM-RFE*). SVM-RFE refines the optimum feature set by using a support vector machine in the context of microarray data (42). The idea of SVM-RFE is that if the orientation of the separating hyperplane found by the SVM is orthogonal to a particular feature dimension, then the feature is informative. SVM-RFE uses the weights of an SVM classifier to produce a feature ranking, and then recursively eliminates the feature with the smallest weight magnitude. Zhang et al. (43) develop and use a similar method called *recursive support vector machine* (R-SVM) to analyze SELDI-TOF mass spectrometry data. However, they use a different evaluation criterion than that used by Guyon et al. (44) to select the most important features. Geurts et al. (45) provide several decision tree–based ensemble methods (bagging, boosting, random forest, and extra-trees) for peak detection and classification for mass spectrometry data.

Both the wrapper method and the embedded techniques for feature selection are computationally demanding compared to the filter methods. However, the embedded methods are lesser computationally intensive than the wrapper method. Also, they are known to be less vulnerable to overfitting than the wrapper method.

3.4. Hybrid Method

This method takes advantage of both the filter method and the embedded or wrapper method to select the features and then uses them to classify the samples with many classification algorithms. One example of such methods is Wu et al. (11). They use a random forest (46) classifier on the full set of features and then select the features on the basis of their importance measures. These selected features or peaks are then used and compared in terms of their classification errors using different classification algorithms. Zhu et al. (32) also prescreen the number of features according to their relative importance with respect to the random forest classifier and use the same number of features they have determined to be significant in terms of *t*-tests on individual features. Consequently, they use them with different classification algorithms.

3.5. Isotopic Distribution

Unlike all the feature selection methods mentioned above, there exists a class of feature selection or peak-picking algorithms based on the isotopic distribution of the peptide molecules. For example, in a MALDI-TOF spectrum, a single peptide can be realized as a series of isotopic peaks. These peaks differ by the number of isotopes of C_{13}, N_{15}, O_{18}, P_{32}, and S_{34}. Note that the peak used most of the time by the peptide mass fingerprinting method is the *monoisotopic peak*. This monoisotopic peak has the unique characteristic of having the lightest mass in an isotopically resolved peak distribution containing only the isotopes C_{12}, N_{14}, O_{16}, P_{31}, and S_{32}.

Note that, in general, the monoisotopic peak is not necessarily the most intense peak. Breen et al. (13, 14) consider a Poisson model to fit an isotopic distribution of peptides. In general, for a relatively large number of atoms n and a relatively small expected proportion p compared to its heavy isotope, it can be modeled as a Poisson distribution with mean $M = np$:

$$P(x;M) = \begin{cases} \frac{e^{-M}M^x}{x!} & \text{if } x = 0, 1, \ldots, \\ 0 & \text{otherwise} \end{cases}$$

However, the values of n and p are not known. Breen et al. (13) use a linear mapping function (least-squares regression line) of the known molecular weight m of a peptide to the mean of the Poisson distribution:

$$M = F(m) = 0.000594m - 0.03091.$$

We anticipate that instead of using this empirical method of predicting the mean of the Poisson distribution from the least-squares line described above, it may be useful to estimate the parameter from the current experimental data. This line was created by taking the database result of the isotopic distribution of a hypothetical average amino acid (13). After a peptide isotopic distribution has been identified, Breen et al. (13, 14) take the leftmost feature of the distribution to be the monoisotopic peak of that peptide. However, there exist added complications to the isotopically resolved distributions due to a process called *deamidation,* where aspartic acid and glutamic acid are converted to aspartate and glutamate, which results in a mass difference of approximately +1 Da. This in turn results in shifted or overlapping isotopic distributions of the peptides. Breen et al. (13) model this as an additive mixture of a Poisson distribution:

$$P(x; F(m)) + P(x-1; F(m+1)).$$

Note that this can be generalized to any number of mixtures.

This model gives us an opportunity to model overlapping isotopic distributions. Breen et al. (13, 14) applied this model to the processed (stick representation) raw mass spectra with mathematical morphology. The stick representation of the data already removes many unwanted features from the mass spectrum. The details of the stick representation of the data can be found in Breen et al. (13). Harvesting monoisotopic peaks following this manner is an efficient way of peak detection without much human intervention.

In the next section, we will consider the process of classification using the selected peaks. Classifying disease and nondisease (case and control) protein spectra has the potential to identify proteomic biomarkers of several diseases/conditions (47). It is to be noted, however, that any classification technique demands an even lower number of important features to classify the samples in the best possible way. So discussions of some feature selection/ reduction techniques are embedded within the classification algorithm as well.

4. Classification

The identification of important biomarkers and the prediction of health outcomes are the two major goals of some mass spectrometry studies. Supervised learning techniques that encompass the whole range of classification algorithms provide a convenient formal framework for building predictive models. If presented with a new MS profile, a classification algorithm should be able to accurately predict the class of a sample (for example, healthy or cancerous) using just the information in that profile. To be able to compare different classification algorithms, we often use common performance measures, such as predictive accuracy, classification error, sensitivity, and specificity. All of them can be computed from a simple 2×2 confusion matrix that tracks the number of correct and incorrect predictions.

Based on the confusion matrix in **Table 11.1**, accuracy is defined as the proportion of correct predictions over all predictions:

$$\text{Accuracy} = \frac{a+d}{a+b+c+d}.$$

Table 11.1
A confusion matrix used to assess the performance of classification algorithms

	Class A	Class B
Predicted Class A	a	b
Predicted Class B	c	d

Classification error is simply equal to 1-accuracy. Sensitivity and specificity are defined as

$$\text{Sensitivity} = \frac{a}{a+c} \quad \text{and} \quad \text{Specificity} = \frac{d}{b+d}.$$

Classification error rates in most cases are sufficient indicators of performance. However, when the cost of misclassifying one class is much greater than the cost of misclassifying another class, sensitivity and specificity measures provide class-specific estimates of predictive accuracy. Since most classification algorithms have one or more tuning parameters, by varying them one obtains the whole range of sensitivity and specificity pairs. They are usually summarized in a graphical plot called the *receiver operating characteristic* (*ROC*) curve of sensitivity versus 1-specificity. A larger area under the curve (AUC) indicates a better performance of the classification algorithm.

Classifiers generally perform better on training data that are used to construct them than the test data. Therefore, to make the estimates of performance measures even more reliable, researchers often use a K-fold cross-validation. The original MS training data set is randomly split into K sets of about equal size. For each $1 \leq k \leq K$, the kth part is regarded as a test and its complement as the training set. The classifier is built using this artificially created training set and its performance measures are computed using the partition that was left out (artificially created test set). When all K classifiers have been built and tested, the estimated error rates are averaged across all partitions.

4.1. Dimension Reduction

High dimensionality of mass spectrometry data can be reduced by applying one of the techniques for feature selection discussed in the previous sections. This essential preprocessing step is usually employed before carrying out classification analysis to remove the "noisy" features (m/z values) to significantly improve the accuracy rates of most classification algorithms (35, 32). In some cases, for example, random forest and penalized-type classification, which will be discussed in detail later, feature selection and classification are fused together in a single algorithm where both goals are achieved simultaneously. Therefore, the dichotomy of feature selection and classification may be somewhat artificial depending on the classification methodology used. Within the context of mass spectrometry data analysis, the feature selection necessity is dictated by both practical considerations from the standpoint of classification accuracy and efficiency as well as the reasonable upper bound on the number of markers that can be used for screening or diagnosis in the future.

In practice, it may happen that the number of selected features can still be relatively large, most of the time much larger than the number of samples in the data. Common classification algorithms such as logistic regression and linear discriminant analysis (LDA) cannot be directly applied when the number of features p is larger than the number of samples N. So when $N << p$, one of the dimension reduction techniques has to be applied first to reduce the number of m/z features even further. The most well-known dimension reduction methods are the principal component analysis (PCA) (48) and the partial least-squares (PLS) (49). Both PCA and PLS effectively reduce the number of dimensions while preserving the structure of the data. They differ in the way they construct latent variables. PCA picks the directions of its principal components along the axis of the largest variability in the data, while PLS maximizes the covariance between the dependent and independent variables, trying to explain as much variability as possible in both the dependent and independent variables. Both PCA and PLS were used in a combination with logistic regression and LDA.

4.2. Common Classification Algorithms

Classification algorithms in both the statistical and machine learning literatures provide researchers with a very broad set of tools for discriminatory analysis. Most of them – sometimes with a bit of extra care – can be used to classify MS samples based on their mass spectra profiles. The first study that used a machine learning approach to discriminate between case and control ovarian cancer proteomic (SELDI-TOF) samples was Petricoin et al. (6). They used a combination of elements from genetic algorithm (GA) (50) and Kohonen's self-organizing maps (SOM) (51) with a classification performance that caught the attention of the bioinformatics community. Sensitivity, specificity, and positive-predictive value were estimated to be 100, 96, and 94%, respectively. However, under further scrutiny, some of these estimates were later questioned. The controversy stirred for some time, but the first step toward a systematic introduction of various classification tools in the analysis of proteomic data was made.

We describe a selected number of classification techniques that have been successfully applied to mass spectrometry data in the past. The list is not exhaustive by any means and is given here to expose the breadth of statistical and machine learning used in the context of proteomic data. Satten et al. (17) and Izmirlian 34) use random forest; Adam et al. (52) use classification trees; Ball et al. (53) use artificial neural networks; Purohit and Rocke (54) use logistic regression with partial least-squares; Hilario et al. (55) use naïve Bayes classifier; Zhu et al. (33) use support vector machines; Lilien et al. (56) use principal component analysis with linear discriminant analysis; Tibshirani et al. (57) use peak probability contrasts (PPC). Wu et al. (11) perform a detailed comparative study of the performance of different classification methods. They apply linear and quadratic discriminant analysis, *K*-nearest-neighbor classifier, support vector machine (SVM), random forest (RF), and bagging and boosting classification trees to ovarian cancer case and control serum samples (MALDI mass spectrometry data set). Their findings suggest that when using multiple *t*-tests for feature selection, SVM has the smallest prediction error, closely followed by RF. When using random forest for feature selection, RF understandably outperforms all other algorithms, while SVM does not perform as well as it did in the first case. Random forest seems to be the most consistent performer among the algorithms considered. Datta and DePadilla (32) study the performance of LDA, QDA (quadratic discriminant analysis), neural networks, 1-nearest-neighbor classifier, SVM, and RF under different feature selection mechanisms. Their results indicate that SVM and RF are the two most consistent classifiers, with error rates of 2.6–7.7%.

Choosing a classification algorithm for particular mass spectrometry data just from the ones mentioned above is not an easy task. Classification algorithms differ in the degree of interpretability of the model, complexity of the model, computation time necessary to build a classifier, applicability, noise tolerance, and many other important aspects. Which algorithm(s) should be chosen at any specific time greatly depends on the data and their intrinsic complexity. A familiarity with the major representatives of different classification approaches is absolutely necessary to understand and weigh the choices that one has when it comes to the practical application of classification techniques to any data, including mass spectrometry data.

In the next several subsections, we will present the most common classification algorithms encountered in the mass spectrometry literature. A comprehensive discussion of different classification algorithms appears in Hastie et al. (58).

4.2.1. Logistic Regression and Penalized Logistic Regression

Logistic regression is perhaps the most widely used model when dealing with binary outcomes. In the context of classification, it applies to a two-class situation. It models the probability of a success (here denoted as class = 1) using the following relationship:

$$P(C = 1|X = x) = \frac{\exp(\beta_0 + \beta^T x)}{1 + \exp(\beta_0 + \beta^T x)},$$

where β_0 and β are the parameters maximizing the log-likelihood function. The model is usually equivalently expressed as a relationship between a linear function of data and the logit transformation of the probability of a success:

$$\log\left(\frac{P(C = 1|X = x)}{1 - P(C = 1|X = x)}\right) = \beta_0 + \beta^T x.$$

Parameters in this model are estimated via the Newton–Raphson algorithm, an iterative numerical technique used for solving nonlinear systems of equations.

As with most classical statistical techniques, the maximum number of parameters that can be reliably estimated should be small when compared to the number of samples in the data. When the number of features is larger than the number of samples, as in the case of mass spectrometry data, feature selection has to be performed to reduce the dimensionality of the data. An alternative approach is to use a penalized logistic regression, where a penalty is imposed on the log-likelihood function $l(\beta)$ corresponding to the logistic regression

$$l^*(\beta) = l(\beta) - \lambda J(\beta),$$

where λ is the tuning parameter controlling how much penalty should be applied, and $J(\beta)$ is the penalty term, which usually takes the two most common forms: ridge penalty $\sum_{i=1}^{p} \beta_i^2$ and lasso penalty $\sum_{i=1}^{p} |\beta_i|$. Due to the lasso penalty term, many of the estimated parameters will be reduced to 0. However, selected variables or features using the lasso penalty are limited by the number of observations, which is much lower than the number of variables. One other problem using the lasso penalty is that it selects only one of the highly correlated variables irrespective of its biological importance. The ridge penalty does not suffer from that problem, and so in the case of the elastic net solution (59), both the lasso penalty and the ridge penalty terms are taken together in the log-likelihood function. This provides a better solution.

Purohit and Rocke (54) use logistic regression coupled with a preliminary PLS dimension reduction step to classify samples based on their mass spectra. They report accuracy rates from 90.2–100% depending on the data transformation used to stabilize the variance. Square-root-transformed data resulted in a perfect classification of samples. Obviously, the results reported are true for the given data.

4.2.2. Linear and Quadratic Discriminant Analysis (LDA and QDA)

Linear discriminant analysis is one of the classical statistical classification techniques originally proposed by Fisher in 1936 (60). As the name suggests, it is a linear classifier, which means that the boundaries between classes are linear (a straight line in a two-dimensional case and a hyperplane in three or more dimensions). The idea behind LDA is very intuitive and relates to the variance decomposition of ANOVA: The more separable the classes, which occurs when the within-class variance is small and the between-class variance is large, the easier it is to correctly classify samples. Suppose that the training mass spectrometry data consist of n samples with p variables (features), which are the intensity values at each m/z value and are denoted by a matrix X with the dimensions of n by p. The LDA seeks the linear transformation of X, Xa, such that when the classes are projected onto the new space, the separation between them is maximized. This can be formally achieved by maximizing the ratio $a^T Ba/a^T Wa$, where B is the between-class covariance matrix, W is the within-class covariance matrix, and a^T stands for the transpose operation. a can always be chosen such that $a^T Wa = 1.$, and the maximization problem can be cast in the form of a constrained maximization problem:

$$\max_a a^T Ba \text{ subject to } a^T Wa = 1.$$

This is a familiar form of a generalized Eigen value problem, and the solution is the eigenvector corresponding to the largest Eigen-value of $W^{-1}B$.

The LDA can also be derived via a probability model by assuming that each class c has a multivariate normal distribution with mean μ_c and a common covariance matrix Σ. Let π_c be the prior probability of class c; then the posterior probability of belonging to class c is given by the Bayes formula

$$p(c|x) = \frac{\pi_c p(x|c)}{p(x)}.$$

We would like to assign samples to classes with the largest posterior probability. By maximizing the logarithm of the posterior distribution with the above assumption of $p(x|c)$ distributed as $N(\mu_c, \Sigma)$, we get

$$L_c = \log(p(x|c)) + \log(\pi_c) = x\Sigma^{-1}\mu_c^T - \frac{\mu_c\Sigma^{-1}\mu_c^T}{2} + \log(\pi_c),$$

which is a linear function in x and directly corresponds to the LDA. When covariance matrices are different for each class (i.e., $\Sigma_i \neq \Sigma_j$), we obtain a quadratic discriminant analysis (QDA), which would be a quadratic function in x. Both LDA and QDA have been extensively used in practice with a fair share of success. When only two classes are being predicted, LDA gives the same results as logistic regression. This correspondence breaks down for more classes.

Wagner et al. (7) considered both LDA and QDA for classification of 41 MS samples, 24 of which were known to come from patients with lung cancer. Both algorithms performed fairly well, particularly when using the top four peaks out of 229 (error rates of 10 and 12%, respectively). It is important to point out here that they observed a significant decline in the performance of LDA and QDA (27 and 34% error rates) when 13 features were used for classification, and the estimates were highly unstable due to covariance matrices being nearly singular. Using PCA or PLS on these 13 features would probably improve the error rates. Lilien et al. (56) propose a classification algorithm Q5, which, in essence, is a PCA dimension-reduced LDA. They test its performance on three ovarian and one prostate cancer SELDI-TOF MS data sets and obtain sensitivity and specificity values in the excess of 97%. Datta (47) use a combination of random forest and LDA, where LDA is used for classification using the top nine features as identified by random forest. The reported estimate of classification error is 14.7%.

4.2.3. Support Vector Machine (SVM)

Support vector machine (SVM) is among the most recent significant developments in the field of discriminatory analysis (61). In its very essence, SVM is a linear classifier (just like logistic regression and LDA), as it directly seeks a separating hyperplane between classes that have the largest possible margin. The margin is defined here as the distance between the hyperplane and the closest sample point. Usually, there are several points called "support vectors" that are exactly one margin away from the hyperplane and on which the hyperplane is constructed. It is clear that, as stated, SVM is of little practical use because most classification problems have no distinct separation between classes and, therefore, no such hyperplane exists. To overcome this problem, two extensions have been proposed in the literature: penalty-based methods and kernel methods.

The first approach relaxes the requirement of a "separating" hyperplane by allowing some sample points to be on the wrong side. It becomes a constrained optimization problem where the constraint is the total distance from all misclassified points to the hyperplane that is smaller than a chosen threshold *c*. The second approach is more elegant and frequently used. Since no linear separation between classes is possible in the original space, the main idea is to project onto a higher-dimensional space, where such a separation usually exists. It turns out that there is no need to specify the transformation $h(x)$ explicitly, and the knowledge of the kernel function is sufficient for optimization:

$$K(x_i, x_j) = h(x_i)^T h(x_j).$$

The most popular choices for the kernel function are the *k*th-degree polynomial

$$K(x_i, x_j) = (1 + x_i^T x_j)^k,$$

radial basis

$$K(x_i, x_j) = e^{\frac{-\|x_i - x_j\|^2}{c}},$$

and the neural network kernel

$$K(x_i, x_j) = \tanh(k_1 x_i^T x_j + k_2),$$

where k, c, k_1, and k_2 are the parameters that need to be specified. The kernel functions involve only the original nontransformed data, which makes them easily computable.

SVM has been successfully applied to mass spectrometry data. It has an advantage in flexibility over most other linear classifiers. The boundaries are linear in a transformed high-dimensional space, but on the original scale they are usually nonlinear, which gives SVM extra flexibility where it is required.

SVM has been extensively applied to MS data. Wagner et al. (7) point out its robustness to different numbers of features used and overall confident classification with low classification error rates of 2% under two of the three settings considered. SVM performed very well in a comparative study of Wu et al. (11) when marginal t-tests were used to identify 15 and 25 markers for classification. In their comparative study on the performance of the most common classifiers under different feature selection schemes, Datta and DePadilla (32) conclude that SVM is the most consistent classification algorithm, with error rates ranging from 2.6–7.7%.

4.2.4. *k*-Nearest-Neighbor Classifier (KNN)

The k-nearest-neighbor algorithm is a good representative of nonparametric classification techniques (62, 63). It is a local classifier in the sense that a class of any given sample is determined by its immediate neighborhood of size k, which is usually much smaller than the number of samples. The algorithm proceeds by finding the k nearest neighbors of each data point and taking a majority vote to determine their classes. A number of distance functions can be used to determine which samples are "close" to each other. The most popular distances are the Euclidean, Mahalanobis, and correlation-based distances.

Appropriately choosing the only parameter k can be a challenge. Some researchers suggest using cross-validation to select the optimal values for k. In practice, however, the most common choices for k are 1 and 3. Since we are usually dealing with two classes, an odd k avoids an issue of ties when predicting a class based on k neighbors.

A major merit of the KNN algorithm is its conceptual simplicity. It is very easy to implement although the computational time required can be intensive and, in some cases, even prohibitive. The interpretability of the results is rather difficult, as no parametric model is fit to the data and classification occurs "behind the scenes." But it turns out that KNN does have a useful interpretation, at least theoretically, as the estimation of the posterior probability $p(c|x)$ by the ratio of the most frequent class over k neighbors.

Zhu et al. (33) successfully applied KNN with $k = 5$ to ovarian cancer data. They report perfect classification accuracy rates based on the independent (testing) data. KNN is known to be quite sensitive to noise, and, in some cases, its performance is clearly affected when applied to noisy MS data. The study by Datta and DePadilla (32) reveals this shortcoming where

1-NN (and any other choice of *k* did not increase the performance) performed rather poorly, with error rates of 7.1–17.2%, while the largest error rate for four other classification algorithms was 7.7%.

4.2.5. Random Forest (RF)

Classification trees are particularly popular among medical researchers due to their interpretability. Given a new sample, it is very easy to classify it by going down the tree until one reaches the terminal node that carries the class assignment. Random forest (64, 46) takes classification trees one step further by building not a single but multiple classification trees using different bootstrap samples (sampled with replacement). A new sample is classified by running it through each tree in the forest. One obtains as many classifications as there are trees. They are then aggregated through a majority voting scheme and a single classification is returned. The idea of bagging, or averaging, multiple classification results, as applied in this context, greatly improves the accuracy of somewhat unstable individual classification trees.

One of the interesting elements of random forest is the ability to compute unbiased estimates of misclassification rates on the fly without explicitly resorting to testing data after building the classifier. By using the samples that were left out of the bootstrap sample when building a new tree, also known as out-of-bag (o-o-b) data, RF runs the o-o-b data through the newly constructed tree and calculates the error estimate. These are later averaged over all trees to obtain a single misclassification error estimate. This combination of bagging and bootstrap is sometimes called 0.632 cross-validation (65) because roughly two thirds of the samples used for building each tree are really $1-1/e$, which is approximately 0.632. This form of cross-validation is arguably very efficient in the way it uses available data.

Variable importance is another element of RF that deserves special attention. Random forest not only classifies samples into classes but also automatically determines the most important features in the data. This ability is exploited quite often when dealing with mass spectrometry data for the feature selection of discriminatory peaks.

Random forest has been applied to MS data (11) and performed well, particularly when RF was also used for feature selection with error rates below 10%. Satten et al. (17) use random forest for the discrimination of bacterial strains based on their MALDI-TOF MS profiles. The estimated error rate in their study is 0%. An extensive and thorough examination of random forest in relationship to the SELDI-TOF proteomic data is undertaken in Izmirlian (34). He pinpoints the key advantages of the algorithm, among which are the efficient use of data for classification and validation, simplicity of the approach, speed of computation, and practically no dependence on the tuning parameters. Classification of SELDI samples is stable due to the bagging approach, which also translates into high noise tolerance and robustness.

5. Software for MS Data Analysis

Software solutions for the preprocessing and analysis of mass spectrometry data are available from a number of different sources. Here, we will concentrate on two open source software applications freely available in a public domain, R (http://www.r-project.org/) and Weka (http://www.cs.waikato.ac.nz/ml/weka/).

5.1. R

R is a popular open source software environment for statistical computing and data visualization available for most mainstream platforms. In the base distribution of R, many statistical tools, input–output capabilities, and a graphics engine are available for immediate use. However, this is not the main reason for its popularity among researchers. R is easily extendable and customizable through user-created libraries, called *packages*, available from the Comprehensive R Archive Network (CRAN) with mirrors around the globe. Packages related to bioinformatics, in particular to microarray data analysis, are being developed under a separate open source project, Bioconductor (http://www.bioconductor.org/).

A number of R packages are available for mass spectrometry data analysis. The CRAN repository contains the *caMassClass* package, which performs preprocessing and classification of SELDI mass spectrometry data. The package provides routines for baseline correction (msc.baseline.subtract), normalization (msc.mass.adjust), peak detection (msc.peaks.find), and alignment (msc.peaks.align), as well as a cross-validation function (msc.classifier.test) for testing several common classification algorithms (SVN, ANN, LDA, QDA, LogitBoost, and recursive partitioning). Two input–output formats are available: CSV and mzXML. Another package for SELDI MS data, *MASDA*, is available from http://bioinformatics.nki.nl/software.php. It performs similar basic preprocessing steps and provides some visualization of results.

The Bioconductor *PROcess* package incorporates a set of functions for baseline correction (bslnoff) and peak detection (isPeak) with very informative color graphical plots. The package can operate in a batch mode, performing baseline removal, normalization, and quality assessment on a number of samples. Three quality parameters are estimated (quality, retain, and peak), which can be used to identify samples of poor quality that should not be used in further analysis. Liquid chromatography (LC/MS) data can be preprocessed using the Bioconductor *xcms* package. Multiple input formats, including NetCDF, mzXML, and zmData, are available for users' convenience. The package performs peak detection (xcmsSet), peak matching (group), peak imputation (fillPeaks), and statistical analysis (diffreport), which reports the most statistically significant differences in analyte intensities.

R provides most, if not all, common classification algorithms. Here we will just list the package names for some of them. Further details about input parameters, implementation, and references can be found in package documentation manuals and/or vignettes. LDA and QDA are available in the *MASS* package, SVM in the *e1071* package, RF in the *randomForest* package, ANN in the *nnet* package, recursive partitioning in the *rpart* package, penalized logistic regression in the *penalized* package, KNN in the *class* package, and peak probability contrasts in the *ppc* package.

5.2. Weka

Weka (Waikato Environment for Knowledge Analysis), developed at the University of Waikato in New Zealand, is an open source, Java-based software package popular among machine learning researchers. It is publicly available online at http://www.cs.waikato. The software provides a comprehensive compilation of machine learning methodologies in both unsupervised (clustering) and supervised (classification) settings. Some data management capabilities are also built in. Many of the classification algorithms discussed above, including additional ones that were not mentioned, have been implemented in Weka. A convenient user interface is perhaps sufficient for beginners, as it is supplemented with a flexible command-line interface for more advanced users.

Advanced classification validation and reporting are built into Weka, providing users with a quick assessment of the performance. Extensive visualization tools are one click away. It is very easy to visualize trees, neural networks, boundaries between classes in the two-dimensional space, and so on. *k*-fold cross-validation with an arbitrary percentage of samples allocated to the training set can also easily be specified. All these features make Weka a very simple and convenient, yet powerful, machine learning tool.

6. Discussion

In spite of several success stories with classifying mass spectra and finding protein biomarkers of diseases, there is no clear consensus among data analysts and statisticians on which classification algorithm should be used for a particular data type. In a recent international competition on classifying mass spectrometry proteomic diagnosis organized at Leiden University Medical Centre (LUMC, the Netherlands, March 2007), various classifiers yielded widely different results when applied to the same data set (e.g., 47, 66). Furthermore, some of the earlier success stories regarding proteomic biomarkers have been questioned due to their lack of reproducibility and the classifying peaks not having biological significance.

It is therefore important to investigate the question of selecting the most suitable classifier for a given data set and try to provide general guidelines. Also, it will be worthwhile to have a data-based way of creating a classifier that performs close to the "best" classifier given a collection of classifiers. Last, but not the least, it is important to identify the features (peaks) that play a main role in the classification process. In a sense, features with a high value of "importance" carry a higher differential signature, which can be studied further for biological understanding of disease etiology.

Acknowledgments

This research was supported in part by NSF grant DMS-0805559, NIH grant 1P30ES014443, and NSF grant MCB-0517135.

References

1. Albrethsen J. (2007) Reproducibility in protein profiling by MALDI-TOF mass spectrometry. *Clin Chem* 53: 852–858.
2. Stühler K., Baessmann C, Sitek B, Jabs W, Lubeck M, Poschmann G, Chamrad DC, Blüggel M, Meyer HE. (2008) Label-free proteomics: a versatile tool for differential proteome, **ABRF 2008, V12-T**: Bruker Daltonics Poster, Salt Lake City, UT.
3. Diamandis EP. (2003) Serum proteomic patterns for detection of prostate cancer. *J Natl Cancer Inst* 95:489–490.
4. Hilario M, Kalousis A, Pellegrini C, Muller M. (2006) Processing and classification of protein mass spectra. *Mass Spectrum Rev* 25:409–449.
5. Baggerly K, Morris J, Coombes K. (2004) Reproducibility of SELDI-TOF protein patterns in serum: comparing datasets from different experiments. *Bioinformatics* 20: 777–785.
6. Petricoin EF, Ardekani AM, Hitt BA, Levine PJ, Fusaro VA, Steinberg SM, Mills GB, Simone C, Fishman DA, Kohn EC, Liotta LA. (2002) Use of proteomic patterns in serum to identify ovarian cancer. *Lancet* 359:527–577.
7. Wagner M, Naik DN, Pothen A, Kasukurti S, Devineni RR, Bao-Ling A, Semmes OJ,

Wright JL. (2004) Computational protein biomarker prediction: a case study for prostate cancer. *BMC Bioinformatics* 5:26.
8. Yasui Y, Pepe M, Thompson ML, Adam BL, Wright GL, Jr., Qu Y, et al. (2003) A data-analytic strategy for protein biomarker discovery: profiling of high-dimensional proteomic data for cancer detection. *Biostatistics* 4:449–463.
9. Coombes KR, Tsavachidis S, Morris JS, Baggerly KA, Hung MC, Kuerer HM. (2005) Improved peak detection and quantification of mass spectrometry data acquired from surface-enhanced laser desorption and ionization by denoising spectra with the undecimated discrete wavelet transform, *Proteomics* 5:4107–4117.
10. Sorace JM, Zhan M. (2003) A data review and re-assessment of ovarian cancer serum proteomic profiling. *BMC Bioinformatics* 4:24.
11. Wu B, Abbott T, Fishman D, McMurray W, Mor G, Stone K, Ward D, Williams K, Zhao H. (2003) Comparison of statistical methods for classification of ovarian cancer using mass spectrometry data, *Bioinformatics* 19: 1636–1643.
12. Baggerly KA, Morris JS, Wang J, Gold D, Xiao LC, Coombes KR. (2003) A comprehensive approach to the analysis of matrix-assisted laser desorption/ionization time of flight proteomics spectra from serum samples. *Proteomics* 3:1667–1672.
13. Breen EJ, Hopwood FG, Williams KL, Wilkins MR. (2000) Automatic Poisson peak harvesting for high throughput protein identification. *Electrophoresis* 21:2243–2251.
14. Breen EJ, Holstein WL, Hopwood FG, Smith PE, Thomas ML, Wilkins MR. (2003) Automated peak harvesting of MALDI-MS spectra for high throughput proteomics. *Spectroscopy* 17:579–596.
15. Sollie P, Breen EJ, Jones R. (1996) Recursive Implementation of Erosions and Dilations Along Discrete Lines at Arbitrary Angles. *IEEE Trans Pattern Anal Mach Intell*, 18: 562–567.
16. Liu H, Li J, Wong L. (2002) A comparative study on feature selection and classification methods using gene expression profiles and proteomic patterns. *Genome Inform* 13: 51–60.
17. Satten GA, Datta S, Moura H, Woolfitt AR, Carvalho MG, Carlone GM, et al. (2004) Standardization and denoising algorithms for mass spectra to classify whole-organism bacterial specimens. *Bioinformatics* 20: 3128–3136.
18. Shao XG, Leung AK, Chau FT. (2003) Wavelet: a new trend in chemistry. *Acc Chem Res* 36:276–283.
19. Saeys Y, Inza I, Larrañaga P. (2007) A review of feature selection techniques in bioinformatics. *Bioinformatics* 23:2507–2517.
20. Kirby M. (2001) *Geometric Data Analysis: An Empirical Approach to Dimensionality Reduction and the Study of Patterns*, John Wiley & Sons, New York.
21. Savitzky A, Golay MJE. (1964) Smoothing and differentiation of data by simplified least squares procedures. *Anal Chem* 36: 1627–1639.
22. Eilers PHC, Marx BD. (1996) Flexible smoothing with B-splines and penalties. *Statist Sci* 11:89–121.
23. Kast J, et al. (2003) Noise filtering techniques for electrospray quadrupole time of fluid mass spectra. *J Am Soc Mass Spectrom* 14:766–776.
24. Morris JS, Coombes KR, Koomen J, Baggerly KA, Kobayashi R. (2005) Feature extraction and quantification for mass spectrometry in biomedical applications using the mean spectrum. *Bioinformatics* 21: 1764–1775.
25. Yasui Y, McLerran D, Adam BL, Winget M, Thornquist M, Feng Z. (2003) An automated peak identification/calibration procedure for high-dimensional protein measures from mass spectrometers. *J Biomed Biotechnol* 2003:242–248.
26. Serra J. (Ed.). (1988) *Image Analysis and Mathematical Morphology, Vol. 2: Theoretical Advances*, Academic Press, New York.
27. Bhanot G, Alexe G, Venkataraghavan B, Levine AJ. (2006) A robust meta classification strategy for cancer detection from MS data. *Proteomics* 6:592–604.
28. Benjamini Y, Hochberg Y. (1995) Controlling the false discovery rate – a practical and powerful approach to multiple testing. *J Roy Statist Soc Ser B* 57:289–300.
29. Westfall P, Young SS. (1993) *Resampling-Based Multiple Testing, Examples and Methods for p-Value Adjustment*, John Wiley & Sons, New York.
30. Dudoit S, Yang YH, Speed TP, Callow MJ. (2002) Statistical methods for identifying differentially expressed genes in replicated cDNA microarray experiments. *Stat Sin* 12:111–139.
31. Datta S, Datta S. (2005) Empirical Bayes screening of many *p*-values with applications to microarray studies. *Bioinformatics* 21:1987–1994.
32. Datta S, DePadilla L. (2006) Feature selection and machine learning with mass

spectrometry data for distinguishing cancer and non-cancer samples. *Stat Methodol*, 3: 79–92.
33. Zhu W, Wang X, Ma Y, Rao M, Glimm J, Kovach JS. (2003) Detection of cancer specific markers amid massive mass spectral data. *Proc Natl Acad Sci USA* 100: 14666–14671.
34. Izmirlian G. (2004) Application of the random forest classification algorithm to a SELDI-TOF proteomics study in the setting of a cancer prevention trial. *Ann NY Acad Sci* 1020:154–174.
35. Yu JS, Ongarello S, Fiedler R, Chen XW, Toffolo G, Cobelli C, Trajanoski Z. (2005) Ovarian cancer identification based on dimensionality reduction for high-throughput mass spectrometry data. *Bioinformatics* 21:2200–2209.
36. Levner I. (2005) Feature selection and nearest centroid classification for protein mass spectrometry. *BMC Bioinformatics* 6:68.
37. Ressom HW, Varghese RS, Abdel-Hamid M, Eissa SA, Saha D, et al. (2005) Analysis of mass spectral serum profiles for biomarker selection. *Bioinformatics* 21:4039–4045.
38. Ressom HW, Varghese RS, Drake SK, Hortin GL, Abdel-Hamid M, Loffredo CA, Goldman R. (2007) Peak selection from MALDI-TOF mass spectra using ant colony optimization. *Bioinformatics* 23:619–626.
39. Dorigo M, Di Caro G, Gambardella LM. (1999) Ant algorithms for discrete optimization. *Artif Life* 5:137–172.
40. Lal TN, Chapelle O, Scholkopf B. (2006) Combining a filter method with SVMs. In *Feature Extraction, Foundations and Applications* (Guyon I, et al., Eds.), Springer-Verlag, New York.
41. Weston J, Elisseeff A, Schölkopf B, Tipping M. (2003) Use of the zero-norm with linear models and kernel methods. *J Mach Learn Res* 3:1439–1461.
42. Guyon I, Weston J, Barnhill S, Vapnik V. (2002) Gene selection for cancer classification using support vector machines. *Mach Learn* 46:389–422.
43. Zhang X, Lu X, Shi Q, Xu XQ, Leung HC, Harris LN, Iglehart JD, Miron A, Liu JS, Wong WH. (2006) Recursive SVM feature selection and sample classification for mass-spectrometry and microarray data. *BMC Bioinformatics* 7:197.
44. Guyon I, Gunn S, Hur AB, Dror G. (2004) Result analysis of the NIPS 2003 feature selection challenge. In *Proceedings of the Neural Information Processing Systems*, Vancouver, Canada, pp. 545–552.
45. Geurts P, Fillet M, de Seny D, Meuwis MA, Malaise M, Merville MP, Wehenkel L. (2005) Proteomic mass spectra classification using decision tree based ensemble methods. *Bioinformatics* 21:3138–3145.
46. Breiman L. (2001) Random forests. *Mach Learn*, 45:5–32.
47. Datta S. (2008) Classification of breast cancer versus normal samples from mass spectrometry profiles using linear discriminant analysis of important features selected by random forest. *Stat Appl Genet Mol Biol* 7:7.
48. Pearson K. (1901) On lines and planes of closest fit to systems of points in space. *Philos Mag*, 2:559–572.
49. Wold S, Martens H, Wold H. (1983) The multivariate calibration problem in chemistry solved by 120 the PLS method. In *Lecture Notes in Mathematics: Matrix Pencils* (Ruhe A, Kaegstroe MB, Eds.), Springer-Verlag, Heidelberg, Germany, pp. 286–293.
50. Holland JH. (1994) *Adaptation in Natural and Artificial Systems: An Introductory Analysis with Applications to Biology, Control, and Artificial Intelligence*, MIT Press, Cambridge, MA, p. 15.
51. Kohonen Y. (1982) Self-organizing formation of topologically correct feature maps. *Biol. Cyber* 43:59–69.
52. Adam BL, Qu Y, Davis JW, Ward MD, Clements MA, Cazares LH, Semmes OJ, Schellhammer PF, Yasui Y, Feng Z, Wright GL. (2002) Serum protein fingerprinting coupled with a pattern-matching algorithm distinguishes prostate cancer from benign prostate hyperplasia and healthy men. *Cancer Res.* 62:3609–3614.
53. Ball G, Mian S, Holding F, Allibone RO, Lowe J, Ali S, Li G, McCardle S, Ellis IO, Creaser C, Rees RC. (2002) An integrated approach utilizing artificial neural networks and SELDI mass spectrometry for the classification of human tumors and rapid identification of potential biomarkers, *Bioinformatics* 18:395–404.
54. Purohit PV, Rocke DM. (2003) Discriminant models for high-throughput proteomics mass spectrometer data. *Proteomics* 3:1699–1703.
55. Hilario M, Kalousis A, Muller M, Pellegrini C. (2003) Machine learning approaches to lung cancer prediction from mass spectra. *Proteomics* 3:1716–1719.
56. Lilien RH, Farid H, Donald BR. (2003) Probabilistic disease classification of expression-dependent proteomic data from mass spectrometry of human serum. *J Comput Biol* 10:925–946.

57. Tibshirani R, Hastie T, Narasimhan B, Soltys S, Shi G, Koong A, Le Q. (2004) Sample classification from protein mass spectrometry, by "peak probability contrasts." *Bioinformatics* 20: 3034–3044.
58. Hastie T, Tibshirani R, Friedman J. (2001) *The Elements of Statistical Learning*, Springer-Verlag, New York.
59. Zou H, Hastie T. (2005) Regularization and variable selection via the elastic net. *J Roy Statist Soc B* 67:301–320.
60. Fisher RA. (1936) The use of multiple measurements in taxonomic problems. *Ann Eugen* 7:179–188.
61. Vapnik VN. (1998) *Statistical Learning Theory*, John Wiley & Sons, New York.
62. Devijiver P, Kittler J. (1982) *Pattern Recognition: A Statistical Approach*, Prentice-Hall, London.
63. Ripley BD. (1996) *Pattern Recognition and Neural Networks*, Cambridge University Press, Cambridge.
64. Breiman L. (1999) Using adaptive bagging to debias regressions. Technical report, 547, Statistics Dept., University of California at Berkeley.
65. Efron B, Tibshirani R. (1995) Cross-validation and the bootstrap: estimating the error rate of a prediction rule. Technical report, TR-477.
66. Strimenopoulou F, Brown PJ. (2008) Empirical Bayes logistic regression. Stanford University *Stat Appl Genet Mol Biol.*, 7:9.

Chapter 12

Computational Methods for Analysis of Two-Dimensional Gels

Gorka Lasso and Rune Matthiesen

Abstract

Two-dimensional gel electrophoresis (2D gels) is an essential quantitative proteomics technique that is frequently used to study differences between samples of clinical relevance. Although considered to have a low throughput, 2D gels can separate thousands of proteins in one gel, making it a good complementary method to MS-based protein quantification. The main drawback of the technique is the tendency of large and hydrophobic proteins such as membrane proteins to precipitate in the isoelectric focusing step. Furthermore, tests using different programs with distinct algorithms for 2D-gel analysis have shown inconsistent ratio values. The aim here is therefore to provide a discussion of algorithms described for the analysis of 2D gels.

Key words: Protein quantitation, 2D gels, algorithms, computational methods.

1. Introduction

1.1. Two-Dimensional Gel Electrophoresis

Two-dimensional gel electrophoresis (1, 2) is an essential quantitative technique in proteomics research. In two-dimensional gels, proteins are separated based on the physical parameters isolectric point (pI) and molecular mass. Thousands of proteins from cells and tissue samples can be separated in the gel. A protein mixture is first loaded onto a non-denaturing polyacrylamide gel and separated by isoelectric focusing in the first dimension. Proteins migrate in a pH gradient until their isoelectric point is reached, where the pI and the pH have identical values and the protein has a total charge of zero. The gel is then equilibrated using sodium dodecylsulfate, which confers a uniform negative charge to all proteins in the gel. In the second dimension, proteins are further

R. Matthiesen (ed.), *Bioinformatics Methods in Clinical Research*, Methods in Molecular Biology 593,
DOI 10.1007/978-1-60327-194-3_12, © Humana Press, a part of Springer Science+Business Media, LLC 2010

separated by their molecular mass. A wide range of protein stains can be used to visualize the gel-contained proteins as spots. Protein dyes can be either colorimetric stains (e.g., Coomassie Blue and silver nitrate) or fluorescent. For a more detailed description of the different dyes, see a recent review by Miller and colleagues (3). Subsequently, the gel is scanned so that protein spots located in the digitalized image can be automatically identified and quantified for a particular sample.

Differential spot patterns contained in samples corresponding to different cellular states (e.g., health and disease state) might relate to changes in protein expression. Therefore, 2D-gel analysis is of invaluable importance in proteome research, where differential spot identification is subsequently analyzed by other methods, such as MS, to identify specific proteins in the gel.

Since it was first described by Margolis and colleagues in 1969 (1), and later reintroduced by O'Farrell in 1975 (2), the 2D-gel technique has been further developed by introducing new equipment, improved protocols, and more sophisticated computational tools. However, two-dimensional gels still suffer from several limitations, such as resolution, sensitivity, and reproducibility. Such limitations result in a high degree of gel-to-gel variation in spot patterns and make it difficult to differentiate between experimentally induced variation and biologically induced variation. One of the major breakthroughs in 2D gels was the development of 2D fluorescence difference gel electrophoresis (DIGE) in the late 1990s (4). With DIGE, two different protein samples can be resolved along with an internal standard on the same gel. Prior to this protein separation technique being applied, each sample is differentially labeled with spectrally resolvable fluorescent dyes, Cy2, Cy3, and Cy5. These fluorescent dyes are structurally similar (they have a similar molecular mass and are positively charged), and they all undergo a nucleophilic substitution with the ϵ-amino group of lysine residues. Along with the fact that different samples are run on the same gel, this ensures that all samples co-migrate under similar conditions (5) and therefore minimizes the experimentally induced variation. Likewise, the use of an internal standard sample enables an accurate comparison of protein amounts between the samples on different gels (**Fig. 12.1**).

The scope of this chapter is to introduce the reader to the computational analysis of 2D gels, an area of computational proteomics that has gained popularity over the last decade and yet is considered the bottleneck in 2D-gel research (6). This chapter describes both classical and novel approaches that are to be applied at different stages of the computational analysis of 2D gels. A comprehensive review of computational methods was recently provided by Dowsey et al. (7). Our aim in this chapter is to provide a more detailed description of a subset of methods that we found interesting.

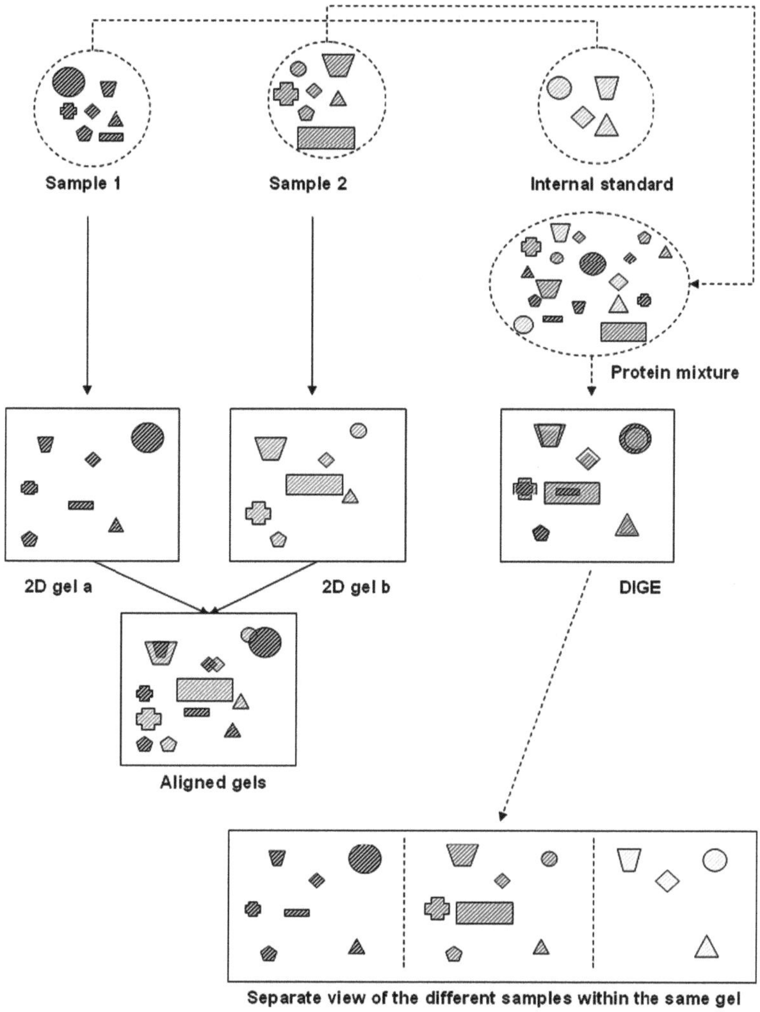

Fig. 12.1. Schematic view of conventional 2D-gel electrophoresis and DIGE workflow. *Dashed arrows* correspond to the DIGE experimental procedure, whereas common *arrows* correspond to the classical 2D-gel protocol where only one sample is loaded onto a single gel.

1.2. Evolution of Computational Methods to Analyze 2D Gels

The very first computational approaches to analyze 2D gels were implemented in the late 1970s (8, 9). In this early period, efforts were made to set up the basis for more sophisticated software. Gaussian curves, previously applied to 1D gels (10), were used to measure the film density distribution of spots and quantitate the amount of protein in each spot (9, 11). Methods were implemented to (i) preprocess digitalized images in order to reduce the streaks and the background intensity of stained gels (12, 13), (ii) facilitate the alignment of multiple gels (14, 15), and (iii) compare patterns of spots in different gels (16). At this stage, the available hardware was not powerful enough to carry out the different tasks required, and computational analyses had to be restricted to particular regions in the gels and required long hours of computer processing.

The introduction of personal computers with more powerful processing and graphical capabilities in the late 1980s along with the World Wide Web revolution in the early 1990s triggered a technological development, which led to the implementation of more sophisticated software and online 2D-gel databases (**Table 12.1**) (17). Such databases give the user the possibility of analyzing reference 2D-gel images where proteins have been identified by other groups, and programs were implemented to integrate different 2D-gel databases and compare such gels (18, 19).

Table 12.1
Commercial and noncommercial software for 2D-gel analysis

Software name	Affiliation	Availability	Year	References
N/A	Vanderbilt University	N/A	1978	(20)
Flicker	National Cancer Institute	http://www.ccrnp.ncifcrf.gov/flicker http://open2dprot.sourceforge.net/Flicker	1979	(14, 19)
N/A	Salk Institute	N/A	1979	(9)
N/A	University of California at San Diego	N/A	1979	(11)
N/A	Roche	N/A	1980	(16)
N/A	University of California at San Diego	N/A	1981	(21)
TYCHO	Argonne National Laboratory	N/A	1981	(12)
N/A	University of Michigan	N/A	1982	(22)
N/A	E. I. du Pont de Nemours & Co.	N/A	1983	(23)
Elsie 4	National Cancer Institute	N/A	1988	(24)
QUEST	Cold Spring Harbor Laboratory	N/A	1989	(25, 26)
LIPS	University of Michigan	N/A	1991	(27)

(continued)

Table 12.1 (continued)

Software name	Affiliation	Availability	Year	References
Phoretix 2D Advanced	Nonlinear Dynamics	http://www.nonlinear.com http://www.phoretix.com	1991	
Melanie series	GeneBio	http://www.2d-gel-analysis.com/index.html	N/A	
PD Quest	Bio-Rad Laboratories	http://www.biorad.com	1998	
AlphaMatch 2-D	Alpha Innotech Corp.	http://alphainnotech.com	1999	
WebGel	National Cancer Institute	http://www.ccrnp.ncifcrf.gov/webgel	1999	(28)
CAROL	Humboldt University, German Heart Institute, Free University Berlin	http://gelmatching.inf.fu-berlin.de/Carol.html	1999	(29)
GELLAB II+	Scanalytics	http://www.scanalytics.com	1989	(30)
HT Analyzer	Genomic Solutions	http://genomicsolutions.com	2000	
Z3	Compugen	http://www.2dgels.com	2000	
Delta 2-D	DECODON	http://www.decodon.com	2000	
Progenesis	Nonlinear Dynamics	http://www.nonlinear.com http://phoretix.com	2001	
DeCyder	Amersham Bioscience	http://www5.gelifesciences.com	N/A	(4, 31)
Image Master 2D	Amersham Bioscience	http://www.apbiotech.com	2001	
Proteome-Weaver	Definiens		2002	
GelScape	University of Alberta	http://www.gelscape.ualberta.ca	2004	(32)
Open2DProt	N/A	http://open2dprot.sourceforge.net	N/A	N/A
SameSpots	Nonlinear Dynamics	http://www.nonlinear.com	2006	(3)
Dymension	Syngene	http://www.syngene.com	N/A	N/A

(continued)

Table 12.1 (continued)

Software name	Affiliation	Availability	Year	References
Pinnacle	The University of Texas MD Anderson Cancer Center	On request	2007	(33)
Rain	Imperial College London & University College Dublin	http://www.proteomegrid.org/rain/	2008	(34)

In the last decade, efforts have been made to refine the sensibility and reproducibility of these methods. Novel and more sophisticated approaches have been introduced to improve tasks such as spot detection, pattern recognition, and interactive analysis of 2D gels.

2. Computational Two-Dimensional Gel Data Analysis

2.1. Preprocessing

In silico analysis of 2D-gel images is performed in order to detect and quantify protein spots and find differences in protein expression levels between two set of samples. This analysis is highly dependent on the quality of the images. Individual 2D-gel images obtained from the same protein sample can vary significantly. The experimentally induced variance can be categorized into three main categories: (i) intensity-related fluctuations, (ii) geometrical distortions, and (iii) spot mismatching. Likewise, intensity-related fluctuations can be further distinguished into background intensity fluctuations and local noise. In order to minimize intensity-related fluctuations, a series of image intensity transformations can be carried out, such as background subtraction and noise filtering methods. **Figure 12.2** shows common artifacts found in 2D gels.

2.1.1. Background Subtraction

Background subtraction methods are applied to eliminate meaningless background intensity level due to nonspecific staining of biological compounds in the gel, which is often evident if the gel has been overexposed during the image development (*see* **Note 1**).

A simple approach to minimize such intensity fluctuations is to compute the average intensity of the lightest and darkest points in the background and replace the entire background intensity

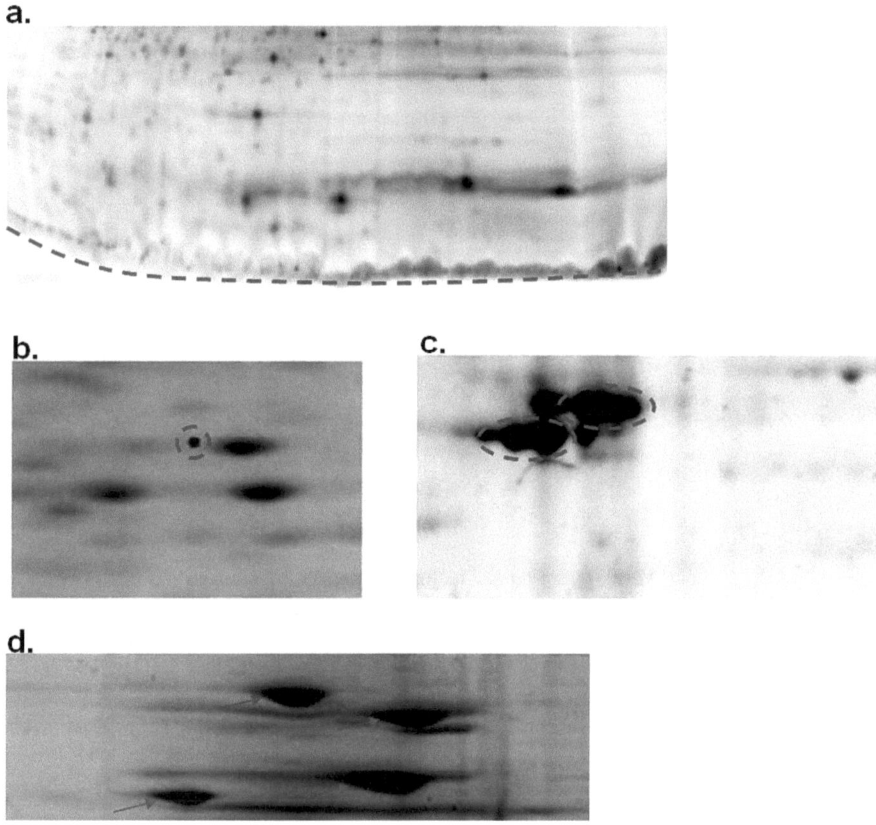

Fig. 12.2. Common artifacts in two-dimensional gels: (**a**) smiley gel; (**b**) noise; (**c**) overlapping spots; (**d**) spot tailing or streaking. Artifacts are to be solved in the different stages carried out during the 2D-gel analysis.

of the gel by the computed average intensity (7). Bossinger and colleagues (11) used a histogram of the relative number of readings at each density level, using averaged density data, to compute the overall background density by taking the histogram's first local maximum from the left as the global background. The TYCHO system (12) applies a local vertical and horizontal segmentation in order to identify the minimum element in the selected region surrounding each pixel. A vertical and a horizontal kernel are passed over the image sequentially, and the minimum value in the kernel is recorded for each pixel. This approach erodes regions of high background by a distance equal to the kernel's arm length. Subsequently, the image is reexpanded and the background is subtracted from the original density-corrected image. Similarly, Tyson and Haralick (35) based their approach on the local minima that represent background depressions and interpolated the background between these minima. Appel and colleagues (36) applied a third-order polynomial function to the background image having all spots previously removed.

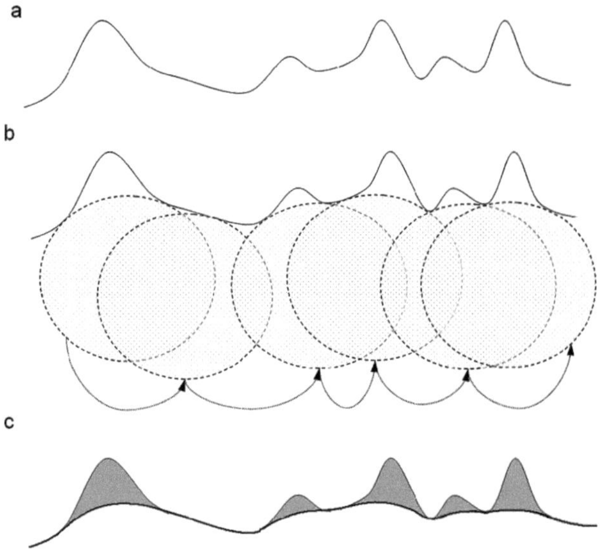

Fig. 12.3. The rolling ball method performing a morphological open transformation (the ball is rolled underneath the background image): (**a**) intensity fluctuations corresponding to the stained background image prior to image preprocessing; (**b**) morphological opening of the background, the a ball is rolled underneath the image; (**c**) the background function after applying the rolling ball, narrow peaks (in *gray*) are lost, thus obtaining a smoother background.

The rolling ball method (**Fig. 12.3**), developed in the early work of Sternberg (37, 38), applies morphological transformations in order to remove smooth, continuous backgrounds from the image. According to this method, a grayscale image is transformed into a 3D image where the intensity value becomes the third dimension. Subsequently, a spherical structural element (whose radius is at least as large as the largest spot) known as the "rolling ball" is rolled over (morphological close transformation) and underneath (morphological open transformation) the 3D image while it is prevented from rolling into spots. While performing background subtraction by opening the image (the ball is rolled underneath the 3D image), the height of each pixel corresponds to the highest point the ball can reach. A similar procedure is performed while closing the image (the ball is rolled over the 3D image), but the method looks for the lowest point at each pixel rather than the highest. According to this method, the rolling ball does not make contact with surface points contained within narrow peaks and depressions of the background image. The obtained background function corresponds to the union of translation paths of the rolling element to every point in the surface. Therefore, these narrow regions are filtered out from the background image. In a new application, a paraboloid is often used rather than a ball object.

The asymmetric least-squares method was first implemented to minimize the background in chromatograms (39) and adapted to two-dimensional space (40, 41). The applied objective function to be minimized is described as

$$Q = \sum_i v_i (y_i - f_i)^2 + \lambda \left(\Delta^d f_i\right)^2, \qquad [1]$$

where y gives the observed data, f denotes the smooth approximation of the data, v are the weights, λ is the penalty coefficient, and Δ^d indicates the derivatives of the dth degree. The first component of the equation describes the signal fit, whereas the second component corresponds to the penalty term used to control the smoothness of the background estimation.

The weights v are given unequally to the data points according to the following definition:

$$v_i = \begin{cases} p & \text{if} \quad y_i > f_i \\ 1 - p & \text{if} \quad y_i \le f_i \end{cases}, \qquad [2]$$

where $0 < p < 1$ and p usually achieves a very small value such as 0.001 (41).

According to this definition, the weights v have high values where the signal analyzed is allowed to affect the estimation of the baseline background, and low values otherwise. The main problem is to simultaneously determine the weights v and the signal approximation f (without the weight, it is not possible to compute the signal approximation, and vice versa). This can be solved iteratively, where all weights v receive equal values, so a first estimate of the approximated signal f can be computed during the first iteration. The subsequent iterations are applied to further refine the initial values given to the signal approximation f and setting weights v. Alternatively, the signal approximation can also be computed using linear combinations of B-spline functions.

2.1.2. Noise Filtering

Noise filtering differentiates from background subtraction in that it does not apply to fluctuations of the background intensity level of gels but to random and locally distributed spots that are due to dust particles, speckling from crystallization of different stains such as SYPRO, and similar artifacts (3). There are two different categories of noise reduction methods: (i) linear and (ii) nonlinear methods. Linear combinations compute the intensity value of a particular pixel in an image by linearly combining the intensity of neighboring pixels (42):

$$\hat{g}(i, j) = \sum \sum_{(m,n) \in O(i,j)} h(g(m, n)), \qquad [3]$$

where $\hat{g}(i, j)$ is the computed intensity of the pixel (i, j), $O(i, j)$ symbolizes its neighboring pixels, and h is the weight value to be applied to the intensity value of a particular neighboring pixel $g(m, n)$. The weight values h are to be described in the convolution matrix where each neighboring pixel is given a particular weight value (11). Different convolution matrices can be applied according to the characteristics of the image to be processed. All neighboring pixels can also be equally weighted, and a mean filter is obtained where an average intensity value is computed based on the intensity value of the neighboring pixels. Alternatively, weights can also be computed using a Gaussian function. The main drawback of using linear methods is that while it reduces the image noise appropriately, it also reduces the intensity values of spots (42). This can therefore have a negative effect for later spot detection and quantification. By using an adaptive filter such as the Wiener filter (43), the main drawback suffered by linear methods can be minimized. Wavelet transform is a nonlinear method that can be applied to nonstationary signals. Following this principle, two-dimensional gel images can be considered as nonstationary signals that contain features of different frequencies (41). Therefore, images can be decomposed into wavelets in order to eliminate noise and yet not modify the significant high-frequency features. In this scenario, 2D gels are decomposed into wavelets whose coefficients are computed from the given gel image. Such wavelets represent scalable frequency-location decomposition in both dimensions of the image. Decomposition can be achieved using either the two-dimensional wavelet basis functions or the one-dimensional wavelets, which are applied along both axes of the decomposed image. The decomposed image can then be denoised by filtering out certain frequencies according to a particular threshold. Finally, in order to reconstruct the image, the inverse wavelet transform is applied. An evaluation of different noise reduction methods was performed by Kaczmarek and colleagues (42). The authors generated 50 synthetic 2D-gel images containing from 300 to 600 spots and a white Gaussian noise with a standard deviation ranking from 10 to 30. The linear methods evaluated differed from each other in the type of filter implemented: (i) mean, (ii) Gaussian, (iii) median, and (iv) Wiener filtering. The different wavelet methods applied were set to use different types of wavelets, different numbers of decomposition levels, different ways of threshold calculation, and different policies of thresholding. Among the linear methods, both the Gaussian filter- and the Wiener filter-based linear methods showed to be the best methods. However, wavelet-based methods were more accurate than linear methods in the different scenarios tested. Among the different parameters set for wavelet methods, the parameters that were found to best suit the given set were the BayesThresh as the threshold estimation method, the Coiflet wavelet with 10 vanishing points as the wavelet type, and a decomposition level of three.

2.2. Spot Detection and Quantification

This task is applied to detect the positions of the spots, identify the boundaries of the protein spots, and estimate the amount of protein in 2D gels. Broadly speaking, the methods applied at this stage are classified into image segmentation techniques and model-based quantification depending on the goals of the applied method. However, in some cases, spot detection and spot quantification are performed by the same mathematical approach, and it is therefore difficult to draw a line that separates both tasks.

2.2.1. Spot Detection

Image segmentation techniques partition the image into nonoverlapping segments, classify each pixel as being a spot pixel or nonspot pixel, and estimate the boundaries of the spot. This process can be performed using different properties of the scanned image such as the raw intensity, slope, and pixels in the surrounding regions (44).

Anderson and colleagues (12) applied a "+"-shape ($x = 21$ pixels- and $y = 15$ pixels) kernel using centered cosine-curves with a 14-pixel period for the x-direction and a 10-pixel period for the y-direction. The kernel was to be applied on a shape window where spots were detected as local maxima whose pixel values were found to be above a particular threshold. This spot detection method produces a sharp spike where the image has a peak or shoulder of shape similar to that of the kernel's central peak.

Conradsen and Pedersen (45) applied a series of median and local maximum filters along with morphological operations to detect spots. While the median filtering removes local noise, the local maximum filtering applies a gray-level morphological erosion. According to this method, a pixel centered in a particular size window is considered a spot pixel if all neighboring pixels are spot pixels; otherwise, it is considered a white pixel. Spot edges are detected by checking the outcomes from second-derivative filters based on four different kernels (**Fig. 12.4**). A particular pixel is considered a spot edge only if all applied kernels give a positive outcome transforming the gel image into a binary image. The authors applied this method using a series of increasing kernel sizes (3×3, 5×5, 7×7, 9×9) in order to consider the variation of spot sizes and applied a more stringent erosion (as the kernel size increases) to separate overlapping and adjacent spots.

The watershed transformation (**Fig. 12.5**) based on immersion simulations (46) was first introduced to detect spots in 2D gels by Pleissner and colleagues (29). The purpose of the watershed is to define the spot contours. This image segmentation technique applies the principle of immersion of the first derivative of the image. According to this principle, holes are drilled in regions where a gray minimum value is observed, and subse-

½	0	0
0	-1	0
0	0	½

0	½	0
0	-1	0
0	½	0

0	0	½
0	-1	0
½	0	0

0	0	0
½	-1	½
0	0	0

Fig. 12.4. Kernels used to detect spot edges.

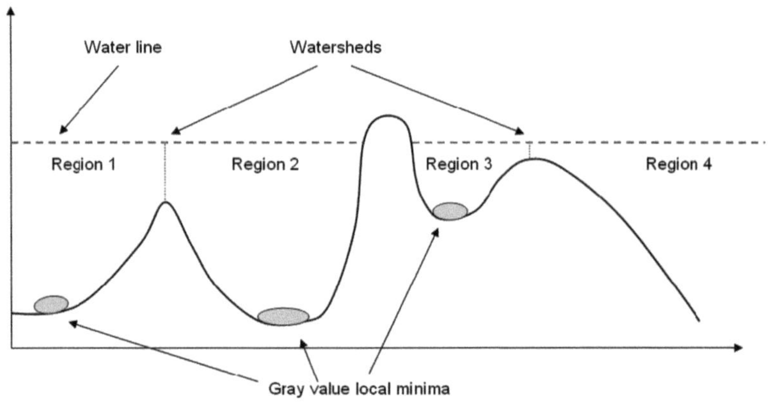

Fig. 12.5. Principle of the watershed transformation [image adapted from (29)].

quently the surface is flooded under a constant water level. Consequently, protein spots correspond to gray-value mountains whose optimal contour is defined by the watersheds. This method results in a string oversegmentation caused by the image noise amplified by the calculation of the gradient image. In order to minimize this effect, a gray-value threshold and a curvature threshold are applied based on the fact that spots have significantly lower gray values than the background and show a convex curvature (29). Another possibility is to use marker-controlled watershed to prevent oversegmentation.

The Phoretix spot detection (Nonlinear Dynamics) algorithm computes the average intensity of pixels located on the edge of an inner window against the average intensity of pixels located on the edge of an outer window (47). When the pixel value of the inner window is higher – by a given ratio – than that of the outer window, the pixel at the center of the inner window is considered a spot pixel. Potential spots are subsequently discarded if their corresponding areas do not cover a threshold determined by the minimum area parameter (not defined by the authors).

Geometric algorithms were also applied in order to detect spots in complex regions where twin spots, streaks, and overlapping spots can be found (48). These regions are usually saturated, and subsequently the gray-level information pertaining to these regions cannot give insights regarding the different spots contained. Elfrat and colleagues developed a spot detection algorithm based on the assumption that each spot has approximately the shape of an axis-parallel ellipse (48). In order to reduce the space of possible solutions, a set of heuristic rules was implemented: (i) Each ellipse must not intersect the set of points located outside the complex region; (ii) the boundaries of any pair of neighboring ellipses can only intersect in at most two points; (iii) the final set of ellipses obtained from a particular complex region must cover a minimum portion of the complex region; and (iv) according to Occams' razor principle, the optimum subset of ellipses must be composed by the lowest number of ellipses that respect the rules described above. The brute-force solution discretizes the parameter space of all possible ellipses combined with a greedy algorithm to select the optimum subset of ellipses. Alternatively, using a logic programming (LP) approach avoids the construction of a large set of ellipses and was found to improve the quality and reduce the complexity of the computation. According to the LP method, a triplet of mutually visible points contained in the complex region are chosen randomly and an axis-parallel ellipse (it must accomplish the conditions described above) that contained the triplet is computed. The ellipse is further extended using the metropolis methodology, by which random neighbors are added into and random points from the covered point set are removed. This process is repeated in a set of rounds: The LP-solver selects the optimum ellipse as the ellipse whose half-axis ratio is closest to 1 and covers the maximum number of points.

2.2.2. Spot Quantification

Model-based quantification methods permit one to estimate the protein expression levels for a particular spot. Generally speaking, once a spot has been detected, there are different methods that can be applied to estimate the original amount of protein loaded onto the gel. The most common parameters extracted from protein spots for quantification are

1. the spot area,

$$A_s = n_s \times A_p, \qquad [4]$$

where A_s is the spot area, n_s is the number of pixels contained within a particular spot, and A_p is the pixel area;

2. the optical density (OD),

$$\mathrm{OD} = \max_{(x,y)\in \mathrm{spot}} (I(x, y)), \qquad [5]$$

where I is the intensity value, and x and y correspond to the coordinates of a particular spot pixel;

3. the integrated optical density (spot volume VOL),

$$\mathrm{VOL} = \sum_{(x,y)\in \mathrm{spot}} I(x, y). \qquad [6]$$

Of these three parameters, OD and VOL are used more commonly for protein quantification. Normalization is a common task in quantitative proteomics. Two-dimensional gels, loaded with the same protein sample, tend to be differentially stained; consequently, the quantification data obtained from a particular spot show differences among different gels. This lack of reproducibility is due to the experimental variance, where several factors are involved (nonlinear dynamics):

i. Differences in sample preparation (e.g., differences in the number of protein samples, pipetting errors, differences in protocols, errors while sample loading).

ii. Differences in sample staining (e.g., inconsistent staining times and differences between stain reagent batches).

iii. Differences in image acquisition (e.g., different exposure times during scanning and images captured under different settings).

Normalization processing minimizes these differences and permits one to compare quantification data from different gels. Several mathematical approaches have been implemented to carry out this task. The most widely used methods consider the total optical density (OD) or spot volume (VOL) of all spots within the image:

4. the relative (normalized by the total over the gel) optical density (%OD),

$$\%\mathrm{OD} = \left(\frac{\mathrm{OD}}{\sum_{s=1}^{n} \mathrm{OD}_s}\right) \times 100; \qquad [7]$$

5. the relative (normalized by the total over the gel) integrated optical density (%VOL),

$$\%\mathrm{VOL} = \left(\frac{\mathrm{VOL}}{\sum_{s=1}^{n} \mathrm{VOL}_s}\right) \times 100. \qquad [8]$$

Of these parameters, %VOL showed to be more accurate for estimating the protein amount in a particular spot using Melanie II (49). All of the above-mentioned parameters do not take into consideration background stain levels and have a limited range of linearity (50). Considering these limitations, Dutt and colleagues described a new parameter, known as the *scaled volume* (SV), that scales an integrated optical density of a particular spot (VOL) by the gel background with secondary signals removed (spots not of interest, such as local artifacts) (50):

$$SV = \frac{VOL_s}{\frac{(VOL_b - VOL_{ns})}{(A_b - A_{ns})}}, \qquad [9]$$

where VOL_s, VOL_b, and VOL_{ns} are the integrated optical densities of the considered spot, background, and not considered spots, respectively. A_b and A_{ns} are the areas of the background and the not-considered spots, respectively. Dutt and Kelvin (50) evaluated OD (Equation [5]), %OD (Equation [7]), %VOL (Equation [8]), and SV (Equation [9]) using three different proteins' standard samples (namely, trypsinogen, trypsin inhibitor, and bovine serum albumin). These three samples were run individually at different concentrations (10^2, 10^3, 10^6 ng/gel) and stained with ammoniacal silver stain. Results showed that the SV parameter correlates silver stain intensity and protein amount better than OD, %OD, and %VOL.

Gaussian functions are undoubtedly the most popular approach to quantitate the spot density and estimate the amount of protein contained in a particular spot (9–12, 25). This is based on the assumptions that (i) the spot density follows a normal distribution and (ii) there is a linear relationship between the spot optical density and the amount of protein (47). Usually, two-dimensional Gaussian functions are used:

$$f(x, y) = Ae^{-\left(\frac{(x-x_0)}{2\sigma_x^2}\right)-\left(\frac{(y-y_0)}{2\sigma_y^2}\right)}, \qquad [10]$$

where A is the amplitude, x_0 and y_0 are the center coordinates of the spot, and σx and σy are the variance on the x- and y-axes, respectively. Gaussian functions are usually fitted to the observed data by applying the least-squares method, which models the observed data by adjusting the parameters of the given Gaussian function.

One of the main drawbacks of applying Gaussian functions is that when the local concentration of protein is high, the corresponding spot appears saturated and is consequently harder to model accurately. In order to improve the spot modeling, Bettens and colleagues (51) developed a diffusion model. This model takes into consideration the diffusion process underlying the formation of protein spots in 2D gels. The authors adapted the fundamental differential equation for diffusion in an isotropic 2D medium by assuming (i) radial symmetry, (ii) two main directions of diffusion, (iii) an anisotropic environment, (iv) the initial distribution of protein is not concentrated on one point but occupies a finite region, and (v) the background is a constant value in the spot region.

The Gaussian and diffusion models described above assume perfect diffusion across the gel. However, protein diffusion is neither regular nor symmetric in practice, therefore forming unpredictable, unusual shapes. Based on this statement, Rogers and colleagues (52) implemented a new spot-modeling method that convolves a shape model with a bivariate Gaussian kernel. According to the authors, the convolved model is flexible enough to appropriately model spots with irregular shape and yet specific enough to discriminate between single protein spots and overlapping spots, which were not further analyzed. The spot shape was modeled using the point distribution model technique (PDM), which uses principal component analysis (PCA) and a set of nonoverlapping spots whose shape is represented by 25 landmark features. The implemented approach uses the watershed algorithm in order to detect spots in 2D gels and applies the Levenberg–Marquardt gradient descent algorithm to determine the best model parameters that fit the detected spot.

Morris and colleagues (33) developed an approach for spot detection and quantification rather different from the classical approaches, where the common steps to follow are (i) spot detection in each individual gel, (ii) spot matching to a reference gel, and, finally, (iii) spot volume computation by summing all pixel values corresponding to the spots. Instead, this method first aligns the corresponding gels and computes an average gel by averaging the intensities of each pixel. The averaged gel is then preprocessed by reducing the white noise using the undecimated discrete wavelet transform (UDWT). Subsequently, spot detection is performed by detecting all pinnacles (where a local maximum in both the horizontal and vertical directions and intensity higher than a certain threshold are observed) in the denoised gel. Pinnacles within a certain range are combined so that only the pinnacles with the highest intensity are kept. Quantification is achieved by taking the maximum intensity within a square whose center corresponds to the pinnacle coordinates. In the final step, the quantification is corrected by applying local background subtraction and normalization using the mean pinnacle intensity of the average denoised gel.

2.3. Gel Alignment

Gel alignment is needed when spots from several gels need to be compared. This is especially the case when DIGE labeling is not used. Furthermore, protein identification by mass spectrometry is normally done using a separate preparative gel. Therefore, comparison of the target 2D-gel image with the preparative reference gels facilitates the protein identification of a particular spot. Multiple gels are usually aligned in a pairwise fashion where the gel image of the best quality is usually selected as the reference gel. Remaining gels are to be aligned with the reference gel (alternatively, the average or sum of intensities in all gels can be used as the reference). Gel alignment of two of more gels can sometimes be a tedious task due to the experimentally induced geometric distortions, such as local translation, rotation, and magnification (53). These geometric distortions are generated by different experimental factors during gel casting, polymerization, running, and gel development. The precise geometry of the gel can vary from cast to cast. Difference in bisacrylamid and polyacrylamide concentration will affect the gel mesh size and therefore how far the proteins run in the gel. Polymerization conditions such as temperature and time will affect the degree of completion of the polymerization reaction, which again affects the mesh size of the gel. The electric field will be influenced by the concentration of electrolytes in the gel buffers, which again affects the protein's migration path. If gel sides are not fully isolated when running a protein sample, a current leakage occurs, which causes a global change in the generated electric field along the gel, resulting in a geometric distortion (54). Likewise, the different buffer solution used during staining procedures has different water activity than that inside the gel, which means that water will diffuse either out or into the gel, leading to shrinkage or swelling of the gel.

The 2D gels can be aligned either by global transformation parameters or by several local transformation parameters. It is currently accepted that global transformation parameters are not accurate enough to deal with the local distortions that are often observed between various 2D gels (55, 23, 56). The alignment process includes (i) image registration, (ii) correspondence analysis, and (iii) optimization of transformation functions by using a specific similarity measure.

Image registration is the process by which parameters associated with the object's location, size, and rotation are determined. The registration can be optimized for area-based matching (ABM) or feature-based matching (FBM). *Correspondence analysis* can be done either automatically or manually. The correspondence analyses are done manually in most commercial programs since automatic correspondence analysis is computationally challenging. "Fuzzy matching" is one example of an algorithm for correspondence analysis. The optimization of transformation functions is the final step in the alignment procedure. The inputs to such an optimization algorithm are two data sets that include the correspondence information about specific data points, transformation functions, and a similarity measure. In the following sections, various procedures used for gel alignment are reviewed.

2.3.1. Image Warping

To handle the problem of gel distortion, it is often necessary to apply a set of local image transformations, known as *image warping processes*, to the corresponding gel image. This is commonly achieved by mapping a local region of the target image onto the geometry of the local region in the reference gel. By adjusting the local geometry of the target gel, the relative spot positions, distances, and angles are considered. Image warping achieves such a transformation while preserving its grayscale values in order to avoid possible artifacts created by this process. This is a classical optimization problem in image alignment where the algorithm searches the space of possible geometrical transformations in order to find an optimum subset of transformations. The optimal subset of transformations is restricted to those subsets that maximize a similarity measure function (*see* **Note 2**) between the target and the reference image while performing a smooth transformation (*see* **Note 3**). Maximal similarity functions ensure maximal efficiency (the ratio between found matches and total matches given a pair of gel images), while smooth transformations guarantee maximal accuracy (the proportion of found matches that are true positives) (7).

Roughly speaking, the image warping method can be categorized into two main classes: feature- and intensity-based methods (57). Feature-based methods commonly extract features from the gel images and compute a geometric transformation based on these elements. These methods are generally robust and can deal with large geometric differences without requiring expensive computations. Classical methods use smooth functions such as low-degree polynomials or thin-plate splines (58) and user-defined landmarks in order to achieve smooth geometric transformations. According to the polynomial approach, both the reference and target gel images can be modeled by linearly combining a set of polynomial functions, each function corresponding to a particular landmark (19). The coefficients for such functions are determined so that the differences between the reference and target models are minimized. The functions employed to create these models are usually low-order monomial functions (*see* **Note 4**), which permits a smooth geometric transformation. Image warping using polynomial functions performs a global transformation, but it is not capable of modeling the complex geometric distortions found in 2D gels (59). An alternative method, which avoids the use of user-defined landmarks, is to use local matches in pairs of 2D gels. Pleissner and colleagues (29) introduced this concept by determining local matches using the Delaunay triangulation principle (*see* **Section 2.4.1** graphs below) and a set of transformations that maps the target point set (a set of intensive points in a window of the target gel) onto the reference point set. The method by Pleissner and colleagues takes into consideration similar geometric patterns and spot intensity relations, is independent of the image resolution, and is also robust in the presence of noise. Kazmarek and colleagues (60, 61) developed the so-called fuzzy matching algorithm, an automatic method for finding corresponding spots. Kazmarek and co-workers used the corresponding spots as inputs to an image-matching algorithm that consists roughly of two different steps: (i) coordinate transformation and (ii) interpolation of image intensity. The coordinate transformation process, recommended by the authors, maps the coordinates corresponding to the reference gel image onto the coordinates of the target image. This process is called the *inverse transform*. The *forward transform* – opposite the inverse transform – is also possible, but it is not recommended by the authors. A bilinear transformation function is applied in order to minimize the differences between both sets of coordinates:

$$\begin{aligned} x2^* &= a_0 + a_1 x1 + a_2 y1 + a_3 x1 y1, \\ y2^* &= b_0 + b_1 x1 + b_2 y1 + b_3 x1 y1, \end{aligned} \quad [11]$$

where $x1$ and $y1$ denote the x- and y-coordinates of the reference image, $x2^*$ and $y2^*$ correspond to continuous coordinates in the transformed image (to which the intensity value can be assigned) mapped onto the target image, and the vectors a and b correspond to transform parameters. These transform parameters can be computed using two different approaches, area-based matching (ABM) and feature-based matching (FBM). According to the ABM approach, transform parameters are optimized according to a particular similarity measure such as the covariance between the intensities of the reference image and the transformed image or the cross-correlation similarity measure. The FBM approach extracts features that are to be paired in the feature space of spots contained in both images to be aligned. This feature correspondence problem is solved by transforming the target features using an iterative method (known as feature-based fuzzy matching): A correspondence matrix M is computed describing the differences between the extracted features (spots) in the target and reference images based on the outputs of the Gaussian functions fitted to each feature. Using the correspondence matrix, a set of weighted transform parameters is computed and feature matching is performed. This process is iteratively repeated at decreasing degrees of fuzziness [i.e., by decreasing the width of the Gaussian function (41)] until reaching convergence. Due to the fact that warped spot coordinates are no longer integer values but continuous values assigned by mapping functions, the intensity value of a spot is estimated by interpolation (e.g., nearest-neighbor and linear and cubic interpolation).

Intensity-based methods perform the appropriate transformations based on the intensity data of the raw image, thus avoiding the feature detection step. This is normally achieved by maximizing the pixelwise correlation between two intensity surfaces. Smilansky (62) implemented an intensity-based method used by Z3 software (Compugen). This method performs a global transformation based on a series of local transformations. The algorithm establishes small rectangular regions that contain spots with unique geometric patterns. The rectangular regions pertaining to a particular section of the reference and target images are compared, and a set of local transformations is computed. Ultimately, the global transformation is achieved by combining the different local transformation vectors by using the Delaunay triangulation transformation. Veeser and colleagues (63) implemented an intensity-based method that applies a multiresolution representation of gel (by applying different degrees of blurring) profiles and decomposes the geometric distortion into its components at each resolution level. At low-resolution levels, the coarse components of the distortion are modeled and it is therefore possible to estimate the optimal geometric transformations, at low resolution, to approximate a target image to its reference image. The misalignment at low resolution is minimized using the computed rough approximation to the optimal geometric transformations and the process is repeated at a higher-resolution level, thus improving the quality of the approximated optimal transformation. A similar approach, based on features rather than intensity, was carried out by Salmi and colleagues (64), who applied a hierarchical grid method along with a set of landmark pairs.

Rohr and colleagues (65) designed a method that combines both landmark information and intensity. By exploiting both types of information, the benefits of each method can be combined in order to obtain optimal geometric transformations that facilitate the alignment of gels. According to Gustafsson and colleagues, the current leakage is the major factor causing geometric distortions (54). The electric field is supposed to be constant along the gel; however, if the gel sides are not fully isolated, a global change in the electric field is generated along the gel, causing a geometric distortion. Following this idea, Gustafsson and colleagues implemented a two-step warping method (54). In the first step, warping transformations are achieved by applying a model that only considers the distortion caused by the current leakage. Such a model estimates the position of a certain protein in an ideal gel where the isolation conditions along the sides are perfect. In the second step, the current leakage-corrected images are aligned to minimize the distortion effects caused by other experimental factors.

2.4. Matching of Protein Profiles

Spot matching is the process of pairing spots corresponding to the same protein in different gels. This is a key process during feature-based 2D-gel image alignment and gel-to-gel comparison. Point pattern recognition is a widely used technique that is applied to identify objects in images. It is based on the fact that any object, at any dimension, can be reduced to a point pattern that can be used to identify similar objects (66). In order to perform point pattern recognition, it is necessary, first, to find an optimal method to represent the space of data points (in our case, all spots in a gel image) and, second, to apply a matching technique that permits one to efficiently match the geometric models generated.

2.4.1. Proximity Graphs

Proximity graphs are widely used to model the space of data points in an image that can subsequently be matched against other images in order to recognize particular objects. These graphs represent neighboring relationships between data points in an *n*-dimensional space based on a proximity definition. The relative neighborhood graph (RNG) (67) and the Gabriel graph (GC) (68) are probably the most common principles to define the proximity between several data points in the space. The relative neighborhood graph (RNG) is based on the "relatively close" neighbor concept described by Lankford (69). According to this approach, two points in the space are relative neighbors if they are at least as close to each other as they are to any other point. This neighborhood relationship between two points i, j is often represented by the intersection of two circles, each centered at one of the pair of points considered, of radius $d(i, j)$, where a neighborhood is considered only if such an intersection does not contain any point in the space (**Fig. 12.6a**). Similarly, the Gabriel graph defines a neighborhood between two points i, j when no other point is contained within a circle, centered at the geometric center of the segment $s(i, j)$, of diameter $D(i, j)$, where $D(i, j) = s(i, j)$ (**Fig. 12.6b**).

The minimum spanning tree (MST) is a subgraph that describes the subset of edges with the lowest weight that connects all nodes in an undirected graph (**Fig. 12.7a**). A graph with interconnected nodes can have several spanning trees. Weights can be given to each edge (e.g., by computing the Euclidean distance between the interconnected nodes), so that the weight of

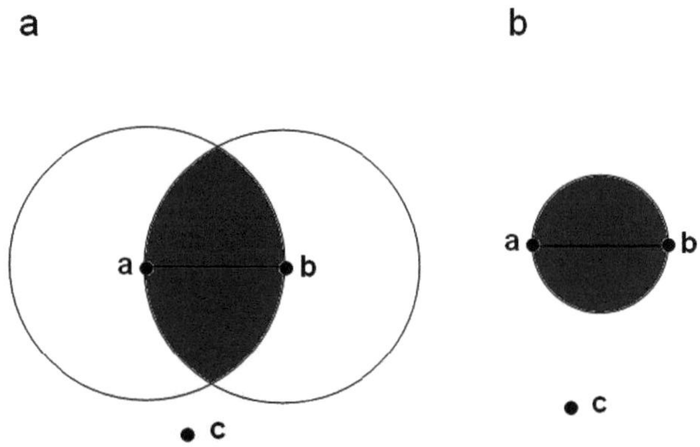

Fig. 12.6. Neighborhood relationships between two points *a* and *b* in the space (**a**) according to the relative neighborhood graph (RNG) and (**b**) according to the Gabriel graph (GC).

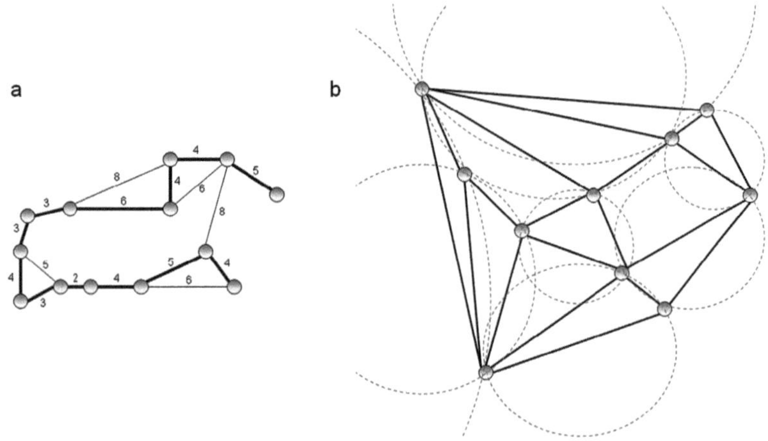

Fig. 12.7. (**a**) Example of a minimum spanning tree (MST). The edges that correspond to the MST have been highlighted in *bold.* (**b**) Delaunay triangulation given a set of data points in the space.

a particular spanning tree corresponds to the sum of the weights of all vertices that connect the edges in the spanning tree. Minimum spanning trees have not been widely used in the field of 2D-gel analysis. Unlike minimum spanning trees, the Delaunay triangulation approach (70) is by far the most common proximity graph implemented in order to represent the space of spots found in a 2D-gel image (**Fig. 12.7b**). According to this method, three data points in the space form a Delaunay triangle if its circumcircle does not contain any other point. Data points can be incrementally considered by this approach, adding (and modifying when necessary) new Delaunay triangles. The net of assembled triangles is called a Delaunay net only if the circumcircles of all triangles contained within the net are "empty."

The following section describes how the obtained minimum spanning tree or Delaunay triangulation can be used to match patterns from different gels.

3. Pattern Matching

The alignment method maps a single arc or segment in the reference pattern onto the targeting image. Once the initial mapping has been performed, a set of transforms is to be applied by mapping the remaining arcs in the pattern onto all arcs in the target image. The candidate with the highest similarity measure is finally selected as the optimum match (7). This approach requires that every arc in the pattern is mapped onto every other arc in the target image, thus making this analysis computationally very expensive.

The Iterative Closest Point (ICP) algorithm (71) performs an iterative loop: (i) The closest points (in the target image) to a model M are computed, and (ii) a set of transformations is computed and applied (e.g., by the least-squares method) so that the distance between the model M and the targeting image is minimized. This loop is iteratively repeated until the mean-square error is below a certain threshold. Standard ICP is based on the Euclidean distance metric. However, this distance metric is not usually appropriate if large deformations are observed between the gels to be aligned (72). Recent research combined the Euclidean distance metric along with the distance metric that takes into account spot shape and intensity (*see* **Note 5**) (72) and the shape context and distributions of neighboring intensities and feature distance metric (*see* **Note 6**) (73) in order to obtain a more accurate and robust ICP.

The geometric hashing (74) technique was originated in an early work by Schwartz and Sharir (75), who focused on the recognition of rotated, translated, and partially occluded two-dimensional objects from their silhouettes. This technique, originally developed in computer vision, was implemented to match geometric features against a database of such features. This approach is composed of two different stages: (i) a preprocessing stage and (ii) a recognition stage. During the preprocessing stage, the model information is encoded and stored in a quantized hash table by means of a reference coordinate system. For each model m, its point features are extracted, and for each ordered pair of features $p_i p_j$ (also known as the basis), the following steps are to be followed (*see* **Fig. 12.8**): (i) The given model is transformed by affine transformation (rotation, scaling, and translation) so that the magnitude of the vector $p_i p_j$ in the reference coordinate system equals 1, the midpoint between the considered pair of points corresponds to the origin of the reference coordinate system, and the corresponding vector $p_i p_j$ has the direction of the positive x-axis; (ii) the coordinates of the remaining features are then computed defined by the basis (the basis acts as the reference); and (iii) the coordinates of the remaining features are used as the index in the corresponding entry (one entry per feature, each entry is also known as a bin) in the hash table, where the model and the basis identifiers are saved. **Figure 12.8** shows the locations of all the hash table entries for model M_1. In the recognition phase, the constructed hash table is accessed by indexing geometric properties of features extracted from an image (e.g., target image) to match the candidate models. An arbitrary pair of points $p_{\mu 1} p_{\mu 2}$ are chosen from the image as the candidate basis and the image is transformed following a similar procedure to that applied to the model. The coordinates of the remaining points are then computed (using the basis as the reference) and mapped onto the hash table as indexes, and all entries in the corresponding hash

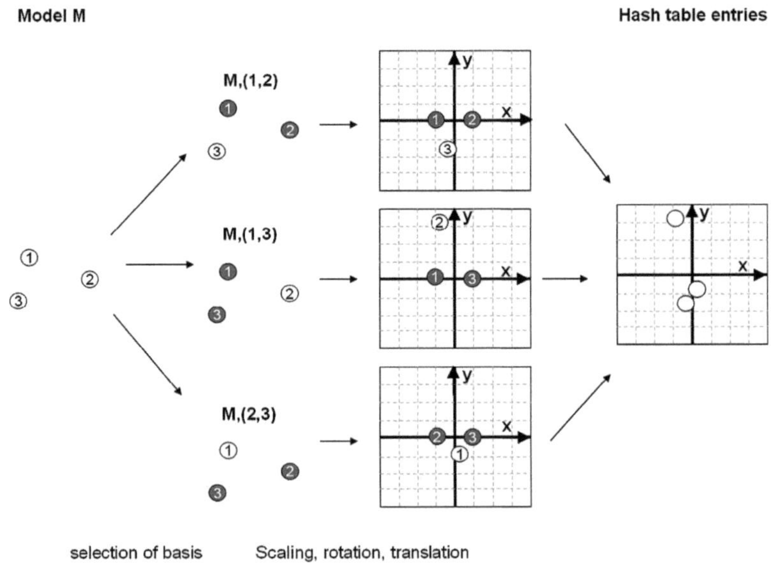

Fig. 12.8. Schematic view of the development of a hash table for a hypothetical model *M* composed of three data points. For the given model, three different bases can be set up (data pairs highlighted in *gray*). For each model *M* basis pair (*a*, *b*), a single entry is stored that corresponds to the data point not used to set up the basis. The hash table contains all entries pertaining to all possible model *M* basis pairs (*a*, *b*).

table bin receive a vote. If there are one or more (model, basis) combinations receiving a support above a certain threshold, a subsequent stage verifies the presence of the model with the given basis matching the chosen basis points in the target image. This is achieved by aligning both the model and the image and applying a set of transformations that results in the best least-squares match between all corresponding features. A further explanation of the geometric hashing technique can be found in Wolfson and Rigoutsos (74).

Pánek and Vohradský (66) developed a point pattern method for matching pair of spots in 2D gels that combines a neighborhood similarity method and a method derived from the directional vectors of candidate match and directional vectors of the closest landmarks. The neighborhood similarity method compares point patterns formed by the closest neighborhood of a candidate-matched spot with the neighborhood of the spot in the reference gel. The underlying idea is that point patterns of matching spot neighborhoods in the reference and target images have to be more similar than the point patterns of other spot pairs. Following this idea, each neighbor spot was given a syntactic identifier describing its position relative to the candidate spot (each spot was defined by its centroid coordinates). The latter method computes (i) the differences between Euclidean lengths of vectors of the candidate match and the Euclidean lengths of vectors of the nearest landmarks and (ii) the differences in absolute angles between the candidate spot and the nearest landmarks. Evaluation of these methods showed that point pattern matching was maximized when the candidate is considered a match by the two methods simultaneously, thus identifying nearly 90% of all spots in a gel pair.

The Carol system (48, 76) implemented a variation of the Delaunay triangulation that consists of an incremental Delaunay triangulation with decreasing spot intensities. Following this approach, it is possible to construct a data structure that represents all intensive edges (a triple of spots is intensive when its circumcircle does not contain any other spot that is more intensive, and edge connecting two spots is intensive if a third spot exists that forms an intensive triple with the connected pair of spots), together with their length and slopes, occurring during the history of the incremental Delaunay triangulation along with flipped diagonals (edges connecting opposite points in neighboring triangles). For the point pattern matching, the authors developed a two-step variant of geometric hashing. First, the approach computes all locations within the target image where a good matching is likely to occur, and subsequently the actual local matching and their evaluation are computed. Good matching is computed via Delaunay edges, where two edges match (i) if their absolute slope difference is smaller than α and the length ratio is equal to or higher than a threshold $1 - \lambda$ and equal to or lower than $1 + \lambda$; (ii) the difference between discrete intensity values of the corresponding pattern spots does not differ by more than two points. The obtained good matches are subsequently scored by means of a translation vector scoring function; all matches with scores above a certain threshold are considered potential locations of matching pattern centers. These potential matches are further evaluated by a voting procedure that computes a partial (λ, α)-matching between the pattern P in the reference image and a pattern P' in the subimage of the target image corresponding to a potential match.

4. Online 2D-Gel Databases

The World Wide Web has allowed scientists all around the world to share data and knowledge. The 2D-gel community is not an exception and has benefited for over a decade from this fact thanks to the different online repositories. Currently, there are more than 50 different online 2D-gel databases (a detailed list of online 2D-gel databases can be found at http://www.expasy.org/ch2d/2d-index.html). These databases are repositories of well-annotated 2D-gel images known as *reference maps*. Such maps are obtained by combining 2D-gel experiments along with other experimental techniques (e.g., mass spectroscopy), which permits the identification and characterization of protein spots. Consequently, reference maps can contain not only descriptive information of particular protein spots but also experimental information such as the isoelectric point pI, molecular mass Mr, amino acid composition, peptide masses, and quantitative data. Furthermore, links to other databases and literature references can also be provided (77). Online 2D-gel databases permit the comparison and exchange of 2D-gel images between laboratories, and reference maps are used for different tasks such as rapidly identify a particular protein spot found to be differentially expressed in different samples, detect those spots corresponding to proteins of no known function, and compare tissue-specific 2D-gel images in order to identify those proteins common to different tissue and proteins that are only expressed in certain tissues.

5. Discussion

Current commercial 2D-gel analysis programs claim to be automatic with high-throughput capabilities. A recent study that compared five commercial programs concluded that, in general, less than 3% of the total processing time was automatic (the "myth" of automated 2D-gel analysis). The authors also conclude that program accuracy decreases with increased 2D-gel complexity and number of 2D gels included in the study. It is clear that current programs are not fully developed to meet the experimentalists' recruitments. However, it is unclear whether this problem is merely an implementation problem rather than an algorithmic problem. In this chapter, we reviewed a fairly large number of algorithms (*see* **Table 12.2**).

Each of them claims to be optimal for specific tasks or specialized cases. It would be very interesting to systematically compare the different algorithmic methods and implement the optimal algorithms for each step in one excellent program. We believe that such an implementation currently should have higher priority than proposing new algorithms for a specific task since the currently available methods have not been sufficiently compared.

Another interesting aspect that has currently not been addressed is the possibility of making search algorithms that can identify similar 2D-gel images in a database using only the image information. Such methods are interesting in the cases where one wants to match a 2D gel from a sample of a specific patient against 2D gels of similar samples of patients having various diseases. Another possibility is that the search algorithms use a simple approach based on a number of extracted features to find simi-

Table 12.2
Specification of algorithms used in publicly available tools

Software	Aim	Specification
Flicker (Open2Dprot)	Image alignment	Spot quantification: VOL Image warping: 1. affine transform 2. polynomial transform 3. pseudo-3D transform
Seg2Dgel (Open2Dprot)	Image segmentation	Background subtraction: zonal notch filter Noise reduction: minimum size threshold Spot detection: Gaussian smooth – Laplacian method Spot quantification: VOL
CAROL	Spot detection and matching	Spot detection: Watershed – gray-value thresholding – convex curvature thresholding Pattern matching: Delaunay triangulation – geometric hashing
Pinnacle	Spot detection and quantification	Noise reduction: wavelet transform Normalization: pinnacle intensity/mean pinnacle intensity Spot detection: local intensity maximum above threshold Spot quantification: max OD in a neighborhood
Rain	Image alignment	Background subtraction: bias field correction Image warping: smooth volume – invariant B-spline Similarity measure: Sum of squared residuals

lar 2D gels in databases. To our knowledge, such a search functionality has not been properly explored. It is therefore unclear whether or not such a search functionality will require improved reproducibility of the 2D-gel technology or if current standards are adequate.

6. Notes

1. Nonspecific staining of protein and other biological compounds can often be observed in practice. For example, silver staining also stains DNA.
2. Several image similarity measures have been implemented. Some of the most popular measures are the sum of absolute

differences (SAD), the sum of squared differences (SSD), the normalized cross-correlation (NCC), and the Hausdorff distance d_h:

$$\mathrm{SAD} = \sum_{i=1}^{n} |A_i - B_i|,$$

$$\mathrm{SSD} = \sum_{i=1}^{n} (A_i - B_i)^2,$$

$$\mathrm{NCC} = \frac{\sum_{i=1}^{n} (A_i - \bar{A})(B_i + \bar{B})}{\sqrt{\left[\sum_{i=1}^{n} (A_i - \bar{A})^2 \sum_{i=1}^{n} (B_i - \bar{B})^2\right]}},$$

$$d_H = \max_{a \in A} \left\{ \min_{b \in B} \{d(a, b)\} \right\},$$

where A and B correspond to the intensities in two windows (the image is segmented in n windows), $\bar{A}$ and $\bar{B}$ denote the sample means of the corresponding windows, a and b are point sets of the corresponding windows, and $d(a, b)$ corresponds to a distance metric between two points in the space (e.g., Euclidean distance).

3. Smooth transformations are usually constrained by a space of allowable transformations and ensure that no discontinuous jumps, undesirable changes, and distorted intermediate stages occur (78).
4. A monomial function is a particular type of polynomial function containing only one term: $g(x) = ax^b$.
5. The IP-Euc distance metric was implemented based on the fact that corresponding spots usually have similar shapes and intensity values after normalization, and it is therefore possible to measure spot similarity by computing the information potential between pair spots P_i , Q_j. This function combines a distance d_{euc} based on the Euclidean distance between the center of a pair of spots and a distance d_{ip} based on information about potential energy (IPE):

$$\mathrm{IPE}(P_i, Q_j) = (1 - \lambda) d_{\mathrm{euc}} - \lambda d_{ip} \quad \lambda \in [0, 1]$$

$$d_{\mathrm{euc}} = \frac{\| P_i, Q_j \|}{\max\limits_{\substack{i \in \{1,\ldots,N_1\} \\ j \in \{1,\ldots,N_2\}}} \left(\| P_i, Q_j \| \right)}$$

$$d_{ip} = \frac{M_{ij}^{ip}}{\max(M_{ij}^{ip})}$$

$$i \in \{1, .., N_1\}$$
$$j \in \{1, .., N_2\}$$

$$M_{ij}^{ip} = \frac{\Delta\text{shape}}{\text{CEF}} = \frac{e^{\frac{2(|P_i|-|Q_j|)}{|P_i|+|Q_j|}}}{\text{CEF}}$$

where $\|P_i, Q_j\|$ corresponds to the Euclidean distance between the center of spots P_i, Q_j, $|\cdot|$ indicates the area size of a spot, and CEF corresponds to the clustering evaluation function (79).

6. Rogers and colleagues (73) formulated a distance metric d that combines the Euclidean distance along with a distance metric d_{SC} that measures the spatial distribution of neighboring points. Furthermore, two other distance metrics, d_{IC} and d_{FD}, were also included: (i) d_{IC} uses the robust least median of squares (LMedS) to calculate the distance between the distribution of intensities within a radial region surrounding the corresponding spots, whereas d_{FD} uses the LMedS to calculate the distance between the distributions of features within a radial region surrounding the corresponding spots:

$$d = \alpha d_{\text{euc}} + \frac{(1-\alpha)(d_{sc} + d_{IC} + d_{FC})}{3}$$

where α is a weighting factor between two measures.

Acknowledgments

Support for GL was provided by the Spanish Ministry of Education and Science Grant BFU2006-09648. Support for RM was provided from Ramon y Cajal (RYC-2006-001446) and the Department of Industry, Tourism and Trade of the Government of the Autonomous Community of the Basque Country (Etortek Research Programs 2005/2006) and from the Innovation Technology Department of the Bizkaia County.

References

1. Margolis J, Kenrick KG. (1969) Two-dimensional resolution of plasma proteins by combination of polyacrylamide disc and gradient gel electrophoresis. *Nature* 221:1056–1057.
2. O'Farrell PH. (1975) High resolution two-dimensional electrophoresis of proteins. *J Biol Chem* 250:4007–4021.
3. Miller I, Crawford J, Gianazza E. (2006) Protein stains for proteomic applica-

tions: which, when, why? *Proteomics* 6: 5385–5408.
4. Unlu M, Morgan ME, Minden JS. (1997) Difference gel electrophoresis: a single gel method for detecting changes in protein extracts. *Electrophoresis* 18:2071–2077.
5. Alban A, David SO, Bjorkesten L, Andersson C, Sloge E, Lewis S, Currie I. (2003) A novel experimental design for comparative two-dimensional gel analysis: two-dimensional difference gel electrophoresis incorporating a pooled internal standard. *Proteomics* 3:36–44.
6. Clark BN, Gutstein HB. (2008) The myth of automated, high-throughput two-dimensional gel analysis. *Proteomics* 8:1197–1203.
7. Dowsey AW, Dunn MJ, Yang GZ. (2003) The role of bioinformatics in two-dimensional gel electrophoresis. *Proteomics* 3:1567–1596.
8. Capel M, Redman B, Bourque DP. (1979) Quantitative comparative analysis of complex two-dimensional electropherograms. *Anal Biochem* 97:210–228.
9. Garrels JI. (1979) Two dimensional gel electrophoresis and computer analysis of proteins synthesized by clonal cell lines. *J Biol Chem* 254:7961–7977.
10. Schumaker MF. (1978) A program which automatically quantitates gel electrophoretic autoradiograms. *Anal Biochem* 91:375–393.
11. Bossinger J, Miller MJ, Vo KP, Geiduschek EP, Xuong NH. (1979) Quantitative analysis of two-dimensional electrophoretograms. *J Biol Chem* 254:7986–7998.
12. Anderson NL, Taylor J, Scandora AE, Coulter BP, Anderson NG. (1981) The TYCHO system for computer analysis of two-dimensional gel electrophoresis patterns. *Clin Chem* 27:1807–1820.
13. Fox SH. (1982) Some relatively simple steps toward a computer system for the analysis of two-dimensional gel-electrophoresis autoradiographs. *Clin Chem* 28:932–934.
14. Lemkin P, Merril C, Lipkin L, Van Keuren M, Oertel W, Shapiro B, Wade M, Schultz M, Smith E. (1979) Software aids for the analysis of 2D gel electrophoresis images. *Comput Biomed Res* 12:517–544.
15. Lipkin LE, Lemkin PF. (1980) Data-base techniques for multiple two-dimensional polyacrylamide gel electrophoresis analyses. *Clin Chem* 26:1403–1412.
16. Alexander A, Cullen B, Emigholz K, Norgard MV, Monahan JJ. (1980) A computer program for displaying two-dimensional gel electrophoresis data. *Anal Biochem* 103:176–183.
17. Sanchez JC, Appel RD, Golaz O, Pasquali C, Ravier F, Bairoch A, Hochstrasser DF. (1995) Inside SWISS-2DPAGE database. *Electrophoresis* 16:1131–1151.
18. Wilkins MR, Hochstrasser DF, Sanchez JC, Bairoch A, Appel RD. (1996) Integrating two-dimensional gel databases using the Melanie II software. *Trends Biochem Sci* 21:496–497.
19. Lemkin PF. (1997) Comparing two-dimensional electrophoretic gel images across the Internet. *Electrophoresis* 18:461–470.
20. Lutin WA, Kyle CF, Freeman JA. (1978) Quantitation of brain proteins by computer-analyzed two dimensional electrophoresis. In *Developments in Biochemistry* (Catsimpoolas N, Ed.), Elsevier North-Holland, Amsterdam, pp. 93–106.
21. Vo KP, Miller MJ, Geiduschek EP, Nielsen C, Olson A, Xuong NH. (1981) Computer analysis of two-dimensional gels. *Anal Biochem* 112:258–271.
22. Skolnick MM. (1982) An approach to completely automatic comparison of two-dimensional electrophoresis gels. *Clin Chem* 28:979–986.
23. Jansson PA, Grim LB, Elias JG, Bagley EA, Longerg-Holm KK. (1983) Implementation and application of a method to quantitate 2-D gel electrophoresis patterns. *Electrophoresis* 4:82–91.
24. Olson AD, Miller MJ. (1988) Elsie 4: quantitative computer analysis of sets of two-dimensional gel electrophoretograms. *Anal Biochem* 169:49–70.
25. Garrels JI. (1989) The QUEST system for quantitative analysis of two-dimensional gels. *J Biol Chem* 264:5269–5282.
26. Garrels JI, Farrar JT, Burwell CB. (1984) The QUEST system for computer-analyzed two-dimensional electrophoresis of proteins. In *Two-Dimensional Gel Electrophoresis of Proteins* (Celis JE, Bravo R, Eds.), Academic Press, Academic Press, pp. 38–90.
27. Kuick RD, Skolnick MM, Hanash SM, Neel JV. (1991) A two-dimensional electrophoresis-related laboratory information processing system: spot matching. *Electrophoresis* 12:736–746.
28. Lemkin PF, Myrick JM, Lakshmanan Y, Shue MJ, Patrick JL, Hornbeck PV, Thornwal GC, Partin AW. (1999) Exploratory data analysis groupware for qualitative and quantitative electrophoretic gel analysis over the Internet-WebGel. *Electrophoresis* 20: 3492–3507.
29. Pleissner KP, Hoffmann F, Kriegel K, Wenk C, Wegner S, Sahlstrom A, Oswald H, Alt H, Fleck E. (1999) New algorith-

mic approaches to protein spot detection and pattern matching in two-dimensional electrophoresis gel databases. *Electrophoresis* 20:755–765.
30. Lemkin PF, Lester EP. (1989) Database and search techniques for two-dimensional gel protein data: a comparison of paradigms for exploratory data analysis and prospects for biological modeling. *Electrophoresis* 10:122–140.
31. Tonge R, Shaw J, Middleton B, Rowlinson R, Rayner S, Young J, Pognan F, Hawkins E, Currie I, Davison M. (2001) Validation and development of fluorescence two-dimensional differential gel electrophoresis proteomics technology. *Proteomics* 1:377–396.
32. Young N, Chang Z, Wishart DS. (2004) GelScape: a web-based server for interactively annotating, manipulating, comparing and archiving 1D and 2D gel images. *Bioinformatics* 20:976–978.
33. Morris JS, Clark BN, Gutstein HB. (2008) Pinnacle: a fast, automatic and accurate method for detecting and quantifying protein spots in 2-dimensional gel electrophoresis data. *Bioinformatics* 24:529–536.
34. Dowsey AW, Dunn MJ, Yang GZ. (2008) Automated image alignment for 2D gel electrophoresis in a high-throughput proteomics pipeline. *Bioinformatics* 24:950–957.
35. Tyson JJ, Haralick RH. (1986) Computer analysis of two-dimensional gels by a general image processing system. *Electrophoresis* 7:107–113.
36. Appel RD, Vargas JR, Palagi PM, Walther D, Hochstrasser DF. (1997) Melanie II – a third-generation software package for analysis of two-dimensional electrophoresis images: II. Algorithms. *Electrophoresis* 18:2735–2748.
37. Sternberg SR. (1983) Biomedical image processing. *Computer* 16:22–34.
38. Sternberg SR. (1986) Grayscale morphology. *Comput Vis Graph Image Process* 35:333–355.
39. Eilers PH. (2004) Parametric time warping. *Anal Chem* 76:404–411.
40. Kaczmarek K, Walczak B, De Jong S, Vandeginste BG. (2005) Baseline reduction in two dimensional gel electrophoresis images. *Acta Chroma* 15:82–96.
41. Daszykowski M, Stanimirova I, Bodzon-Kulakowska A, Silberring J, Lubec G, Walczak B. (2007) Start-to-end processing of two-dimensional gel electrophoretic images. *J Chromatogr A* 1158:306–317.
42. Kaczmarek K, Walczak B, de Jong S, Vandeginste BG. (2004) Preprocessing of two-dimensional gel electrophoresis images. *Proteomics* 4:2377–2389.
43. Wiener N. (1949) *Extrapolation, Interpolation, and Smoothing of Stationary Time Series*, MIT Press, Cambridge, MA
44. Berth M, Moser FM, Kolbe M, Bernhardt J. (2007) The state of the art in the analysis of two-dimensional gel electrophoresis images. *Appl Microbiol Biotechnol* 76: 1223–1243.
45. Conradsen K, Pedersen J. (1992) Analysis of two-dimensional electrophoretic gels. *Biometrics* 48:1273–1287.
46. Vincent L, Soille P. (1991) Watersheds in digital spaces: an efficient algorithm based on immersion simulations. *Trans Patt Anal Mach Intell* 16:583–598.
47. Mahon P, Dupree P. (2001) Quantitative and reproducible two-dimensional gel analysis using Phoretix 2D Full. *Electrophoresis* 22:2075–2085.
48. Efrat A, Hoffmann F, Kriegel K, Schultz C, Wenk C. (2002) Geometric algorithms for the analysis of 2D-electrophoresis gels. *J Comput Biol* 9:299–315.
49. Yan JX, Sanchez JC, Tonella L, Williams KL, Hochstrasser DF. (1999) Studies of quantitative analysis of protein expression in *Saccharomyces cerevisiae*. *Electrophoresis* 20:738–742.
50. Dutt MJ, Lee KH. (2001) The scaled volume as an image analysis variable for detecting changes in protein expression levels by silver stain. *Electrophoresis* 22:1627–1632.
51. Bettens E, Scheunders P, Van Dyck D, Moens L, Van Osta P. (1997) Computer analysis of two-dimensional electrophoresis gels: a new segmentation and modeling algorithm. *Electrophoresis* 18:792–798.
52. Rogers M, Graham J, Tonge RP. (2003) Statistical models of shape for the analysis of protein spots in two-dimensional electrophoresis gel images. *Proteomics* 3:887–896.
53. Lemkin P, Thornwal GC, Evans J. (2005) Comparing 2-D electrophoretic gels across Internet databases. In *The Proteomics Protocols Handbook*, 5th ed (Walker JM, Ed.), Humana Press, Totowa, NJ.
54. Gustafsson JS, Blomberg A, Rudemo M. (2002) Warping two-dimensional electrophoresis gel images to correct for geometric distortions of the spot pattern. *Electrophoresis* 23:1731–1744.
55. Goshtasby, A. (1986) Piecewise linear mapping functions for image registration. *Patt Recog* 19:459–466.
56. Goshtasby, A. (1987) Piecewise cubic mapping functions for image registration. *Patt Recog* 20:525–533.

57. Rohr K, Cathier P, Wörz S. (2003) Elastic registration of electrophoresis images using intensity information and point landmarks. *Patt Recog* 3:1035–1048.
58. Bookstein FL. (1989) Principal warps: thin plate splines and the decomposition of deformations. *IEEE Trans Patt Anal Mach Intell* 11:567–585.
59. Aittokallio T, Salmi J, Nyman TA, Nevalainen OS. (2005) Geometrical distortions in two-dimensional gels: applicable correction methods. *J Chromatogr B Analyt Technol Biomed Life Sci* 815:25–37.
60. Kaczmarek K, Walczak B, de Jong S, Vandeginste BG. (2002) Feature based fuzzy matching of 2D gel electrophoresis images. *J Chem Inf Comput Sci* 42:1431–1442.
61. Kaczmarek K, Walczak B, de Jong S, Vandeginste BG. (2003) Matching 2D gel electrophoresis images. *J Chem Inf Comput Sci* 43:978–986.
62. Smilansky Z. (2001) Automatic registration for images of two-dimensional protein gels. *Electrophoresis* 22:1616–1626.
63. Veeser S, Dunn MJ, Yang GZ. (2001) Multiresolution image registration for two-dimensional gel electrophoresis. *Proteomics* 1:856–870.
64. Salmi J, Aittokallio T, Westerholm J, Griese M, Rosengren A, Nyman TA, Lahesmaa R, Nevalainen O. (2002) Hierarchical grid transformation for image warping in the analysis of two-dimensional electrophoresis gels. *Proteomics* 2:1504–1515.
65. Rohr K, Cathier P, Wörz S. (2004) Elastic registration of electrophoresis images using intensity information and point landmarks. *Patt Recog* 3:1035–1048.
66. Panek J, Vohradsky J. (1999) Point pattern matching in the analysis of two-dimensional gel electropherograms. *Electrophoresis* 20:3483–3491.
67. Toussaint GT. (1980) The relative neighbourhood graph of a finite planar set. *Patt Recog* 12:261–268.
68. Gabriel KR, Sokal RR. (1969) A new statistical approach to geographic variation analysis. *Syst Zool* 18:259–270.
69. Lankford PM. (1969) Regionalization: theory and alternative algorithms. *Geogr Anal* 1:196–212.
70. Delaunay B. (1934) Sur la sphère vide. *Izvestia Akademii Nauk SSR, Otdelenie Matematicheskikh i Estestvennykh Nauk* 7: 793–800.
71. Besl PJ. (1992) A method for registration of 3-D shapes. *IEEE Trans Patt Anal Mach Intell* 14:239–256.
72. Shi G, Jiang T, Zhu W, Liu B, Zhao H. (2007) Alignment of two-dimensional electrophoresis gels. *Biochem Biophys Res Commun* 357:427–432.
73. Rogers M, Graham J, Tonge R. (2004) Two-dimensional electrophoresis gel registration using point matching and local image-based refinement. In *British Machine Vision Conference*, Kingston University, London
74. Wolfson HJ, Rigoutsos I. (1997) Geometric hashing: an overview. *IEEE Comput Sci Eng* 4:10–21.
75. Schwartz J, Sharir M. (1986) Identification of partially obscured objects in two and three dimensions by matching noisy characteristic curves. *Int J Robot Res* 6:29–44.
76. Hoffmann F, Kriegel K, Wenk C. (1999) *An Applied Point Pattern Matching Problem: Comparing 2D Patterns of Protein Spots,* Vol. 93, Elsevier Science Publishers, Amsterdam, pp. 75–88.
77. Hoogland C, Mostaguir K, Sanchez JC, Hochstrasser DF, Appel RD. (2004) SWISS-2DPAGE, ten years later. *Proteomics* 4:2352–2356.
78. Douglas D, Jean G. (1996) Topological evolution of surfaces. In *Proceedings of the Conference on Graphics Interface '96,* Toronto, Ontario, Canada.
79. Gokcay E, Principe JC. (2002) Information theoretic clustering. *IEEE Trans Patt Anal Mach Intell* 24:158–171.

Chapter 13

Mass Spectrometry in Epigenetic Research

Hans Christian Beck

Abstract

The inhibition of the histone deacetylase enzymes induces hyperacetylation of the histone proteins. This hyperacetylation causes cell cycle arrest and cell death in cancer cells but not in normal cells. Therefore, the development of histone deacetylase inhibitors for the treatment of various cancers has gained tremendous interest in recent years, and many of these inhibitors are currently undergoing clinical trials. Despite intense research, however, the exact molecular mechanisms of action of these molecules remain, to a wide extent, unclear. The recent application of mass spectrometry-based proteomics techniques to histone biology has gained new insight into the function of the nucleosome: Novel posttranslational modifications have been discovered at the lateral surface of the nucleosome. These modifications regulate histone–DNA interactions, adding a new dimension to the epigenetic regulation of nucleosome mobility.

Key words: Mass spectrometry, VEMS, HDAC inhibitors, quantitative proteomics, epigenetics.

1. Introduction

The fundamental unit of eukaryote chromatin is the nucleosome core particle. This particle consists of 147 base pairs (bp) of DNA wrapped around an octamer of the four core histones, H2A, H2B, H3, and H4 (two heterodimers of H2A and H2B and a heterotetramer of histones H3 and H4). Nucleosomes are joined together by linker DNA and histone H1 to form chromatin. Nucleosomes are equally spaced along the genome and form nucleofilament that can adopt higher levels of compaction. A central mechanism for regulating chromatin activity is the covalent modifications of histones catalyzed by enzymes. A complex interplay between these posttranslational modifications dictates the chromatin structure and function by activating or repressing gene transcription.

R. Matthiesen (ed.), *Bioinformatics Methods in Clinical Research*, Methods in Molecular Biology 593,
DOI 10.1007/978-1-60327-194-3_13, © Humana Press, a part of Springer Science+Business Media, LLC 2010

The N- and C-terminals of the core histone proteins are subjected to a variety of posttranslational modifications, including methylation of lysines and arginines, acetylation of lysine, phosphorylation of threonines and serines, ADP-ribosylation, ubiquitination (1, 2), formylation (3), and sumoylation (4). These modifications dictate the chromatin structure and function by activating or repressing gene transcription.

A histone code was hypothesized (5–7) where a specific combinatorial code of these modifications regulates specific genomic functions such as protein–protein and protein–DNA interactions for the recruitment of transcription factor interactions; the code thus functions as a regulator of gene expression (8).

The acetylation of lysine residues is the major mediator of the histone code. The acetylation and deacetylation of the core histones are balanced by the activity of histone acetyl transferases (HATs) and histone deacetylases (HDACs). The hyperactylation of histone tails is usually associated with high transcriptional activity (**Fig. 13.1**), whereas the hypoacetylation of the N-termini of the core histone proteins is associated with transcriptional silencing (9). The inhibition of HDAC enzymes by HDAC inhibitors leads to a hyperacetylation of lysine residues in the N-terminus of histones H2A, H2B, H3, and H4, which results in the expression of genes that induce growth arrest, cell differentiation, and apoptotic cell death in cultured tumor cells (10–12). Furthermore, HDAC inhibitors (HDACis) have shown promising pharmacological properties in animal models and clinical trials, where tumor growth inhibition has been observed. Therefore, the inhibition of HDACs has great potential in the combat against cancer, and a number of HDACis are currently being evaluated as cancer drugs in preclinical studies.

Over the last decade, it has become increasingly evident that histones are not the only targets of acetylation and deacetylation. Since the discovery of the tumor suppressor and sequence-specific DNA-binding transcription factor p53 a decade ago (13), the list of nonhistone DNA-binding proteins has continuously grown. Nonhistone protein transcription factors comprise the largest group, and most importantly many of these proteins are modulated by acetylation and play key roles in oncogenesis and cancer progression (14). Furthermore, many of these transcription factors are also regulated at the transcription level by histone acetylation. The treatment of various cancers with HDAC inhibitors induces the hyperacetylation of many transcription factors, including p53 – with cellular outcomes such as the induction of cell cycle arrest and apoptosis.

Mass spectrometry-based proteomics has, at present, the centere stage in protein research and molecular biology and is playing an increasing role in the elucidation of the primary structure of histones. The first of the myriads of modifications identified in the histone proteins was discovered four decades ago by Murray (15), who discovered an ε-*N*-methyl lysine in histone proteins using liquid-gas chromatography coupled to electron impact quadrupole mass spectrometry. Shortly after, histone acetylation (16), phosphorylation (17, 18), and ADP-ribosylation (19) were discovered using the same MS technique. The major limitation with this MS technique is, however, that peptides with more than approximately 10 amino acid residues could not be analyzed due to thermal breakdown caused by the thermal gradient of the GC during the analysis. Since then, mass spectrometry technology has undergone tremendous developments, starting with the invention of soft-ionization techniques, such as matrix-assisted laser desorption and application of electrospray ionization, combined with fragmentation methods (known as tandem mass spectrometry – MS/MS), such as collision-induced fragmentation (CID), electron capture dissociation (ECD), and electron transfer dissociation (ETD) in combination with quadrupole time-of-flight (q-TOF), ion-traps, orbitraps, or Fourier transformation (FT) mass analyzers allowing the analysis of larger biomolecules (>1000 Da) and the achievement of detailed structural information of amino acid sequence and posttranslational modifications.

These developments in mass spectrometry allowed the detailed structural analysis of histone peptides as well as that of intact histone proteins and their posttranslational modifications. Furthermore, labeling techniques using chemical labeling methods or metabolic labeling allowed the relative and absolute quantification of histone modifications (*see* **Chapter 10**). These developments in mass spectrometry make this technique ideal to study histone modifications in disease and in the developments of drugs directed against histone-modifying enzymes such as histone deacetylases.

In this chapter, we will highlight the recent achievements in the role of mass spectrometry-based proteomics in histone biology, with a special emphasis on histone modifications and their functional role in cancer and HDAC inhibitor–induced cell death in cancer treatment.

2. Histone Modifications Identified by Mass Spectrometry-Based Proteomics Technologies

The recent developments in mass spectrometry-based proteomics revealed exciting findings in histone posttranslational modifications. These technologies have overcome the major disadvantages of the immunochemical methods traditionally used for the characterization of histone modifications. The main disadvantage of the immunochemical techniques is that the modification studied included a priori knowledge of the specific histone modification to be studied. Another drawback is the difficulty in interpreting the obtained results due to the potential cross-reactivity or epitope masking by a neighboring modification (20). In contrast, mass spectrometry-based proteomics technologies have several advantages over the immunochemical methods traditionally used for the characterization of posttranslational histone modifications. These include the selectivity and specificity as well as the speed of analysis. Also, the recent developments of quantitative mass spectrometry-based proteomics technologies have enabled the analysis of novel modifications in a quantitative manner in a single experiment (21). **Table 13.1** summarizes the histone modifications identified by mass spectrometry.

2.1. Analytical Strategies for Histone Proteomics

Several MS-based approaches have been used to analyze and discover histone modifications. This include classic approaches such as peptide mass mapping using MALDI MS, peptide sequencing, and the identification and annotation of location of modification site by nanospray and capillary liquid chromatography coupled to q-TOF tandem mass spectrometry (MS/MS). Less commonly used mass spectrometry technologies include Orbitrap and FT MS in combination with ECD and ETD ionization techniques.

Basically, all approaches can be divided into two different strategies: the "bottom-up" strategies that rely on enzymatic digestion of the histone proteins prior to analysis, and the "top-down" approach that analyzes intact histone proteins. The "histone code" theory correlates distinct patterns of histone modifications with a distinct genetic readout, leading to molecular events such as cell cycle arrest and apoptosis. A particular challenge in the MS analysis of histone modifications is therefore to analyze the complete modification state of individual histone proteins. The optimal strategy would include the measurement of intact histone proteins combined with structural information on amino acid sequence and the type and location of modifications. This requires mass spectrometers with a very high resolution combined with a fragmentation technique that enables the fragmentation of large peptides. The introduction of high-resolution instruments, such as Fourier transform ion cyclotron resonance (FT-ICR), MS combined with electron capture dissociation (ECD) fragmentation, or Orbitrap MS combined with electron transfer dissociation (ETD), currently forms the basis for the future development of the ideal "top-down" approach, and such instruments have already been applied for histone analysis (22, 27, 43). CID is based on the fragmentation of selected peptides through collision with an inert gas, typically argon, in a gas cell. Upon collision with the gas molecules, the kinetic (or "translational") energy of the peptide is converted into internal energy, leading to peptide bond breakages and fragmentation of the peptides into smaller fragments. One of the main disadvantages with CID, however, is the limitation in energy available for fragmentation of the peptide, thus limiting the size of the peptide that can be fragmented.

In contrast, ECD fragmentation relies on the capture of electrons emitted from a hot filament in the FT cell of the FT-ICR instrument that is captured by a multiprotonated specimen such as large peptides. Upon electron capture, an odd-electron ion is formed, rapidly leading to fragmentation of the peptide backbone. This MS technique using ECD ionization was applied to characterize histone H2B variants and their "populations" of posttranslational modifications, where two major variants of this protein were found, each of them present in at least six posttranslationally modified forms. However, conventional LC-MS/MS using CID was also applied to gain more detailed structural information on the exact location of specific modifications.

ETD, an ionization technique similar to ECD based on the transfer of electrons from gas-phase radical ions to the positively charged peptide ions, was recently introduced. Like ECD, the ETD ionization technique cleaves the amide groups along the peptide backbone, yielding a ladder of sequence ions leaving labile modifications such as phosphorylations intact. In a recent study, the 1-N-terminal tail of histone H3.1 (residues 1–50) was analyzed using ETD-PTR mass spectrometry (22).

MS/MS spectra of histone-derived peptides containing posttranslational modifications often contain sufficient structural information for the unambiguous identification of the modified histone residue. Collision-induced fragmentation of the peptide often produces fragment ions resulting from the fragmentation of the modified amino acid side chain. These ions are often unique and may therefore serve as diagnostic ions for a given modification. For example, peptides containing acetylated lysine residue have a mass shift of 42.011 Da per acetylated residue. Fragmentation of the acetylated peptide produces diagnostic ions at *m/z* 143.118 (immonium ion) and at *m/z* 84 and 126.091 (elimination of ammonia from the immonium ion and rearrangement) (23). MS/MS analyses of peptides containing trimethylated lysine (causes a mass increment of 42.047) also produce a diagnostic ion at *m/z* 84.081 and 143.1179. In addition, CID fragmentation of trimethylated lysine residues also produces diagnostic neutral losses of 59.073 and at 60.081 Da (MH^+ –59 or –60). Therefore, acetylated and trimethylated peptides can be differentiated on the basis of diagnostic fragment ions and neutral losses. Furthermore, unique fragment ions can be exploited in survey scans during MS/MS analyses for the search for acetylated peptides. In fact, a detailed investigation of the fragmentation of peptides containing acetylated lysine residues was recently performed by Trelle and Jensen (24). A thorough investigation of spectral data from MS/MS analysis of 172 acetylated tryptic peptides showed that the immonium ion derivative at *m/z* 126 is highly specific for the fragmentation of acetylated peptides. In fact, more than 98% of the acetylated peptides investigated produced this ion upon CID fragmentation, making this ion an important and indispensable feature of tandem mass spectrometry when analyzing for unknown lysine acetylations.

The first comprehensive analysis of histones by MS was done by Zhang et al. (25), who characterized the level of histone H4 acetylation by matrix-assisted laser desorption ionization time-of-flight mass spectrometry and annotated the exact acetylation sites by nano-electrospray tandem mass spectrometry. It was found that the acetylation of H4 at lysines 5, 8, 12, and 16 proceeds in the direction from lysine 16 to lysine 5 and that deacetylation occurs in the opposite direction, leading to the proposal of a "zip" model that was confirmed in a study of mouse lymphosarcoma cells treated with the HDACis trichostatin A (TSA) or depsipeptide (26) and also in human small cell lung cancer cells treated with the HDACi PXD101 (21).

In another study by Zhang and co-workers (27), Fourier transform ion cyclotron resonance mass spectrometry (FT-ICR MS) was employed to analyze histone modifications in mammalian histones. FT-ICR MS offers a resolution power of $>10^6$, which, in some cases, may be sufficient to analyze each peptide in a digestion mixture without chromatographic separation prior to MS analysis. Utilizing the resolving power of this technique, it was possible to determine whether a peptide is tri-methylated histone or acetylated (mass shift of 42.0470 Da vs. 42.0106) based on the measured mass alone, thereby circumventing the need for confirmative MS/MS analysis. Utilizing these features of FT-ICR MS in a screen of proteolytic digest of histone proteins, Zhang et al. annotated more than 20 novel modification sites, most of which were located in the core region and COOH-tail of the histone proteins. Functional analysis of the methylation on lysine 59 in histone H4 showed that this modification – consistent with the role of lysine 79 in histone H3 – is essential for transcriptional silencing at the yeast silent mating loci and telomers.

A range of other studies have contributed to MS-based identification of histone modifications and are summarized in **Table 13.1**.

Table 13.1
Summary of histone modifications identified by mass spectrometry

Residue	Modification	References	Residue	Modification	References
Histone H2A			H3 - continued		
Ser 1	Phosphorylation	(28)	Thr 6	Phosphorylation	(27)
Lys 5	Acetylation	(27, 28)	Lys 9	Me/Ac	(21, 27, 29, 30)
Lys 9	Acetylation	(27)	Ser 10	Phosphorylation	(27, 31)
Lys 13	Acetylation	(27)	Thr 11	Phosphorylation	(27, 29, 30)
Lys 15	Acetylation	(27)	Lys 14	Me/Ac	(21, 27, 29, 30)
Lys 36	Acetylation	(27)	Arg 17	Methylation	(27)
Lys 74	Methylation	(27)	Lys 18	Me/Ac	(21, 27, 29, 30)
Lys 75	Methylation	(27)	Lys 23	Me/Ac	(21, 27, 29, 30)
Arg 77	Methylation	(27)	Arg 26	Methylation	(27)
Lys 99	Methylation	(27)	Lys 27	Me/Ac	(21, 27, 29, 30)
Lys 119	Ac/Ub	(27)	Ser 28	Phosphorylation	(27, 31)
Lys 124	Methylation	(27)	Ser 31	Phosphorylation	(31)
Lys 126	Methylation	(27)	Lys 36	Me/Ac	(21, 27, 29, 30, 32)
Histone H2B			Arg 52	Methylation	(27)
Lys 5	Me/Ac	(21, 27)	Arg 53	Methylation	(27)
Lys 11	Acetylation	(21, 33)	Lys 56	Me/Ac	(27, 29, 34)
Lys 12	Acetylation	(21, 27, 28, 33)	Lys 64	Me/Ac	(29, 30)
Lys 15	Acetylation	(21, 27, 28)	Lys 79	Me/Ac	(21, 27, 29, 30, 35–37)
Lys 16	Acetylation	(21)	Lys 115	Acetylation	(27)

(continued)

Table 13.1 (continued)

Residue	Modification	References	Residue	Modification	References
Lys 20	Acetylation	(21, 27, 28)	Thr 118	Phosphorylation	(27)
Lys 23	Methylation	(27)	Lys 122	Me/Ac	(27, 29, 30, 37)
Lys 34	Methylation	(27)	Arg 128	Methylation	(37)
Ser 36	Phosphorylation	(38)	Arg 129	Methylation	(37)
Lys 43	Methylation	(27)			
Lys 46	Me/Ac	(21, 33)	**Histone H4**		
Lys 57	Methylation	(21)	Lys 5	Acetylation	(21, 25, 27, 29, 35, 39)
Arg 79	Methylation	(27)	Lys 8	Acetylation	(21, 25, 27, 29, 35, 39)
Lys 85	Acetylation	(27)	Lys 12	Me/Ac	(21, 25, 27, 29, 35, 39)
Arg 86	Methylation	(27)	Lys 16	Acetylation	(21, 25, 27, 29, 35, 39)
Arg 92	Methylation	(27)	Lys 20	Me/Ac	(21, 25, 27, 29, 35, 39)
Arg 99	Methylation	(27)	Lys 31	Me/Ac	(21, 27, 29)
lys 108	Me/Ac	(21, 27)	Ser 47	Phosphorylation	(27)
Lys 116	Me/Ac	(27, 33)	Arg 55	Methylation	(21)
Lys 120	Ac/Ub	(21, 27)	Lys 59	Methylation	(27)
			Lys 77	Me/Ac	(21, 27)
Histone H3			Lys 79	Me/Ac	(27)
Arg 2	Methylation	(27)	Lys 91	Acetylation	(27)
Thr 3	Phosphorylation	(31)	Arg 92	Methylation	(27)
Lys 4	Me/Ac	(27, 29–31, 35, 36)			

3. Quantification of Histone Modifications by Mass Spectrometry

The prerequisite for linking specific patterns of histone modifications with regulatory events leading to cancer initiation and progression is the quantitative measurement of the spatial and temporal distributions of the histone modification. Traditionally, the quantitative analysis of histone modifications has been carried out by immunochemical methods, but the recent achievements in quantitative mass spectrometry-based proteomic methods allowing the multisided analysis of protein modifications are becoming the method of choice for the quantitative analysis of histone modifications.

Quantitative mass spectrometry-based methods are normally based on the incorporation of stable isotopes by in vivo (biochemical) or ex vivo (enzymatic or chemical) approaches or may be based on peptide intensity profiling (label-free). Several of these strategies have been applied in quantitative studies of histone modifications. In vivo labeling is based on the incorporation of stable isotopes during cell growth; one of the most commonly used such methods is SILAC (*s*table *i*sotope *l*abeling with *a*mino acids in *c*ell culture) (40). Ex vivo labeling includes methods such as the enzymatic incorporation of ^{18}O in the C-terminus of the resulting peptides during proteolytic cleavage of the proteins by endonucleases prior to analysis by mass spectrometry (41), or chemical tagging of reactive groups of the amino acid side chains, such as acetylation of the ε-amino group side of the side-chain lysine residues using deuterated anhydrides (39), or tagging the cystein sulfhydryl groups with isotope tags for relative and absolute quantitation (iTRAQ) (42).

For example, Smith and co-workers (39) used a protocol based on chemical acetylation of unacetylated lysine residues in histone H4 with deuterated acetic anhydride followed by trypsination and concomitant LC-MS and MS/MS analysis. The mass shift caused by the exogenous introduction of deuterated acetyl groups was exploited to determine the fraction of in vivo acetylation at lysine residues 5, 8, 12, and 14 of histone H4 from yeast. They found that lysine 16 was the preferred site of acetylation, followed by lysines 12, 5, and 8. In another study, the SILAC approach was applied to monitor the de novo synthesis of H2A during cell cycle progression (43).

In a recent study, Beck et al. used a stable isotope-free method in the quantitative proteomic study of the dose-response effect of the HDAC inhibitor Belinostat (formerly named PXD101) (44) on histone acetylation in human cancer cells in an unbiased manner (21). Histone fractions from small cell lung cancer cells exposed to increased doses of Belinostat were isolated by acid extraction followed by in-solution trypsination. The resulting peptide mixtures were analyzed by LC-MS/MS (six samples run in triplicate) for protein identification and assignment of posttranscriptional modifications. Coefficient of variance (CV) analysis was used to pinpoint nonvarying (unmodified) "internal standard" peptides for data set normalization. After normalization of the data set (six samples analyzed in triplicate), the relative changes in intensity of each of the identified (modified) peptides were calculated. Statistically significant changes in peptide (modified peptides) abundance were determined by Tukey comparison test and SAM (statistical analysis of microarray) analysis. This method revealed a series of posttranslational modified peptides from all four core histones, which exhibit a dose-response effect upon HDACi treatment of small cell lung cancer cells.

4. Covalent Histone Modifications and Cancer

Generally, two different forms of chromatin exist: An open state (euchromatin) is associated with transcriptional activation, whereas tightly packed chromatin (heterochromatin) is associated with transcriptionally silent genomic regions. The fundamental unit of the chromatin molecule is the *nucleosome*, which consists of approximately 147 DNA base pairs wrapped around two units of each of the core histones H2A, H2B, H3, and H4. The N-termini of the core histones protrude from this structure and are in contact with adjacent nucleosomes in a higher-order structure whose structure still remains elusive.

It is evident that specific combinations of posttranslational modifications of the histone proteins achieved by histone-modifying enzymes such as HDACs and HATs alter the structure of these domains and thus affect binding of effector molecules, which, in turn, affect gene expression patterns. Histone acetylations occur at the ε-amino groups of conserved lysine residues of histone proteins, most often in their N-tail domains (**Table 13.1**).

4.1. Histone Deacetylases and Cancer

A defect in the life cycle of a cell may result in the development of cancer or uncontrolled growth of the cell. It is evident that histone deacetylases play a crucial role in the development of cancer, since the inhibition of these enzymes can cause the morphological reversion of the transformed cell phenotype (45–47). Inhibition of HDACs also leads to the blockage of cell proliferation, the promotion of differentiation, and the induction of apoptosis of cancer cells (10–12). As a consequence, there is significant interest in the development of HDAC inhibitors as anticancer medicine. At least 18 different forms of human HDACs have been identified.

They are grouped into four different classes, classes I – IV. Class I is comprised of HDAC1–3 and -8, class II of HDAC4–7, -9, and -10; class III of the NAD+-dependent SIR1–7; and class IV of HDAC11 (48). Currently, several HDACis are being evaluated in clinical trials to treat a variety of malignancies.

Histone acyltransferases catalyze the acetylation of the ε-amino group of lysine residues in N-termini of the core histones, thereby blocking the positive charge of the side chain of this amino acid residue and diminishing the interaction with the negatively charged DNA. This leads to the disruption of the higher-order chromatin structure (euchromatin). Histone deacetylases catalyze the reverse reaction, restoring the positively charged ε-amino group of lysine, which allows the compact chromatin form to be restored (heterochromatin). The open state chromatin form provides accessibility to transcription factors and the enzymes related to transcription processes. Specific patterns of histone modifications are affected by other histone modifications and generate a "histone code" (49); that is, specific readouts control specific transcriptional events. For example, histone H3 acetylation at lysine 9 and methylation of lysine 4 are associated with active transcription, whereas the loss of histone H3 acetylation at K9 and gain of histone H3 K9 and K27 methylation is indicative of heterochromatin and thus transcriptional silencing (50).

Histone proteins from cancer cells are characterized by very specific changes in the modification patterns. For example, the loss of acetylation at lysine 16 and trimethylation at lysine 20 of histone H4 appeared early and accumulated during the tumorigenic process and was associated with the hypomethylation of DNA repetitive sequences, a well-known characteristic of cancer cells (51). Specific histone modifications can also be used to predict cancer recurrence. Seligson and co-workers found that histone H3 acetylation coupled with H3-K4 dimethylation confers a lowered risk for prostate cancer recurrence (52), and it is likely that these findings can be extended to other cancer types.

4.2. Histone Deacetylase Inhibitors

A large number of structurally diverse HDAC inhibitors have been developed, purified from natural sources or developed in chemical laboratories, several of which have progressed to clinical development. The first HDACi discovered was butyrate. First, this molecule was discovered to have anticancer activity by the induction of cellular differentiation (53). It was subsequently discovered that butyrate was able to induce histone hyperacetylation (54), still without recognizing that HDACs were the target. The HDACis can be divided into six distinct classes based on their chemical structure. Butyrate, valproate, and AN-9 (prodrug) are short-chain fatty acid HDACis. The largest group of HDACis is the hydroxamates; these include the potent HDACis trichstatin (TSA), PXD101, LAQ824, and benzamides, cyclic tetrapeptides, electrophilic ketones, and a miscellaneous group (12). Many of these HDACis are currently being evaluated in clinical trials. These agents inhibit the enzymatic activity of HDACs with varying efficiency. All hydroxymate-based HDACis induce hyperactylation of the histones H3 and H4 and also alpha-tubulin.

4.3. Anticancer Mechanism of HDAC Inhibitors

HDACs are recruited to promoters to repress transcription and thereby counteract the transcription activation actions of the HATs. The general mechanism of HDAC activity is that HDAC recruitment leads to reduced levels of acetylation of the histone and as a consequence to compact chromatin. This precludes the access of the transcriptional machinery and, consequently, repression of transcription.

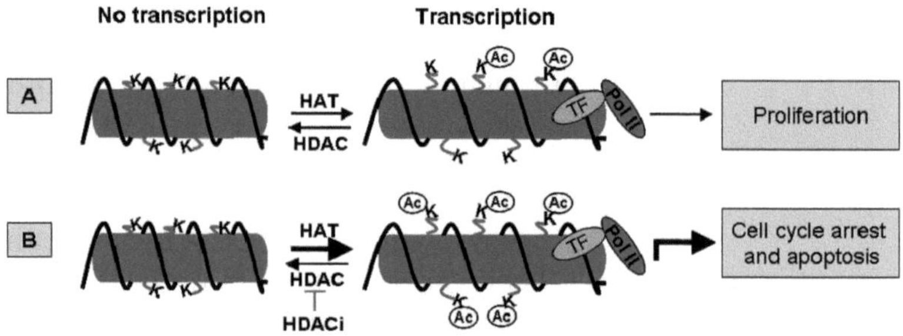

Fig. 13.1. The opposed activities of HAT and HDAC regulate the level at which a gene is transcribed. Reduced acetylation levels lead to repression of tumor suppressor genes and unlimited growth of cancer cells (**A**). Treatment with HDAC inhibitors leads to hyperacetylation of histone proteins. This activates expression of genes, leading to cell cycle arrest and apoptosis [adapted from (14)].

Examples of this mechanism include the repression of the Mad-Max target genes through the recruitment of HDAC1 and HDAC2 by a complex of the transcriptional repressor Mad, the transcription factor Max, and the mSin3 scaffold protein (55). A similar mechanism is active in the repression of E2F-mediated transcription, where HDAC1 recruitment coincides with decreased histone acetylation (56). Many of the repressed genes in these studies are related to tumor suppression, cell cycle inhibition, and cell differentiation or inducers of apoptosis, and the loss of repression of these favors the development and growth of cancerous cells. Most importantly, the treatment with an HDAC inhibitor de-represses the promoter in question and offers a clue as to why HDAC inhibitors may be useful for cancer treatment.

Histone deacetylation by HDAC influences the expression of genes that are involved in both cancer initiation and progression. As a consequence, treatment with HDACis has a major impact on the expression of many of these genes. Numerous studies have investigated the influence of HDAC inhibitor treatment with gene expression and provide a basic understanding of the mechanism by which HDACis modulate gene expression.

The protein p21 is an antiproliferative, cyclin-dependent kinase inhibitor that associates with cyclin-dependent kinases such as cyclin A2 and inhibits their kinase activities, leading to cell cycle arrest (57). In general, treatment of cancer cells with an HDACi leads to *p21* upregulation, which correlates with the hyperacetylation of histones H3 and H4 in their promoter region (58).

5. Nonhistone HDAC Targets in Cancer

It is evident that histones are not the only proteins regulated by the reversible action of HATs and HDACs, and the number of proteins identified as targets for these enzymes is continuously increasing, many of these playing a role in cancer. The acetylation and deacetylation of nonhistone proteins may have multiple effects on protein functions, including modulation of protein–protein interactions, protein stability, and localization (59). The largest group of these proteins is comprised of transcription factors. Like the histones, nonhistone proteins are acetylated at the ε-amino group of lysine residues. In contrast to N-terminal acetylation, this acetylation is highly reversible, and treatment of cancer cells with HDACis leads to hyperacetylation of these proteins at specific lysine residues. Thus, treatment with HDACis leads to altered protein function that affects the DNA binding and activation of transcription of genes involved in cancer-related processes such as apoptosis and cell death (**Fig. 13.2**).

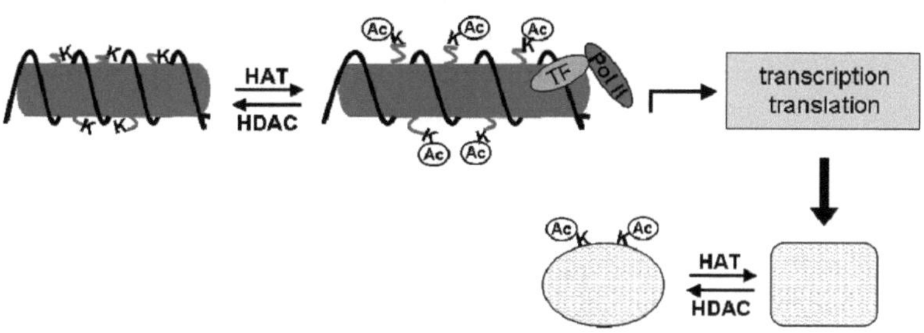

Fig. 13.2. Hyperacetylation of nonhistone proteins leads to altered functional properties of the proteins, such as protein–protein interactions, transcriptional activation, DNA binding ability, and subcellular location [adapted from (14)].

The tumor suppressor protein p53 is one of the first nonhistone proteins discovered as targets for acetylation and deacetylation reactions. Treatment of lung cancer cells with the HDACi depsipeptide causes specific acetylation of this protein at lysines 373 and 382. This recruits p300, a multifunctional protein with

HAT activity, and increases the expression of p21 (60). Treatment of prostate cancer cells by the HDACi TSA stabilizes the acetylation of lysine 382, whereas CG-1521 treatment of the same cells stabilizes the acetylation of lysine 373. Here, only the acetylation of lysine 373 was sufficient to increase p21 expression (61).

A few proteins other than transcription factors with a role in cancer development and progression, thus being potential targets for intervention strategies, include Ku70, a multifunctional protein involved in DNA repair. In the cytoplasm, Ku70 is kept in an unacetylated state by the action of HDACs and/or sirtuin deacetylases, thus ensuring the binding of the proapoptotic protein Bax. Upon treatment of cancer cells with HDACi, Ku70 is hyperacetylated at lysines 539 and 542. This releases Bax from Ku70, permitting this protein to translocate to the mitochondria, where it initiates apoptosis (62).

6. Clinical Development of HDAC Inhibitors

Since the discovery of the anticancer effects of the small molecule HDAC inhibition, numerous molecules have entered clinical trials, such as the pan-inhibitors Belinostat (PXD101) Vorinostat (SAHA, Zolinza™), and LBH589, and more selective agents, such as Romidepsin (depsipeptide, FK228), MS-275, and MGCD103. The first HDACi to be approved for cancer treatment was Vorinostat, which was approved by the U.S. Food and Drug Administration in October 2006 (http://www.fda.gov/bbs/topics/NEWS/2006/NEW01484.html). Many other HDACis are currently being tested in clinical trials, and the future will undoubtedly lead to the approval of other HDACis for cancer treatment, either alone or in combination with other, synergizing therapeutics (63).

In fact, combining these epigenetic modulators with therapeutics commonly used for cancer treatment has shown great promise. For example, a common reason for treatment failure when treating colorectal cancer with 5-fluorouracil (5-FU) is resistance to 5-FU (64). Resistance to a chemotherapeutic agent may occur by several mechanisms, such as the upregulation of efflux pumps or metabolizing enzymes, downstream effectors of the target proteins, or the target protein itself. In the case of 5-FU, resistance to 5-FU is found to be related to the upregulation of Thymidylate synthase (TS) – the target protein in 5-FU treatment of colorectal cancer (64). An improved response was observed in 5-FU-treated patients with a low tumoral TS expression, whereas only a weak response to 5-FU treatment was observed in patients with a high TS expression. Gene expression studies in various carcinoma cells revealed that the TS gene was targeted by HDACi, leading to repression of the TS gene, and, in fact, combining 5-FU and HDACi treatment increased the chemosensitivity of 5-FU (65–67). The HDACi Belinostat is currently being tested in clinical trials alone or in combination with 5-FU in patients with solid tumors (http://clinicaltrials.gov/ct2/show/NCT00413322).

Histone acetylation is a defining event for any of these HDACis and can be used as an indicator of HDAC inhibitor activity in both tumor cells and normal cells. it has therefore led to the widespread use of histone acetylation in peripheral blood mononuclear cells as surrogate markers in clinical phase I trials (68–70). A more accurate and direct measurement of the efficacy of HDACi treatment was obtained by measuring histone H4 acetylation in fine-needle biopsies of solid tumors using specific anti-H4 histone antibody. Acetylated H4 was monitored in vivo with immunochemical methods during treatment with Belinostat (PXD101) and compared with pharmacokinetics in plasma and tumor tissue. It was found that the acetylation level correlated well with the Belinostat levels in both plasma and tumors, indicating that this method is useful for monitoring HDACi efficacy in clinical trials involving humans with solid tumors (71).

The prediction of the clinical response of various drugs is a challenging task in the development of cancer drugs. This requires in-depth knowledge of the molecular mechanism action of the specific drug in question. Despite the intense research in the field of epigenetics and mechanisms of HDACis, their precise molecular mechanisms are still rather unknown. However, in a recent study, Dejligbjerg et al. (72) identified 16 potential genes that in the literature were proposed to be involved in HDACi sensitivity. Four of these genes, ornithine decarboxylase (ODC1), v-ski sarcoma viral oncogene homologue (SKI), signal transducer and activator of transcription 1 (STAT1), and thymidylate synthetase (TYMS), showed a correlation in expression levels with Belinostat sensitivity, indicating their usefulness as markers for the clinical outcome of HDACi sensitivity. Unfortunately, whether or not the regulation of these genes is reflected at the protein level was not investigated. Furthermore, the study was performed in human cancer cell lines and needs further validation in human trials of various cancers.

7. Conclusion and Perspectives

Despite intense research during the last decade, we are only in the beginning of understanding the regulatory role of the variety of modifications in epigenetics. It is clear that specific patterns of histone modifications regulate specific gene readouts leading to cellular events such as cell cycle arrest and apoptosis. In addition, recent research has led to the realization that nonhistone proteins and transcription factors are also targets of HATs and that HDACs play a crucial role in controlling the epigenetic gene readout leading to cell cycle arrest and apoptosis, specifically in cancer cells. These cellular events are induced by the inhibition of HDACs and form the basis for the development of HDACis as anticancer agents. These agents also provide an excellent basis for epigenetic research – and unraveling their exact molecular "mechanism of action" will undoubtedly provide a basis for the development of more efficient drugs, provide markers for monitoring drug efficacy in clinical trials, and form the basis for individualized cancer treatment.

Historically, immunochemical methods have been the methods of choice in epigenetic research. Recent developments in mass spectrometry have enabled the discovery and quantification of multiple site protein modifications and have led to the characterization of myriads of histone posttranslational modifications. By contrast, immunochemical methods are limited by antibody specificity and are therefore primarily used for the quantification of single modification sites at once.

Conversely, chromatin immunoprecipitation (ChIP) allows the assessment of gene-specific variances in histone modification patterns. Basically, the principles of this technique rely on in vivo cross-linking of protein-DNA complexes using formaldehyde or UV-radiation followed by extraction of the cross-linked chromatin, disruption by ultra sonication, and immunoprecipitation of DNA-cross-linked protein using highly specific antibodies raised against the protein of interest or – in histone research – the histone modification of interest. After reversal of the cross-linking, the DNA fragment is purified and subjected to quantitative PCR or DNA microarray (ChIP-on-chip) analysis. In principle, these approaches allow the investigation of gene-specific patterns of histone modifications for specific DNA elements (ChIP) or the investigation of the histone modification pattern at nucleosomal resolution on entire genomes (ChIP-on-chip). ChIP has been used for the characterization of the binding of histone proteins to the *HSP70* gene during heat shock (73) and ChIP-on-chip experiments using site-specific antibodies. In another study, ChIP was used to demonstrate the link between hypoacetylated histones and silent chromatin (74). High-resolution genome-wide mapping of histone acetylation using ChIP-on-chip analyses revealed significant localized differences, where both acetylation and methylation was found to be associated with transcriptional activation but with significant local differences in modification abundances. Acetylations occur predominantly in the beginning of the genes, whereas methylations can occur throughout transcribed regions (75).

Recent achievements in histone biology and epigenetics are clearly linked to the recent developments in mass spectrometry-based proteomics and immunochemical methods, and these technologies will undoubtedly dominate future epigenetic research. The protein output of ChIP experiments is most often assessed by specific antibodies targeting known proteins and protein modifications, leaving a source of unexploited information such as unknown proteins and protein modifications behind. The integration of mass spectrometry-based technologies with ChIP methods will allow the discovery of not only novel histone modification patterns linked to specific genomic regions but also modification-specific roles of novel nonhistone proteins involved in diseases such as cancer. A major obstacle in combining these technologies is, however, the limited amount of protein available for MS analysis, as the quantity of material required for DNA analysis is much lower (orders of magnitude) than that required for MS analysis. The future challenge is the refinement of sample preparation and enrichment procedures prior to MS analysis.

References

1. Strahl BD, Allis CD. (2000) The language of covalent histone modifications. *Nature* 403(6765):41–45.
2. Berger SL. (2002) Histone modifications in transcriptional regulation. *Curr Opin Genet Dev* 12(2):142–148.
3. Wisniewski JR, Zougman A, Mann M. (2008) Nepsilon-formylation of lysine is a widespread post-translational modification of nuclear proteins occurring at residues involved in regulation of chromatin function. *Nucleic Acids Res* 36(2):570–577.
4. Shiio Y, Eisenman RN. (2003) Histone sumoylation is associated with transcriptional repression. *Proc Natl Acad Sci USA* 100(23):13225–13230.
5. Fischle W, Wang Y, Allis CD. (2003) Binary switches and modification cassettes in histone biology and beyond. *Nature* 425(6957):475–479.
6. Turner BM. (2000) Histone acetylation and an epigenetic code. *Bioessays* 22(9):836–845.
7. Fischle W, Tseng BS, Dormann HL, Ueberheide BM, Garcia BA, Shabanowitz J, Hunt DF, Funabiki H, Allis CD. (2005) Regulation of HP1-chromatin binding by histone H3 methylation and phosphorylation. *Nature* 438(7071):1116–1122.
8. Agalioti T, Chen G, Thanos D. (2002) Deciphering the transcriptional histone acetylation code for a human gene. *Cell* 111(3): 381–392.
9. Turner BM. (2002) Cellular memory and the histone code. *Cell* 111(3):285–291.
10. Johnstone RW. (2002) Histone-deacetylase inhibitors: novel drugs for the treatment of cancer. *Nat Rev Drug Discov* 1(4):287–299.
11. Marks PA, Richon VM, Rifkind RA. (2000) Histone deacetylase inhibitors: inducers of differentiation or apoptosis of transformed cells. *J Natl Cancer Inst* 92(15):1210–1216.
12. Bolden JE, Peart MJ, Johnstone RW. (2006) Anticancer activities of histone deacetylase inhibitors. *Nat Rev Drug Discov* 5(9):769–784.
13. Gu W, Roeder RG. (1997) Activation of p53 sequence-specific DNA binding by acetylation of the p53 C-terminal domain. *Cell* 90(4):595–606.
14. Glozak MA, Seto E. (2007) Histone deacetylases and cancer. *Oncogene* 26(37):5420–5432.
15. Murray K. (1964) The occurrence of epsilon-n-methyl lysine in histones. *Biochemistry* 3:10–15.
16. Allfrey VG, Faulkner R, Mirsky AE. (1964) Acetylation and methylation of histones and their possible role in the regulation of RNA synthesis. *Proc Natl Acad Sci USA* 51: 786–794.

17. Ord MG, Stocken LA. (1966) Metabolic properties of histones from rat liver and thymus gland. *Biochem J* 98(3):888–897.
18. Stevely WS, Stocken LA. (1966) Phosphorylation of rat-thymus histone. *Biochem J* 100(2):20C–21C.
19. Ueda K, Omachi A, Kawaichi M, Hayaishi O. (1975) Natural occurrence of poly(ADP-ribosyl) histones in rat liver. *Proc Natl Acad Sci USA* 72(1):205–209.
20. Cheung P. (2004) Generation and characterization of antibodies directed against di-modified histones, and comments on antibody and epitope recognition. *Methods Enzymol* 376:221–234.
21. Beck HC, Nielsen EC, Matthiesen R, Jensen LH, Sehested M, Finn P, Grauslund M, Hansen AM, Jensen ON. (2006) Quantitative proteomic analysis of post-translational modifications of human histones. *Mol Cell Proteomics* 5(7):1314–1325.
22. Mikesh LM, Ueberheide B, Chi A, Coon JJ, Syka JE, Shabanowitz J, Hunt DF. (2006) The utility of ETD mass spectrometry in proteomic analysis. *Biochim Biophys Acta* 1764(12):1811–1822.
23. Zhang K, Yau PM, Chandrasekhar B, New R, Kondrat R, Imai BS, Bradbury ME. (2004) Differentiation between peptides containing acetylated or tri-methylated lysines by mass spectrometry: an application for determining lysine 9 acetylation and methylation of histone H3. *Proteomics* 4(1):1–10.
24. Trelle MB, Jensen ON. (2008) Utility of immonium ions for assignment of epsilon-N-acetyllysine-containing peptides by tandem mass spectrometry. *Anal Chem* 80(9):3422–3430.
25. Zhang K, Williams KE, Huang L, Yau P, Siino JS, Bradbury EM, Jones PR, Minch MJ, Burlingame AL. (2002) Histone acetylation and deacetylation: identification of acetylation and methylation sites of HeLa histone H4 by mass spectrometry. *Mol Cell Proteomics* 1(7):500–508.
26. Ren C, Zhang L, Freitas MA, Ghoshal K, Parthun MR, Jacob ST. (2005) Peptide mass mapping of acetylated isoforms of histone H4 from mouse lymphosarcoma cells treated with histone deacetylase (HDACs) inhibitors. *J Am Soc Mass Spectrom* 16(10):1641–1653.
27. Zhang L, Eugeni EE, Parthun MR, Freitas MA. (2003) Identification of novel histone post-translational modifications by peptide mass fingerprinting. *Chromosoma* 112(2): 77–86.
28. Bonenfant D, Coulot M, Towbin H, Schindler P, van Oostrum J. (2006) Characterization of histone H2A and H2B variants and their post-translational modifications by mass spectrometry. *Mol Cell Proteomics* 5(3):541–552.
29. Garcia BA, Hake SB, Diaz RL, Kauer M, Morris SA, Recht J, Shabanowitz J, Mishra N, Strahl BD, Allis CD, Hunt DF. (2007) Organismal differences in post-translational modifications in histones H3 and H4. *J Biol Chem* 282(10):7641–7655.
30. Hake SB, Garcia BA, Duncan EM, Kauer M, Dellaire G, Shabanowitz J, Bazett-Jones DP, Allis CD, Hunt DF. (2006) Expression patterns and post-translational modifications associated with mammalian histone H3 variants. *J Biol Chem* 281(1):559–568.
31. Garcia BA, Barber CM, Hake SB, Ptak C, Turner FB, Busby SA, Shabanowitz J, Moran RG, Allis CD, Hunt DF. (2005) Modifications of human histone H3 variants during mitosis. *Biochemistry* 44(39):13202–13213.
32. Morris SA, Rao B, Garcia BA, Hake SB, Diaz RL, Shabanowitz J, Hunt DF, Allis CD, Lieb JD, Strahl BD. (2007) Identification of histone H3 lysine 36 acetylation as a highly conserved histone modification. *J Biol Chem* 282(10):7632–7640.
33. Medzihradszky KF, Zhang X, Chalkley RJ, Guan S, McFarland MA, Chalmers MJ, Marshall AG, Diaz RL, Allis CD, Burlingame AL. (2004) Characterization of Tetrahymena histone H2B variants and posttranslational populations by electron capture dissociation (ECD) Fourier transform ion cyclotron mass spectrometry (FT-ICR MS). *Mol Cell Proteomics* 3(9):872–886.
34. Xu F, Zhang K, Grunstein M. (2005) Acetylation in histone H3 globular domain regulates gene expression in yeast. *Cell* 121(3):375–385.
35. Zhang K, Tang H. (2003) Analysis of core histones by liquid chromatography-mass spectrometry and peptide mapping. *J Chromatogr B Anal Technol Biomed Life Sci* 783(1):173–179.
36. Zhang K, Tang H, Huang L, Blankenship JW, Jones PR, Xiang F, Yau PM, Burlingame AL. (2002) Identification of acetylation and methylation sites of histone H3 from chicken erythrocytes by high-accuracy matrix-assisted laser desorption ionization-time-of-flight, matrix-assisted laser desorption ionization-postsource decay, and nanoelectrospray ionization tandem mass spectrometry. *Anal Biochem* 306(2): 259–269.
37. Cocklin RR, Wang M. (2003) Identification of methylation and acetylation sites on mouse histone H3 using matrix-assisted

laser desorption/ionization time-of-flight and nanoelectrospray ionization tandem mass spectrometry. *J Protein Chem* 22(4): 327–334.
38. Maile T, Kwoczynski S, Katzenberger RJ, Wassarman DA, Sauer F. (2004) TAF1 activates transcription by phosphorylation of serine 33 in histone H2B. *Science* 304(5673):1010–1014.
39. Smith CM, Gafken PR, Zhang Z, Gottschling DE, Smith JB, Smith DL. (2003) Mass spectrometric quantification of acetylation at specific lysines within the amino-terminal tail of histone H4. *Anal Biochem* 316(1):23–33.
40. Ong SE, Blagoev B, Kratchmarova I, Kristensen DB, Steen H, Pandey A, Mann M. (2002) Stable isotope labeling by amino acids in cell culture, SILAC, as a simple and accurate approach to expression proteomics. *Mol Cell Proteomics* 1(5):376–386.
41. Bantscheff M, Dumpelfeld B, Kuster B. (2004) Femtomol sensitivity post-digest ^{18}O labeling for relative quantification of differential protein complex composition. *Rapid Commun Mass Spectrom* 18(8): 869–876.
42. Ross PL, Huang YN, Marchese JN, Williamson B, Parker K, Hattan S, Khainovski N, Pillai S, Dey S, Daniels S, Purkayastha S, Juhasz P, Martin S, Bartlet-Jones M, He F, Jacobson A, Pappin DJ. (2004) Multiplexed protein quantitation in *Saccharomyces cerevisiae* using amine-reactive isobaric tagging reagents. *Mol Cell Proteomics* 3(12):1154–1169.
43. Boyne MT, 2nd, Pesavento JJ, Mizzen CA, Kelleher NL. (2006) Precise characterization of human histones in the H2A gene family by top down mass spectrometry. *J Proteome Res* 5(2):248–253.
44. Plumb JA, Finn PW, Williams RJ, Bandara MJ, Romero MR, Watkins CJ, La Thangue NB, Brown R. (2003) Pharmacodynamic response and inhibition of growth of human tumor xenografts by the novel histone deacetylase inhibitor PXD101. *Mol Cancer Ther* 2(8):721–728.
45. Ginsburg E, Salomon D, Sreevalsan T, Freese E. (1973) Growth inhibition and morphological changes caused by lipophilic acids in mammalian cells. *Proc Natl Acad Sci USA* 70(8):2457–2461.
46. Altenburg BC, Via DP, Steiner SH. (1976) Modification of the phenotype of murine sarcoma virus-transformed cells by sodium butyrate. Effects on morphology and cytoskeletal elements. *Exp Cell Res* 102(2):223–231.
47. Boffa LC, Vidali G, Mann RS, Allfrey VG. (1978) Suppression of histone deacetylation in vivo and in vitro by sodium butyrate. *J Biol Chem* 253(10):3364–3366.
48. Gregoretti IV, Lee YM, Goodson HV. (2004) Molecular evolution of the histone deacetylase family: functional implications of phylogenetic analysis. *J Mol Biol* 338(1): 17–31.
49. Jenuwein T, Allis CD. (2001) Translating the histone code. *Science* 293(5532): 1074–1080.
50. Maison C, Bailly D, Peters AH, Quivy JP, Roche D, Taddei A, LachnerM, Jenuwein T, Almouzni G. (2002) Higher-order structure in pericentric heterochromatin involves a distinct pattern of histone modification and an RNA component. *Nat Genet* 30(3): 329–334.
51. Fraga MF, Ballestar E, Villar-Garea A, Boix-Chornet M, Espada J, Schotta G, Bonaldi T, Haydon C, Ropero S, Petrie K, Iyer NG, Perez-Rosado A, Calvo E, Lopez JA, Cano A, Calasanz MJ, Colomer D, Piris MA, Ahn N, Imhof A, Caldas C, Jenuwein T, Esteller M. (2005) Loss of acetylation at Lys16 and trimethylation at Lys20 of histone H4 is a common hallmark of human cancer. *Nat Genet* 37(4):391–400.
52. Seligson DB, Horvath S, Shi T, Yu H, Tze S, Grunstein M, Kurdistani SK. (2005) Global histone modification patterns predict risk of prostate cancer recurrence. *Nature* 435(7046):1262–1266.
53. Leder A, Orkin S, Leder P. (1975) Differentiation of erythroleukemic cells in the presence of inhibitors of DNA synthesis. *Science* 190(4217):893–894.
54. Riggs MG, Whittaker RG, Neumann JR, Ingram VM. (1977) n-Butyrate causes histone modification in HeLa and Friend erythroleukaemia cells. *Nature* 268(5619): 462–464.
55. Hassig CA, Fleischer TC, Billin AN, Schreiber SL, Ayer DE. (1997) Histone deacetylase activity is required for full transcriptional repression by mSin3A. *Cell* 89(3):341–347.
56. Brehm A, Miska EA, McCance DJ, Reid JL, Bannister AJ, Kouzarides T. (1998) Retinoblastoma protein recruits histone deacetylase to repress transcription. *Nature* 391(6667):597–601.
57. Boulaire J, Fotedar A, Fotedar R. (2000) The functions of the cdk-cyclin kinase inhibitor p21WAF1. *Pathol Biol (Paris)* 48(3):190–202.
58. Richon VM, Sandhoff TW, Rifkind RA, Marks PA. (2000) Histone deacetylase

inhibitor selectively induces p21WAF1 expression and gene-associated histone acetylation. *Proc Natl Acad Sci USA* 97(18):10014–10019.
59. Glozak MA, Sengupta N, Zhang X, Seto E. (2005) Acetylation and deacetylation of non-histone proteins. *Gene* 363: 15–23.
60. Zhao Y, Lu S, Wu L, Chai G, Wang H, Chen Y, Sun J, Yu Y, Zhou W, Zheng Q, Wu M, Otterson GA, Zhu WG. (2006) Acetylation of p53 at lysine 373/382 by the histone deacetylase inhibitor depsipeptide induces expression of p21(Waf1/Cip1). *Mol Cell Biol* 26(7):2782–2790.
61. Roy S, Tenniswood M. (2007) Site-specific acetylation of p53 directs selective transcription complex assembly. *J Biol Chem* 282(7):4765–4771.
62. Cohen HY, Lavu S, Bitterman KJ, Hekking B, Imahiyerobo TA, Miller C, Frye R, Ploegh H, Kessler BM, Sinclair DA. (2004) Acetylation of the C terminus of Ku70 by CBP and PCAF controls Bax-mediated apoptosis. *Mol Cell* 13(5):627–638.
63. Glaser KB. (2007) HDAC inhibitors: clinical update and mechanism-based potential. *Biochem Pharmacol* 74(5):659–671.
64. Longley DB, Harkin DP, Johnston PG. (2003) 5-fluorouracil: mechanisms of action and clinical strategies. *Nat Rev Cancer* 3(5):330–338.
65. Tumber A, Collins LS, Petersen KD, Thougaard A, Christiansen SJ, Dejligbjerg M, Jensen PB, Sehested M, Ritchie JW. (2007) The histone deacetylase inhibitor PXD101 synergises with 5-fluorouracil to inhibit colon cancer cell growth in vitro and in vivo. *Cancer Chemother Pharmacol* 60(2):275–283.
66. Ocker M, Alajati A, Ganslmayer M, Zopf S, Luders M, Neureiter D, Hahn EG, Schuppan D, Herold C. (2005) The histone-deacetylase inhibitor SAHA potentiates proapoptotic effects of 5-fluorouracil and irinotecan in hepatoma cells. *J Cancer Res Clin Oncol* 131(6): 385–394.
67. Zhang X, Yashiro M, Ren J, Hirakawa K. (2006) Histone deacetylase inhibitor, trichostatin A, increases the chemosensitivity of anticancer drugs in gastric cancer cell lines. *Oncol Rep* 16(3):563–568.
68. Byrd JC, Marcucci G, Parthun MR, Xiao JJ, Klisovic RB, Moran M, Lin TS, Liu S, Sklenar AR, Davis ME, Lucas DM, Fischer B, Shank R, Tejaswi SL, Binkley P, Wright J, Chan KK, Grever MR. (2005) A phase 1 and pharmacodynamic study of depsipeptide (FK228) in chronic lymphocytic leukemia and acute myeloid leukemia. *Blood* 105(3)959–967.
69. Ryan QC, Headlee D, Acharya M, Sparreboom A, Trepel JB, Ye J, Figg WD, Hwang K, Chung EJ, Murgo A, Melillo G, Elsayed Y, Monga M, Kalnitskiy M, Zwiebel J, Sausville EA. (2005) Phase I and pharmacokinetic study of MS-275, a histone deacetylase inhibitor, in patients with advanced and refractory solid tumors or lymphoma. *J Clin Oncol* 23(17):3912–3922.
70. Munster P, Marchion D, Bicaku E, Schmitt M, Lee JH, DeConti R, Simon G, Fishman M, Minton S, Garrett C, Chiappori A, Lush R, Sullivan D, Daud A. (2007) Phase I trial of histone deacetylase inhibition by valproic acid followed by the topoisomerase II inhibitor epirubicin in advanced solid tumors: a clinical and translational study. *J Clin Oncol* 25(15):1979–1985.
71. Marquard L, Petersen KD, Persson M, Hoff KD, Jensen PB, Sehested M. (2008) Monitoring the effect of Belinostat in solid tumors by H4 acetylation. *APMIS* 116(5):382–392.
72. Dejligbjerg M, Grauslund M, Christensen IJ, Tjornelund J, Buhl Jensen P, Sehested M. (2008) Identification of predictive biomarkers for the histone deacetylase inhibitor Belinostat in a panel of human cancer cell lines. *Cancer Biomark* 4(2):101–109.
73. Solomon MJ, Larsen PL, Varshavsky A. (1988) Mapping protein-DNA interactions in vivo with formaldehyde: evidence that histone H4 is retained on a highly transcribed gene. *Cell* 53(6):937–947.
74. Braunstein M, Rose AB, Holmes SG, Allis CD, Broach JR. (1993) Transcriptional silencing in yeast is associated with reduced nucleosome acetylation. *Genes Dev* 7(4):592–604.
75. Pokholok DK, Harbison CT, Levine S, Cole M, Hannett NM, Lee TI, Bell GW, Walker K, Rolfe PA, Herbolsheimer E, Zeitlinger J, Lewitter F, Gifford DK, Young RA. (2005) Genome-wide map of nucleosome acetylation and methylation in yeast. *Cell* 122(4):517–527.

Chapter 14

Computational Approaches to Metabolomics

David S. Wishart

Abstract

This chapter is intended to familiarize readers with the field of metabolomics and some of the algorithms, data analysis strategies, and computer programs used to analyze or interpret metabolomic data. Specifically, this chapter provides a brief overview of the experimental approaches and applications of metabolomics followed by a description of the spectral and statistical analysis tools for metabolomics. The chapter concludes with a discussion of the resources that can be used to interpret and analyze metabolomic data at a biological or clinical level. Emerging needs, challenges, and recent progress being made in these areas are also discussed.

Key words: Metabolomics, bioinformatics, cheminformatics, data analysis.

1. Introduction

Metabolomics (also known as *metabonomics* or *metabolic profiling*) is a newly emerging field of omics research concerned with the high-throughput identification and quantification of the small molecule metabolites in the metabolome (1). The metabolome can be defined as the complete collection of all small molecule (<1500-Da) metabolites (amino acids, sugars, organic acids, bases, vitamins, lipids, etc.) found in a specific cell, organ, or organism (2). It is a close counterpart to the genome, the transcriptome, and the proteome. Because of its unique focus on small molecules and small molecule interactions, metabolomics is finding widespread applications in a variety of clinically important areas, including diabetes (3), osteoarthritis (4), genetic disease diagnosis and monitoring (5, 6), infectious disease diagnosis (7), clinical toxicology (8), and organ transplant monitoring (9). Indeed, metabolomics is evolving into a very important player in the clinical "omics" field.

R. Matthiesen (ed.), *Bioinformatics Methods in Clinical Research*, Methods in Molecular Biology 593,
DOI 10.1007/978-1-60327-194-3_14, © Humana Press, a part of Springer Science+Business Media, LLC 2010

Because metabolomics is a relatively new addition to the "omics" sciences, it is still developing some of its basic computational infrastructure (10). Whereas most data in the field of proteomics, genomics, or transcriptomics are readily available and easily analyzed through online electronic databases, most metabolomic data are still housed in books, journals, and other paper archives. Metabolomics also differs from the other "omics" sciences because of its strong emphasis on chemicals and analytical chemistry techniques such as mass spectrometry (MS), nuclear magnetic resonance (NMR) spectroscopy, and chromatography. As a result, some of the analytical software used in metabolomics, particularly as it relates to metabolite identification, is often a little different than the software used in genomics, proteomics, or transcriptomics (10).

The field of metabolomics is not only concerned with the identification and quantification of metabolites, but it is also concerned with relating metabolite data to genes, proteins, pathways, physiology, and phenotypes. As a result, metabolomics requires that whatever chemical information it generates must be linked to both biochemical causes and physiological consequences. This means that computational approaches to metabolomics must combine two very different informatics disciplines: bioinformatics and cheminformatics.

This chapter is intended to familiarize readers with the field of metabolomics along with some of the algorithms, data analysis strategies, and computer programs used to analyze or interpret metabolomic data. The chapter is divided into five sections. **Section 1** is intended to define metabolomics in the context of other "omics" technologies. **Section 2** provides a brief overview of the experimental approaches used in clinical metabolomics. **Section 3** gives a detailed description of the spectral and statistical analysis tools commonly used in collecting and processing metabolomic data. The last two sections describe some of the resources (databases and modeling software) that can be used to interpret, visualize, and analyze metabolomic data at a biological or clinical level.

2. Clinical Metabolomics: Experimental Methods

Metabolic profiling, in one form or another, has been a part of clinical practice for thousands of years. As far back as the fifth century BC, Hippocrates described the diagnosis and detection of diseases through a simple sensory analysis of urine (color, taste, smell). The analysis of biofluids eventually became more quantitative in the mid-19th century, when clinicians began to identify and quantify biofluid constituents and associate them with various medical conditions (11). However, it wasn't until the early 20th century that clinical chemistry and metabolic profiling became a part of routine medical practice with the development of simple colorimetric tests and dedicated instruments to quantify metabolites in blood and urine (12). Nowadays, blood and urine tests, which offer from 5 to 15 different chemical readouts, are routinely performed by multicomponent clinical analyzers or by simple paper strip tests (13). However, what distinguishes metabolomics from clinical chemistry is the fact that in metabolomics one is not attempting to characterize a few compounds at a time, but literally dozens or even hundreds of compounds at a time. By being able to measure so many metabolites at once, it is possible to get a far more comprehensive picture of what is happening to a patient's physiology or metabolism. Indeed, metabolomics provides a metabolic profile or "signature" that is as potentially as informative as the genetic signature from a gene chip. What's more, because metabolic responses are often measured in seconds or minutes (whereas other types of physiological responses are typically measured in days or weeks), metabolomic measurements can potentially yield important physiological information that is not normally accessible with gene chips, 2D gels, or tissue biopsies (9).

High-throughput metabolomics only became possible in the late 1990s as a result of technological breakthroughs in small molecule separation and identification. These include the development of robust, very high-resolution mass spectrometry instruments for precise mass determination, the widespread deployment of high-resolution, high-throughput NMR spectrometers, improvements in capillary electrophoresis (CE), the invention of ultra-high-pressure liquid chromatography (UPLC), and the development of multidimensional chromatographic systems for rapid compound separation (14). Equally important to the rise of metabolomics has been the creation of new software programs to rapidly process spectral or chromatographic data along with the development of dedicated electronic databases containing detailed descriptive and spectral information on the constituent chemicals found in different metabolomes (15–17). These computer programs and databases will be discussed in more detail in **Sections 3** and **4** of this chapter.

In clinical metabolomics, one is almost always working with a biofluid [urine, saliva, serum, sinovial fluid, cerebrospinal fluid (CSF)] or a fluidized tissue extract. The preference of working with biofluids over tissues is primarily dictated by the fact that fluids are far easier to process and analyze with today's NMR, MS, or HPLC instruments. Likewise, the collection of biofluids is generally much less invasive than the collection of tissues. Biofluid analysis is always done with the assumption that the chemicals found in different biofluids are largely reflective of the physiological state of the organ that produces, or is bathed in, that fluid. Hence, urine reflects processes going on the kidney, bile – the liver, cerebrospinal fluid – the brain, and so on. Blood is a special biofluid, as it potentially reflects all processes going on in all organs. This can be both a blessing and a curse, as metabolite perturbations in the blood, while easily detectable, cannot be easily traced to a specific organ or a specific cause.

Some biofluids, such as urine and CSF, are essentially protein-free and can be used almost immediately without any further workup. Other fluids contain high-molecular-weight proteins (saliva, serum, plasma, and tissue extracts) and lipoprotein particles (serum and plasma). These fluids usually require further extraction or filtration to remove the high-molecular-weight components, especially for MS- and NMR-based metabolomic studies. For many MS-based metabolomic studies, the removal of inorganic salts (sodium, potassium, phosphate) via chromatographic separation is also critical to obtaining interpretable spectra.

Regardless of which biofluid is being used, care must be taken to avoid bacterial contamination or the addition of any adulterating chemical additives or preservatives, such as EDTA, heparin, isopropyl alcohol, glycerol, or organic buffers such as Tris, MOPS, or HEPES. Signals from these additives can mask important metabolite resonances or can be mistaken as "unknown" endogenous metabolites. As a general rule, biological samples should be stored at –80 °C in sterile glass or plastic containers with 0.02% azide added as a bacteriocide. Leaving samples out at room temperature or even in refrigerators for extended periods of time can lead to noticeable changes in metabolite concentrations (18). Several excellent papers have appeared recently that elaborate on the best practices for handling a number of different types of biological samples (18–20).

Once a clinical sample is prepared, it may then be split into several aliquots and subject to detailed metabolomic analysis using a variety of instruments. In general, there are two experimental approaches to metabolite analysis: (1) global metabolic profiling and (2) targeted metabolic profiling. Global metabolic profiling is an experimental technique that attempts to measure all detectable metabolites in a sample without selective enrichment or concentration (21). On the other hand, targeted metabolic profiling or targeted metabolomics attempts to measure certain classes of metabolites using selective enrichment or selective concentration via solid-phase extraction, liquid–liquid extraction, chemical derivatization, or chromatographic partitioning (22). Global metabolic profiling is generally a much simpler and a much higher throughput technique, but it lacks the sensitivity of targeted metabolic profiling. Typically, the lower limit of detection for global metabolomic techniques is about 100–200 nM, while for targeted methods metabolite concentrations, as low as 1 pM can be detected (20–22).

Regardless of whether a global or targeted approach is chosen, the vast majority of metabolomic analyses are eventually performed on NMR, GC-MS, and/or LC-MS instruments. For GC-MS analyses, the samples must be derivatized via trimethylsilylation prior to injection into the gas chromatograph. On the other hand, for NMR and LC-MS studies, the samples are usually injected (for MS) or inserted (for NMR) into the instrument with almost no further workup. However, the use of selective isotopic labeling techniques for LC-MS methods (which roughly parallels the concept of trimethylsilyl derivatization in GC-MS) appears to be a promising new approach to improve the separation, identification, and quantification of metabolites by LC-MS (23).

Table 14.1 provides a brief description of the advantages and disadvantages of the three major technologies (NMR, GC-MS, and LC-MS) used in modern metabolomic studies. Hybrid systems, such as LC-MS-NMR platforms, also exist. These hybrid systems, while relatively rare, can often take advantage of the strengths of each of the component technologies. As a general rule, NMR is typically capable of detecting 50 and 75 compounds in a given human biofluid, with a lower sensitivity limit of about 1 μM (24). Most of the compounds detected by NMR are intrinsically polar molecules, such as organic acids, sugars, amino acids, and small amines. GC-MS is also capable of detecting between 50 and 150 compounds (depending on the biofluid), with a lower sensitivity limit of about 100 nM (20). GC-MS provides relatively broad metabolite coverage, with amino acids, sugars, organic acids, phosphorylated compounds, fatty acids, and even cholesterol being routinely detected. Because of the exquisite sensitivity of today's MS instruments, LC-MS methods can detect hundreds or even thousands of "features" (14). However, the number of compounds that can be positively identified is typically much less (~100). LC-MS methods are particularly useful in targeted metabolomic studies of human serum lipids, where up to 400 different lipids and fatty acids can be detected and quantified (22). Recent studies have shown that the combination of multiple detection technologies (GC-MS plus NMR plus LC-MS) gives a far more complete picture of the metabolome than just a single detection technology (25). Indeed, the number of "shared" compounds identified or detected by one method (LC-MS) versus another (GC-MS) is often less than 50%, and in some cases can be as little as 20%.

Additional details pertaining to the data collection conditions and instrument operation protocols for LC-MS, GC-MS, and NMR are beyond the scope of this chapter, but several excellent technical reviews have been published that cover these details (14, 19, 20). Regardless of which technology is used, the net output

Table 14.1
A comparison of different metabolomics technologies

Technology	Advantages	Disadvantages
GC-mass spectrometry	– Robust, mature technology – Relatively inexpensive – Quantitative (with calibration) – Modest sample size need – Good sensitivity – Large body of software and databases for metabolite ID – Detects most organic and some inorganic molecules – Excellent separation reproducibility	– Sample not recoverable – Requires sample derivatization – Requires separation – Slow (20–30 min/sample) – Cannot be used in imaging – Novel compound ID is difficult
LC-mass spectrometry	– Superb sensitivity – Very flexible technology – Detects most organic and some inorganic molecules – Minimal sample size requirement – Can be used in metabolite imaging (MALDI) – Can be done without separation (direct injection) – Has potential for detecting largest portion of metabolome	– Sample not recoverable – Not very quantitative – Expensive instrumentation – Slow (20–30 min/sample) – Poor separation resolution and reproducibility (vs. GC) – Less robust instrumentation than NMR or GC-MS – Limited body of software and databases for metabolite ID – Novel compound ID is difficult
NMR Spectroscopy	– Quantitative – Nondestructive – Fast (2–3 min/sample) – Requires no derivatization – Requires no separation – Detects all organic classes – Allows ID of novel chemicals – Robust, mature technology – Can be used for metabolite imaging (fMRI) – Large body of software and databases for metabolite ID – Compatible with liquids and solids	– Not very sensitive – Expensive instrumentation – Large instrument footprint – Cannot detect or ID salts and inorganic ions – Cannot detect nonprotonated compounds – Requires larger (0.5-ml) samples

from almost any metabolomics run is a series of spectra with hundreds of different peaks or spectral features. The computational challenge is to figure out what the detected peaks are and exactly what they mean in a biological context.

3. Spectral and Statistical Analysis Tools for Metabolomics

There are two very distinct schools-of-thought about how metabolomic data should be processed and interpreted (**Fig. 14.1**). In one version (called *chemometric* or *nonquantitative* metabolomics), the compounds are not initially identified – only their spectral patterns and intensities are recorded, statistically compared, clustered, or correlated, and used to make diagnoses, identify phenotypes, or draw conclusions (26). In the other version (called *quantitative* metabolomics), the compounds are actually identified and quantified. The resulting list of compounds and concentrations is then used to make diagnoses, identify phenotypes, or draw conclusions (2, 27). The principle difference between the two approaches lies at the point at which the metabolite identifications are made. In quantitative metabolomics, one identifies metabolites in the first step, while in chemometric approaches to metabolomics, one typically identifies metabolites in the last step (if at all). The second difference lies in their emphasis on metabolite identification, with quantitative metabolomics being entirely dependent upon it, while chemometric methods treat metabolite identification more as an afterthought. Both methods have their advantages and disadvantages.

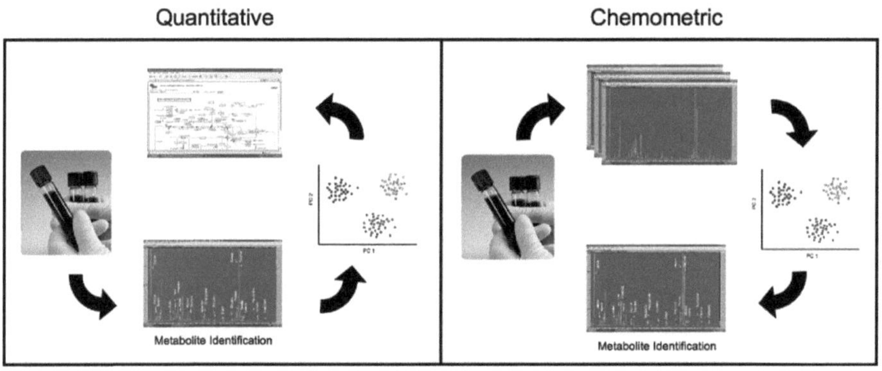

Fig. 14.1. A schematic illustration of the differences and similarities between quantitative and chemometric approaches to metabolomics.

In particular, the key strengths of chemometric profiling lie in its amenability to automation and its nonbiased assessment of metabolite data. However, for chemometric methods to work, it is critical to have a large number of spectra from many different samples collected and processed identically. It is also important to ensure careful experimental design and strict sample uniformity (i.e., uniform diet, uniform environment, uniform sample collection, uniform sample workup). For instance, inbred (genetically identical) animals maintained in metabolic cages or plants grown in identical laboratory environmental chambers are frequently used. The use of a large number of spectra, in combination with the strict sample uniformity, helps reduce the statistical problems arising from improper spectral alignment, instrumental variations, baseline distortion, line-width differences, or natural instrument drift (26).

In contrast to chemometric metabolomics, a key advantage of quantitative metabolomics is that it does not require the collection of identical sets of cells, tissues, or lab animals, and so it is more amenable to human studies or studies that require less day-to-day monitoring (27). In other words, there is no need for specially designed metabolic chambers. On the other hand, two notable disadvantages of quantitative metabolomics are (1) its relative lack of automation and (2) the fact that technologies do not yet exist to identify/quantify all detectable metabolites in any given biological sample. Without comprehensive metabolite identification, it is possible to introduce some bias into subsequent biological or clinical interpretations.

In the following sections, we will discuss both chemometric and quantitative metabolomics in more detail and illustrate how the two approaches may be combined to generate even more useful interpretations of metabolomic data.

3.1. Chemometric Methods in Metabolomics

Chemometrics can be defined as the application of mathematical, statistical, graphical, or symbolic methods to maximize the information that can be extracted from chemical or spectral data (26, 28). Chemometric approaches for spectral analysis emerged in the 1980s and are primarily used to extract useful information from complex spectra consisting of many hard-to-identify or unknown components (28). Chemometric approaches can also used to identify statistically significant differences between large groups of spectra or large groups of chromatograms collected on different samples or under different conditions. As useful as chemometrics is, it is important to remember that chemometric software is not designed to identify or quantify chemical compounds from MS, HPLC, or NMR traces. Instead, a completely different type of software (called *spectral fitting* or *spectral deconvolution* software – *see* **Section 3.2**) must be used for this purpose.

For chemometric methods to be successful, they generally need a relatively large number ($N > 40$) of samples. Therefore, to facilitate metabolomic analysis via chemometric approaches, it is essential to collect a number of spectra (HPLC traces, total ion chromatograms, GC-MS retention profiles, NMR spectra – depending on the technology available) from a number of different biological or clinical samples. These spectral traces should contain "features" that have both intensity data and peak position data (i.e., x, y-coordinates). Once these spectra are collected, it is critical to scale, align, and/or center them to permit proper statistical comparisons. This scaling can be done either through spectral or chromatographic alignment (29, 30) or through a technique called *spectral binning* (27). The general concept behind spectral alignment is illustrated in **Fig. 14.2**. In this approach, peaks are computationally matched and then digitally shifted into alignment to create "synthetic" spectra that can all be superimposed. Programs such as MZmine and XCMS are particularly useful for this process (30). In contrast to spectral alignment, spectral binning involves dividing each input spectrum into smaller regions or bins. This spectral partitioning or spectral digitizing process, like spectral alignment, allows specific features, peaks, or peak clusters in a multipeak spectrum or multipeak chromatogram to be systematically compared. Once the peaks are binned or aligned, the peak intensities (or total area under the curve) in each bin or under each peak can be tabulated and analyzed using multivariate statistical analysis. This "divide-and-conquer" approach allows spectral components to be quantitatively compared within a single spectrum or between multiple spectra.

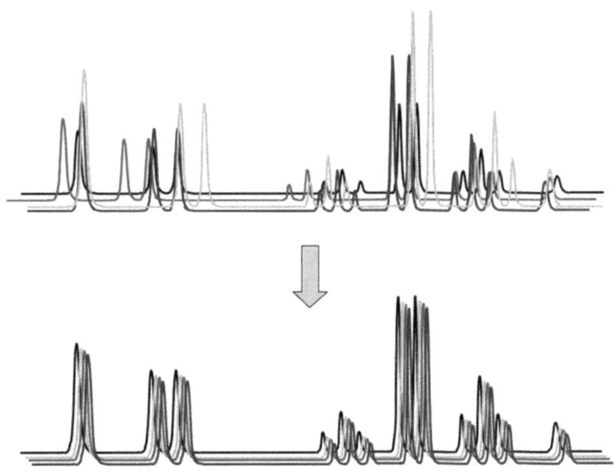

Fig. 14.2. An illustrative example of how spectral or chromatogram alignment works. Note that the peaks all have slightly different migration times or resonance positions due to instrument drift or other external effects.

Of course, the number of peaks, features, or "dimensions" that a given spectrum may represent could number in the hundreds or even thousands. To reduce the complexity or the number of parameters, chemometricians (and statisticians) use a class of statistical techniques called *dimensional reduction* to identify the key components that seem to contain the maximum amount of information or that are responsible for the greatest differences. The most common form of dimensional reduction is known as *principal component analysis*, or PCA.

PCA is not a classification technique but rather an unsupervised clustering technique. It is also known as singular-value decomposition (SVD) or eigenvector analysis. Recently, PCA has also been shown to be related to *k*-means clustering (31). PCA can be easily performed using a variety of free or nearly free software programs such as Matlab or the statistical package R (http://www.r-project.org) using their *prcomp* or *princomp* commands. More sophisticated (and expensive) commercial software tools with high-quality graphical displays and simplified interfaces are also available. A particularly popular choice in the metabolomics community is the Umetrics (Sweden) package called SIMCA-P.

Formally, principal component analysis is a statistical technique that determines an optimal linear transformation for a collection of data points such that the properties of that sample are most clearly displayed along the coordinate (or principal) axes. In other words, PCA allows one to easily plot, visualize, and cluster multiple metabolomic data sets based on linear combinations of their shared features. A somewhat simplified visual explanation of PCA is given in **Fig. 14.3**. Here we use the analogy of projecting shadows on a wall using a flashlight to find a "maximally informative projection" for a particular object. More precisely, we are trying to reduce a three-dimensional object into a series of maximally informative two-dimensional projections that would allow us to reconstruct a proper model of the original object. If the object of interest is a five-pointed star, then by shining the flashlight directly on the face of the star, one would generate the tell-tale "star" shadow. On the other hand, if the flashlight was directed at the edge of the star, the resulting shadow would be a less informative "rectangle" shape. This rectangular shadow, if used alone, would likely lead the observer to the wrong conclusion about what the object was. However, by combining the star shadow with the rectangular shadow (i.e., the two principal

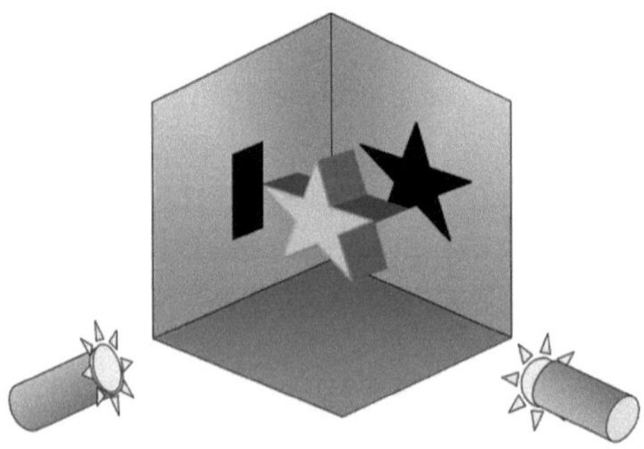

Fig. 14.3. A simplified picture of principal component analysis (PCA) where a three-dimensional object is reduced to a two-dimensional representation by prudent selection of one or more projection planes.

components or the two orthogonal projections), it is possible to reconstruct the shape and thickness of the original 3D star. While this example shows how a 3D object can be projected or have its key components reduced to two dimensions, the strength of PCA is that it can do the same with a hyperdimensional object just as easily.

In practice, PCA is most commonly used in metabolomics to identify how one or more samples are different from another, which variables contribute most to this difference, and whether those variables contribute in the same way (i.e., are correlated) or independently (i.e., uncorrelated) from each other. As a data reduction technique, PCA is particularly appealing because it allows one to visually or graphically detect sample clusters or groupings. This is most easily seen in **Fig. 14.4**. **Figure 14.4a** illustrates an example of a so-called PCA scores plot, where three well-defined clusters have been identified using just two principal components (PC 1 and PC 2). These two principal components account for >99 % of the variation in the samples. **Figure 14.4b** illustrates an example where separation or clustering is not achievable using the two largest principal components. In this latter case, the use of additional principal components or different combinations of principal components (i.e., different models) may achieve better separation. However, in some cases PCA will not succeed in identifying any clear clusters or obvious groupings no matter how many components are used. If this is the case, it is wise to accept the result and assume that the presumptive classes or groups cannot be distinguished. As a general rule, if a PCA analysis fails to achieve even a modest separation of classes or if the noise in the data set is too great, then it is unwise to attempt to separate classes using more complex models. One will only end up overfitting the model and introducing errors into the interpretation.

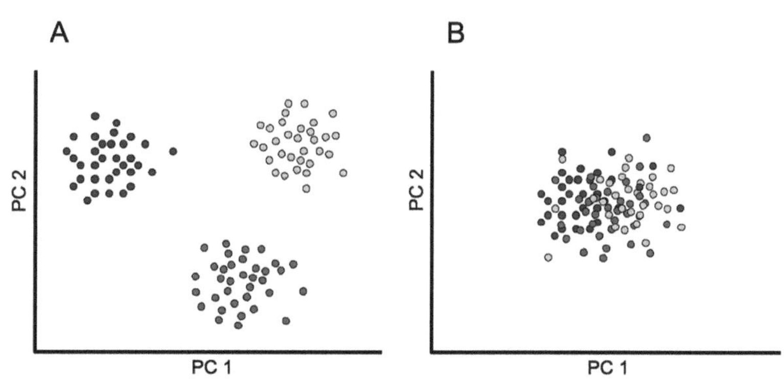

Fig. 14.4. A PCA scores plot. **(a)** A PCA scores plot where three well-defined clusters have been identified using just two principal components (PC 1 and PC 2). **(b)** An example where separation or clustering is not achievable using the two largest principal components.

The performance of a PCA model can be quantitatively evaluated in terms of an R^2 and/or a Q^2 value. R^2 is the correlation index and refers to the goodness of fit or the explained variation. On the other hand, Q^2 refers to the predicted variation or quality of prediction. R^2 is a quantitative measure (with a maximum value of 1) that indicates how well the PCA model is able to mathematically reproduce the data in the data set. A poorly fit model will have an R^2 of 0.2 or 0.3, while a well-fit model will have an R^2 of 0.7 or 0.8. If too many principal components are used, it is possible to overfit the data or to create clusters where clusters don't really exist. To guard against overfitting, the value Q^2 is commonly determined. Q^2 is usually estimated by cross-validation or permutation testing to assess the predictive ability of the model relative to the number of principal components used in the model. Cross-validation is a process that involves partitioning a sample of data into subsets such that the analysis is initially performed on a single subset (the training set), while the other subsets (the test sets) are retained to confirm and validate the initial analysis. In practice, Q^2 typically tracks very closely to R^2 as the number of components in the PCA model rises. However, if the PCA model begins to become overfit, Q^2 reaches a maximum value and then begins to fall. Generally, a $Q^2 > 0.5$ if considered good while a Q^2 of 0.9 is outstanding. A good rule of thumb is that the difference between Q^2 and R^2 should not exceed 0.2 or 0.3.

PCA is also a very useful technique for quantifying the amount of useful information or signal that is contained in the data. This is typically done by plotting the "weightings" of the individual components in what is called a *PCA loadings plot.* **Figure 14.5** provides an example of a hypothetical loadings plot for the first two principal components from a data set comparing urine samples from patients with cystinuria with urine samples from a normal patient pool. This data set shows, not unexpectedly, that patients with cystinuria have higher concentrations of the amino acids cysteine, lysine, arginine, and ornithine, along with lower concentrations of creatinine and citrate. Note that this kind of loadings plot is only possible if the compounds have been identified and quantified using quantitative metabolomic methods (*see* **Section 3.2**). If the compounds are not identified, this kind of plot, and hence this kind of interpretation, is not possible. This particular example serves to emphasize the fact that PCA (along with many other statistical methods) can be used in both nonquantitative and quantitative metabolomics.

PCA is not the only chemometric or statistical approach that can be applied to spectral analysis in metabolomics. A second class of chemometric methods is known as *supervised learning* or *supervised classification.* Supervised classifiers require that information about the class identities must be provided by the user in advance

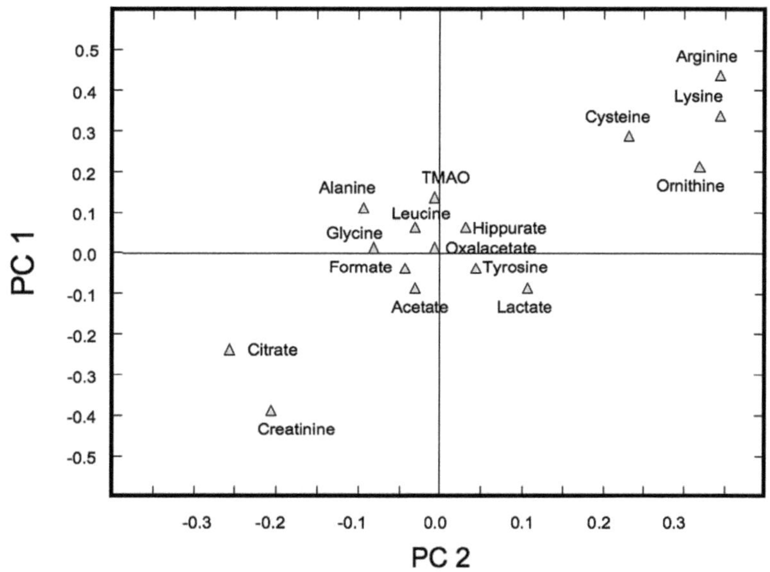

Fig. 14.5. An example of a hypothetical loadings plot for the first two principal components from a data set comparing urine samples from patients with cystinuria with urine samples from a normal patient pool.

of running the analysis. In other words, prior knowledge or prior clinical diagnoses are used to identify one group of spectra as being normal and the other group of spectra as being diseased. Examples of supervised classifiers include SIMCA (soft independent modeling of class analogy), PLS-DA (partial least-squares discriminant analysis), and *k*-means clustering (26–28). All of these techniques have been used to interpret NMR, MS/MS, and infrared (FTIR) spectral patterns in a variety of metabolomic or metabonomic applications (32–34).

PLS-DA can be used to enhance the separation between groups of observations by rotating PCA components such that a maximum separation among classes is obtained. This separation enhancement allows one to better understand which variables are most responsible for separating the observed (or apparent) classes. The basic principles behind PLS (partial least-squares) are similar to that of PCA. However, in PLS a second piece of information is used, namely, the labeled set of class identities (say "cystinuria" and "normal"). PLS-DA, which is a particular form of PLS, is a regression or categorical extension of PCA that takes advantage of a priori or user-assigned class information to attempt to maximize the covariance between the "test" or predictor variables and the training variable(s). PLS-DA is typically used after a relatively clear separation between two or more groups has been obtained through an unsupervised (PCA) analysis. Care must be taken in using PLS-DA methods, as it is easy to create convincing clusters or classes that have no statistical meaning (i.e., they overfit the data). The best way of avoiding these problems is to use *N*-fold cross-validation methods, bootstrapping, or resubstitution approaches to ensure that the data clusters derived by PLS-DA or other supervised methods are real and robust (35).

As seen in **Fig. 14.1**, statistical methods such as PCA, PLS-DA, or other techniques (*k*-means clustering, hierarchical clustering, artificial neural networks, ANOVA – analysis of variance, MANOVA – multivariate analysis of variance, etc.) can be used either at the beginning or toward the end of a metabolomic analysis. In chemometric approaches to metabolomics, these techniques are used at the beginning of the analysis process. In quantitative metabolomics, they are used at the end. One of the strengths of using chemometric methods at the beginning of the analysis process is that it allows one to look at all metabolites or all spectral features (both known and unknown) in an unbiased way. The advantage of this kind of holistic approach lies in the fact that one is not selectively ignoring or including key metabolic data in making a phenotypic classification or diagnosis. Furthermore, once the scores plots and loadings plots are generated from a chemometric analysis, it is possible to use this information to direct most of one's effort at identifying or quantifying only the most important or informative metabolites. This process certainly reduces, although it does not eliminate, the burden of metabolite identification.

3.2. Metabolite Identification and Quantification in Metabolomics

Whether one chooses to use quantitative metabolomics or chemometric methods in metabolomic analysis, eventually all paths lead to the need to identify (and quantify) metabolites (*see* **Fig. 14.1**). In metabolomics, most metabolite identification and quantification are done by comparing the sample spectrum or sample chromatogram to a library of reference spectra derived from pure compounds (5, 27, 36, 37). This can be done manually, semi-automatically, or automatically. In all cases, the basic premise is that the spectra obtained for a clinical sample of interest (which is a mixture of metabolites) are a linear combination of individual spectra for each of the pure metabolites in the mixture (*see* **Fig. 14.6**). This approach to compound identification is commonly done with both NMR and GC-MS data. It is also possible to use this concept, albeit to a much more limited extent, with LC-MS data (38).

Computer-aided GC-MS metabolite identification is typically performed by comparing GC retention times or retention indices (RI) with known compounds or by comparing against pregenerated retention index/mass spectral library databases. A very large GC-MS spectral library, covering tens of thousands of compounds, is available through the National Institute of Standards (NIST). While this GC-MS library is quite large, it actually contains a relatively modest number of

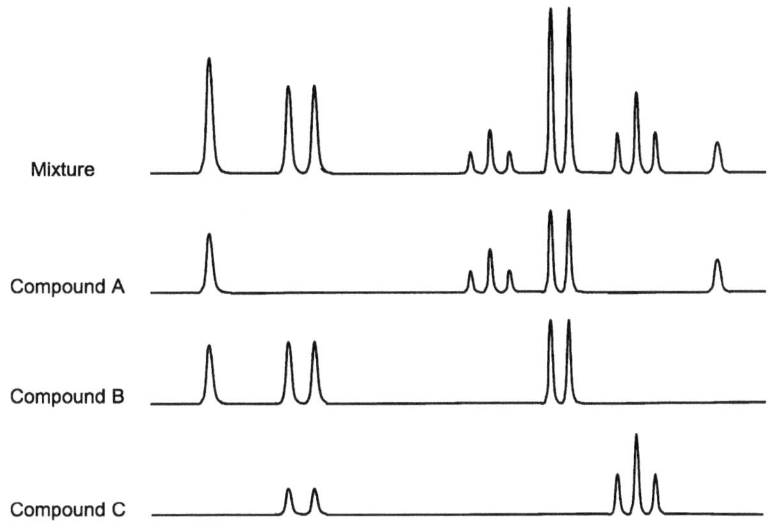

Fig. 14.6. A simplified illustration of spectral deconvolution showing how the top spectrum (obtained from a biofluid mixture) is actually a linear combination of three other spectra (A, B, and C) collected from pure compounds.

mammalian metabolites. Somewhat more metabolite-specific GC-MS libraries (that are NIST02 and AMDIS formatted) are available through the Golm metabolite database (39) and the Human Metabolome Database (16). In addition to the requirement for properly formatted databases containing a large number of mass spectral tags (MSTs), GC-MS metabolite identification also requires specialized GC deconvolution software such as AMDIS (http://chemdata.nist.gov/mass-spc/amdis/) or other commercial tools such as ChromaTof that support GC peak detection, peak area calculation, and mass spectral deconvolution. These programs identify and score possible metabolite matches by computing the similarity (using a normalized Euclidean or Hamming distance) between the observed electron-impact (EI) mass spectrum and the observed retention index with the database's collection of EI spectra and retention indices (MSTs). The resulting matches are typically presented as a sorted table containing the similarity score, the retention index, the spectral identifier, and the compound name (if known). Using well-resolved GC-MS spectra and an appropriate combination of databases, it is possible to identify 50 to 100 metabolites in a given sample. Quantification must typically be done by spiking in authentic chemical standards and creating standardized concentration curves that are specific to the instrument and running conditions.

Somewhat similar concepts used in GC-MS metabolite identification are also used in NMR-based metabolite identification. Just as with GC-MS metabolite identification, a database of reference one-dimensional ^{1}H NMR spectra is required, as is a set of software tools for comparing and calculating spectral matches between the observed NMR spectrum and the database spectra. A number of freely available, metabolite-specific ^{1}H and ^{13}C NMR spectral libraries have recently been described, including the BioMagResBank spectral library (40) and the HMDB spectral library (16). These web-enabled NMR databases also support (albeit somewhat limited) compound identification through spectral matching. Several excellent commercial software packages also exist for NMR-based compound identification, including Chenomx Inc.'s Eclipse (27) and Bruker's AMIX software. These user-friendly software tools, both of which support both compound identification and quantification, are widely used in the NMR metabolomics community.

Chenomx's Eclipse software uses a patented spectral deconvolution process that fits observed ^{1}H NMR spectra against a specially constructed library of 260 reference metabolite spectra. These spectra were collected over a wide range of pHs (4–9) and a wide range of spectrometer frequencies (300–900 MHz), allowing the software to be used on almost any ^{1}H NMR data set collected under almost any solution condition. They were also calibrated with quantification standards (imidazole and DSS) to permit semiautomated concentration determinations. Historically, the Eclipse software only supported semiautomated (i.e., user-assisted) compound identification and/or quantification. This is a relatively slow process (30–45 minutes per spectrum) that can yield inconsistent compound identification results. Upcoming releases of the software are expected to support fully automated compound identification and quantification. This enhancement should greatly accelerate the compound identification/quantification process and significantly improve the consistency in metabolite identification.

Bruker's AMIX software is another commercial product that offers support for compound identification and quantification for 1D and 2D NMR as well as LC-MS spectra. It used a method called AutoDROP to facilitate compound ID and structure verification (41). The key idea behind AutoDROP is the systematic decomposition of reference spectra into spectral patterns of molecular fragments. Compound identification is based on the recognition of such patterns in the target spectra. Like Eclipse, the AMIX approach is also semiautomated. While AMIX's support for compound identification and quantification is not quite as extensive or simple as with Eclipse, the AMIX software is quite unique in its support of 2D NMR spectral analysis.

One of the strengths of the NMR curve-fitting approaches is the fact that the NMR spectra for many individual metabolites are often composed of multiple peaks covering a wide range of chemical shifts. This means that most metabolites have unique or characteristic "chemical shift" fingerprints. This particular characteristic of NMR spectra helps reduce the problem of spectral (or chromatographic) redundancy, as it is unlikely that any two compounds will have identical numbers of peaks with identical chemical shifts, identical intensities, identical spin couplings, or identical peak shapes. Likewise, with higher magnetic fields (>600 MHz), the chemical shift separation among different peaks and different compounds is often good enough to allow the unambiguous identification of up to 100 compounds at a time – through simple curve fitting (5, 24, 27).

As noted earlier, automated or semiautomated metabolite identification is not restricted to NMR or GC-MS methods. It is also possible to apply the same techniques to LC-MS systems (38). Because liquid chromatography (LC) is not as reproducible as gas chromatography, the use of LC retention times or LC retention indices in metabolite identification is generally not feasible. Consequently, compound identification via LC-MS systems must depend almost exclusively on acquired mass data. In particular, if the resolution of the mass spectrometer is sufficiently high [as with Fourier transform mass spectrometers (FT-MS) or OrbiTrap mass spectrometers], it is possible to determine the chemical formula and often the identity of the compound directly from the parent ion masses and their isotope intensity patterns. A very effective and freely available program, called "Seven Golden Rules," was recently described by Kind and Fiehn that permits rapid chemical formula extraction and compound identification (or ranking) from high-resolution MS spectra (42). The performance of the algorithm improves substantially if one restricts the database search to known metabolites and/or drugs.

In addition to using FT-MS techniques, it is also possible to use soft-ionization tandem mass spectrometry or MS/MS methods to determine compound identity. In this approach, the MS/MS spectra must be collected at reasonably similar collision energies and on similar kinds of instruments (43). Query MS/MS spectra are compared to a library of MS/MS spectra collected for pure compounds and scored in a manner similar to the way EI spectra are scored in GC-MS methods. The Human Metabolome Database maintains a library of more than 600 pure metabolite MS/MS spectra and also supports this kind of MS/MS-based metabolite identification through a web-based query tool (16). Quantification of metabolites by LC-MS is somewhat more difficult than by GC-MS or NMR. Typically, quantification requires the addition or spiking of isotopically labeled derivatives of the metabolites of interest to the biofluid or tissue sample (23). The intensity of the isotopic derivative can then be used to quantify the metabolite of interest.

As has been pointed out earlier, the identification and quantification of compounds do not preclude the use of statistical or machine learning approaches to interpret the data. In fact, the same statistical techniques used in chemometric or non-quantitative metabolomic studies – PCA, SIMCA, PLS-DA, *k*-means clustering – can still be used to analyze metabolite profiles. Indeed, the added information (i.e., compound name and concentration) seems to significantly improve the discriminatory capabilities of most statistical techniques over what is possible for unlabeled or binned spectral data (34). Quantitative metabolomics also seems to be particularly amenable to other, more powerful, classification techniques such as artificial neural networks (ANNs), support vector machines (SVMs), and decision trees (DTs).

4. Biological Interpretation and Visualization of Metabolomic Data

The clinical or biological interpretation of metabolomic data is generally a much different process than the methods associated with metabolite identification or spectral discrimination. In particular, once a researcher or clinician has identified a key set of metabolites or biomarkers that have changed in a significant way, the next challenge is to provide some biological or clinical context to these changes. In making these interpretations, it is important to remember that metabolites are normally associated with specific pathways and processes, just as genes and proteins are. As might be expected, most of the small molecule metabolites measured by today's metabolomic techniques are associated with generic metabolic processes (glycolysis, gluconogenesis, lipid metabolism) found in all living cells. Changes in the relative concentrations of certain "universal" metabolites such as glucose, citrate, lactate, alpha-ketoglutarate, and others can reflect changes in cell viability (apoptosis), levels of oxygenation (anoxia or ischemia), local pH, general homeostasis, and so on. Often these metabolites can provide useful information about cell function or cell stress and organ function (9). Other kinds of metabolites are specifically associated with tissue remodeling, muscle atrophy, and muscle breakdown, such as methyl-histidine, creatine, tuarine, and glycine. By noting changes in the levels of these metabolites, it is possible to determine the extent of tissue repair or tissue damage (9). Still other compounds, such as malondialdehyde, 8-isoprostane F2, glutathione, and hydrogen peroxide are used as markers of oxidative stress (44). Increased amounts of these compounds in either blood or urine are indicative of poor redox status. Inflammation is also detectable through the monitoring of plasma levels of eicosanoids such as thromboxane B2, leukotriene B4, prostaglandin E2, or metabolic end products such as uric acid (45). Finally, plasma levels of homocyteine, triacylglycerol, cholesterol, and lipoprotein particles (LDL, HDL) have long been used to assess individuals for increased risk of cardiovascular disease (46). In short, each metabolite tells a unique story. The challenge for the physician and the scientist is to accurately interpret each one.

Key to the proper biological or clinical interpretation of metabolomic data is the availability of high-quality metabolism databases and metabolic pathway visualization tools. There are two types of metabolomics databases: (1) metabolic pathway databases and (2) metabolomic databases. Metabolic pathway databases are designed to house and display biochemical pathways or metabolite–gene–protein interactions. They are fundamentally visual aids, designed to facilitate the exploration of metabolism, metabolites, pathways, genes, and enzymes (often across many species). Metabolomic databases, on the other hand, contain much more information about metabolites, their chemical or spectral properties, as well as their physiological, biological, or clinical roles. Metabolomic databases are also somewhat more species-specific. Here we will review three metabolic pathway databases (KEGG, HumanCyc, and the Reactome database), and, one metabolomics database (the Human Metabolome Database) and provide a brief description of their respective features that may be useful for clinical metabolomics and the interpretation of metabolomic data. Additional databases, along with their URLs and some brief comments, are listed in **Table 14.2**.

4.1. The KEGG Database

Perhaps the most comprehensive metabolic pathway database on the web is the Kyoto Encyclopedia of Genes and Genomes, or KEGG (47). KEGG has been under development at the Kanehisa lab at the Institute for Chemical Research in Kyoto, Japan, since 1995. This particular resource brings a very broad, multi-organism view to metabolism, as it contains genomic, chemical, and network/pathway information for more than 360 organisms, including 72,171 pathways, 15,050 chemical compounds, and 7,342 reactions (at last count). KEGG is actually composed of four smaller databases (BRITE, GENES, LIGAND, and PATHWAY), with the LIGAND and PATHWAY databases being most relevant to those interested in metabolism.

KEGG's LIGAND or chemical compound database contains chemical structures of most known metabolites and sugars (glycans) as well as a growing number of pharmaceutical and environmental compounds. This database may be queried by KEGG compound identifiers, formal names, synonyms, chemical formulas, masses, associated enzyme names, and reactions. Similar compound structures may also be searched using KEGG's SIMCOMP and SUBCOMP utilities via KEGG compound identifiers or manually uploaded MOL files. These queries return synoptic "compound cards," which provide information about the compound (formula, molecular weight, chemical structure), its connection to different reactions, pathways, and enzymes, as well as hyperlinks to external databases.

Table 14.2
A summary of metabolomic and metabolic pathway databases

Database name	URL or web address	Comments
Human Metabolome Database	http://www.hmdb.ca	– Largest and most complete collection of metabolite data (biophysical, biological, and clinical) – 90+ data fields per metabolite –Specific to humans only
PubChem	http://pubchem.ncbi.nlm.nih.gov	–Largest public collection of chemical substances (includes many metabolites) – Links to PubMed articles
Chemicals Entities of Biological Interest (ChEBI)	http://www.ebi.ac.uk/chebi/	– Covers metabolites and drugs – 10 data fields per metabolite – Primary focus on ontology and nomenclature
HumanCyc (Encyclopedia of Human Metabolic Pathways)	http://humancyc.org/	– Large collection of human metabolite and pathway data – 10 data fields per metabolite –Includes tools for illustration and annotation
KEGG (Kyoto Encyclopedia of Genes and Genomes)	http://www.genome.jp/kegg/	– Best-known and most complete metabolic pathway database – 15 data fields per metabolite – Covers many organisms – Limited biomedical data
LipidMaps	http://www.lipidmaps.org/	– Limited to lipids only (not species-specific) – Nomenclature standard
METLIN Metabolite Database	http://metlin.scripps.edu/	– Human-specific – Mixes drugs, drug metabolites together – 10 data fields per metabolite
Golm Metabolome Database	http://csbdb.mpimp-golm.mpg.de/csbdb/gmd/gmd.html	– Emphasis on MS or GC-MS data only –No biological data – 5 to 10 data fields per chemical – Specific to plants
Reactome (A Curated Knowledgebase of Pathways)	http://www.reactome.org/	– Pathway database with more advanced query features – Not as complete as KEGG or MetaCyc
Roche Applied Sciences Biochemical Pathways Chart	http://www.expasy.org/cgi-bin/search-biochem-index	– The old metabolism standard (online)

In addition to a large collection of metabolites and metabolite structures, KEGG's PATHWAY database also contains 359 manually drawn and fully annotated reference pathways or wiring diagrams for metabolism, gene signaling, and protein interactions. KEGG uses graph-theoretic concepts (i.e., combinations of line graphs and nested graphs) to map and propagate its reference pathways to other organisms. In KEGG's wiring diagrams, the nodes typically represent metabolites and the edges represent the enzymes (identified with an Enzyme Classification, or EC, number) responsible for the metabolite conversion. Both the nodes and edges are hyperlinked to KEGG data cards. To use KEGG's PATHWAY database, users may select from several hundred hierarchically named metabolic and catabolic processes/pathways. Clicking on these hyperlinked names will send the user to a hyperlinked image describing the pathway and containing additional hyperlinks to compounds and protein/enzyme data or structures. KEGG's PATHWAY database has recently been expanded to include more than just hyperlinked metabolic pathways, as it now contains wiring diagrams for DNA/RNA processing, signal transduction, immune responses, cell communication and development, human diseases, and even drug development history. KEGG offers many other features including flat files for FTP downloads, an application programming interface (API), and standalone Java drawing tools (KegDraw and KegArray) for chemical querying and microarray annotation. A much more complete description of KEGG and its contents can be found in an article by Kanehisa et al. (47) and references therein.

Despite its comprehensiveness, KEGG is somewhat limited in its application to human diseases and genetic disorders. First, KEGG's query system only supports browsing or querying of single entries (a single compound, a single pathway) as opposed to large-scale relational queries. This limits users from asking complex questions such as "find all human enzymes regulated by tyrosine or tyrosine metabolites." Second, the vast majority of KEGG pathways and KEGG compounds are not found in humans, but rather in plants or microbes. Third, KEGG presents its pathways as "consensus" pathways combining all reactions known in all species to generate a map of, for example, tyrosine biosynthesis. This makes it difficult to distinguish which metabolic intermediates, pathways and enzymes are specific only to humans. Despite these limitations for certain biomedical applications, the KEGG database still represents one of the most valuable and comprehensive resources for understanding and exploring metabolism.

4.2. The HumanCyc Database

The HumanCyc database is part of the "Cyc" suite of databases (including EcoCyc, BioCyc, and MetaCyc) that have been developed and maintained by Peter Karp's group at the Stanford Research Institute since 1999 (48). HumanCyc (version 10.6) is a web-accessible database containing information on 28,782 human genes, 2,594 human enzymes, 1,296 reactions, 1,004 human metabolites, and 197 human-specific metabolic pathways. HumanCyc contains extensively hyperlinked metabolic pathway diagrams, enzyme reactions, enzyme data, chemical structures, chemical data, and gene information. Likewise, users can query HumanCyc by the name of a protein, gene, reaction, pathway, chemical compound, or EC (enzyme classification number). Just as with KEGG, most HumanCyc queries or browsing operations return a rich and colorful collection of hyperlinked figures, pathways, chemical structures, reactions, enzyme names, references, and protein/gene sequence data.

Unlike most other metabolic pathway databases, HumanCyc provides much more detailed enzyme information, including data on substrate specificity, kinetic properties, activators, inhibitors, cofactor requirements, and links to sequence/structure databases. Additionally, HumanCyc supports sophisticated relational queries, allowing complex searches to be performed and more detailed information to be displayed. These search utilities are supplemented with a very impressive "Omics Viewer" that allows gene expression and metabolite profiling data to be painted onto any organism's metabolic network. HumanCyc also displays metabolic pathway information at varying degrees of resolution, allowing users to interactively zoom into a reaction diagram for more detailed views and more detailed pathway annotations.

4.3. The Reactome Database

A much more recent addition to the collection of metabolic pathway databases is the Reactome database (49). The Reactome project was started in 2002 to develop a curated resource of core pathways and reactions in human biology. The *reactome* is defined as the complete set of possible reactions or pathways that can be found in a living organism, including the reactions involved in intermediary metabolism, regulatory pathways, signal transduction, and cell cycle processes. The Reactome database is a curated resource authored by biological researchers with expertise in their fields. Unlike KEGG or HumanCyc, the Reactome database takes a much more liberal view of what constitutes metabolism (or biochemical reactions) by including such processes as mitosis, DNA-repair, insulin-mediated signaling, translation, transcription, and mRNA processing in addition to the standard metabolic pathways involving amino acids, carbohydrates, nucleotides, and lipids.

The Reactome database (Version 23) currently has 781 human-associated pathways assembled from 2,327 reactions involving 2,293 proteins or protein complexes. Central to the Reactome database is a schematic "Reaction Map," which graphically summarizes all high-level reactions contained in the Reactome database. This map allows users to navigate through the database in an interactive and progressively more detailed fashion. Users may also browse through the database by selecting topics from a table of contents, or they may query the database using a variety of text and keyword searches. The Reactome database also supports complex Boolean text queries for different combinations of reactions, reaction products, organisms, and enzymes. The results from these queries include higher-resolution pathway maps (in PDF, PNG, and SVG formats), SBML (systems biology mark-up language) descriptions, and synoptic Reactome web "cards" on specific proteins or metabolites with hyperlinks to many external databases.

One of the most useful and innovative features of the Reactome database is a tool called the Reactome "skypainter." This allows users to paste in a list of genes or gene identifiers (GenBank, UniProt, RefSeq, EntrezGene, OMIM, InterPro, Affymetrix, Agilent, and Ensembl formats) and to "paint" the Reactome reaction map in a variety of ways. In fact, it is even possible to generate "movies" that can track gene expression changes over different time periods – as might be obtained from a time-series gene or protein expression study. This tool is particularly useful for analyzing microarray data, but it is also useful for visualizing disease genes (say from OMIM) and mapping the roles they play and the pathways in which they participate. In general, the central concepts behind the Reactome database are quite innovative, and it certainly appears that this resource that could play an increasingly important role in many areas of biology, biochemistry, and systems biology.

4.4. The Human Metabolome Database (HMDB)

The HMDB (16) currently contains more than 2,921 human metabolite entries that are linked to more than 28,083 different synonyms. These metabolites are further connected to some 77 nonredundant pathways, 3,364 distinct enzymes, 103,000 SNPs, as well as 862 metabolic diseases (genetic and acquired). Much of this information is gathered manually or semiautomatically from thousands of books, journal articles, and electronic databases. In addition to its comprehensive literature-derived data, the HMDB also contains an extensive collection of experimental metabolite concentration data for plasma, urine, CSF, and/or other biofluids for more than 1,200 compounds. The HMDB also has more than 600 compounds for which experimentally acquired "reference" ^{1}H and ^{13}C NMR and MS/MS spectra have been acquired.

The HMDB is fully searchable, with many built-in tools for viewing, sorting, and extracting metabolites, biofluid concentrations, enzymes, genes, NMR or MS spectra, and disease information. Each metabolite entry in the HMDB contains an average of 90 separate data fields, including a comprehensive compound description, names and synonyms, structural information, physicochemical data, reference NMR and MS spectra, normal and abnormal biofluid concentrations, tissue locations, disease associations, pathway information, enzyme data, gene sequence data, SNP and mutation data, as well as extensive links to images, references, and other public databases. A screen shot montage of the HMDB and some of its data content is given in **Fig. 14.7**. A key feature that distinguishes the HMDB from other metabolic resources is its extensive support for higher-level database searching and selecting functions. In particular, the HMDB offers a chemical structure search utility, a local BLAST search that supports both single- and multiple-sequence queries, a Boolean text search, a relational data extraction tool, an MS spectral matching tool, and an NMR spectral search tool. These spectral query tools are particularly useful for identifying compounds via MS or NMR data from other metabolomic studies.

5. Metabolic Modeling and the Interpretation of Metabolomic Data

As we have already seen, the statistical tools and metabolomics databases described in **Sections 3** and **4** are particularly useful at identifying metabolic differences, finding interesting biomarkers, and discerning relevant biological pathways. However, these approaches provide a relatively static view of metabolism and biology. To gain a more complete understanding of the dynamics of metabolic networks along with their temporal (and spatial) dependencies, it is often necessary to turn to metabolic modeling. Metabolic modeling offers both scientists and clinicians the capacity to predict the consequences of gene knockouts, the effects of gene mutations, or the consequences of metabolite/drug intervention strategies. In other words, metabolic simulation effectively turns biology (and metabolomics) from a purely observational science imto a far more predictive science.

Metabolic modeling or metabolic simulation can be done in a variety of ways. Traditionally, it is done by writing down and solving systems of time-dependent ordinary differential equations (ODEs) that describe the chemical reactions and reaction rates of the metabolic system of interest. There are now a host of metabolic simulation programs that allow very complex, multicomponent simulations to be performed (50, 51). These include programs such as GEPASI (52), CellDesigner (53), SCAMP (54), and Cellerator (55). GEPASI is a good example of a typical metabolic or biochemical pathway simulation package. This program, which has been under development for almost 15 years, uses a simple interface to allow one to build models of metabolic pathways and simulate their dynamics and steady-state behavior for given sets of parameters. GEPASI also generates the coefficients of metabolic control analysis for steady states. In addition,

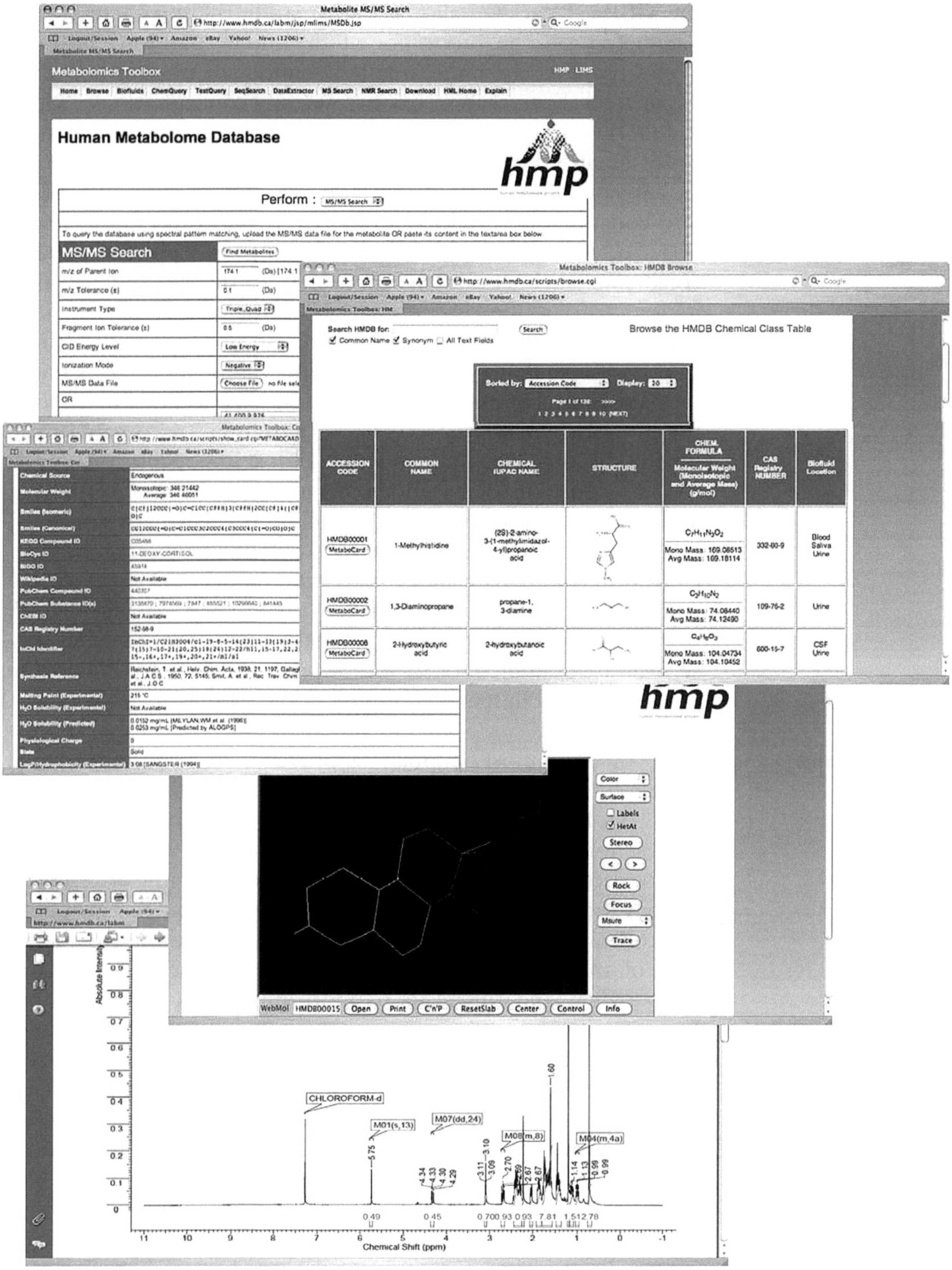

Fig. 14.7. A montage of screen shots from the Human Metabolome Database (HMDB) illustrating some of the data content and query capabilities of the database.

the GEPASI package allows one to study the effects of several parameters on the properties of the model pathway. GEPASI allows users to enter the kinetic equations of interest and their parameters (Km, reaction velocity, starting concentrations), solves the ODEs using an ODE solver, and generates plots that can be easily visualized by the user.

An alternative to solving large systems of time-dependent rate equations is a technique known as *constraint-based modeling* (56, 57). Constraint-based modeling uses physicochemical constraints such as mass balance, energy balance, and flux limitations to describe the potential behavior of a large metabolic system (a cell, an organ, an organism). In this type of modeling, the time dependence and rate constants can be ignored, as one is only interested in finding the steady-state conditions that satisfy the physicochemical constraints. Because cells and organs are so inherently complex and because it is almost impossible to know all the rate constants or instantaneous metabolite concentrations at a given time, constraint-based modeling is particularly appealing to those involved in large-scale metabolomic studies. In particular, through constraint-based modeling, models and experimental data can be more easily reconciled and studied on a whole-cell or genome-scale level (56, 57). Furthermore, experimental data sets can be examined for their consistency against the underlying biology and chemistry represented in the models.

5.1. Flux Balance Analysis

One of the most popular approaches to constraint-based metabolic modeling is known as *flux-balance analysis*, or FBA (58, 59). FBA requires knowledge of the stoichiometry of most of reactions and transport processes that are thought to occur in the metabolic system of interest. This collection of reactions defines the metabolic network. FBA assumes that the metabolic network will reach a steady state constrained by the stoichiometry of the reactions. Normally, the stoichiometric constraints are too few, and this leads to more unknowns than equations (i.e., an underdetermined system). However, possible sets of solutions can be found by including information about all feasible metabolite fluxes (metabolites added or excreted) and by specifying maximum and minimum fluxes through any particular reaction. The model can also be refined or further constrained by adding experimental data – from known physiological or biochemical data obtained from specific metabolomic studies. Once the solution space is defined, the model is refined and its behavior can be studied by optimizing the steady-state behavior with respect to some objective function. Typically, the objective function optimization involves the maximization of biomass, the maximization of growth rate, the maximization of ATP production, the maximization of the production of a particular product, or the maximization of reducing power. Once the model is fully optimized, it is possible to use that FBA model to create predictive models of cellular, organ, or whole organism metabolism. These predictions can be done by changing the network parameters or flux balance, changing the reactants, adding new components to the model, or changing the objective function to be maximized.

Critical to the success of any FBA model is the derivation or compilation of appropriate mass and charge balance (58, 59). Mass balance is defined in terms of both the flux of metabolites through each reaction, the stoichiometry of that reaction, and the conservation of mass and charge. Mass and charge balance considerations give rise to a set of coupled differential equations. This set of equations is often expressed as a matrix equation, which can be solved through simple linear algebra and optimized through linear programming. The goal of FBA is to identify the metabolic fluxes in the steady state (i.e., where the net flux is 0). Because there are always more reactions than metabolites, the steady-state solution is always underdetermined. As a result, additional constraints must be added to determine a unique solution. These constraints can be fluxes measured through metabolomics experiments (such as isotope labeling experiments) or through estimated ranges of allowable (feasible) flux values.

FBA methods have been used in a variety of metabolomic studies, including bacterial metabolism (60), yeast metabolism (61), erythrocyte metabolism (62), myocardial metabolism (63), and most impressively the entire human metabolomic network (64). Certainly, as more detailed flux data is acquired through isotope tracer analysis and more information is obtained from quantitative, targeted metabolic profiling, it is likely that flux balance analysis and other kinds of constraint-based modeling will play an increasingly important role in the interpretation of metabolomic data, especially in clinical metabolomic data.

6. Conclusions

This chapter was written to provide a general-purpose overview of the field of metabolomics along with higher-level descriptions of some of the algorithms, databases, data analysis strategies, and computer programs used to analyze or interpret metabolomic data. As seen in **Section 2**, metabolomics shares many experimental and procedural similarities with proteomics, with requirements for the same types of instrumentation (LC/MS, NMR, HPLC, UPLC, etc.) and similar types of sample preparation protocols. It is also clear from the discussion in **Section 3** that metabolomics shares many of the same computational needs as proteomics and transcriptomics, particularly in terms of the use and analysis of statistical, data reduction, and data visualization tools. All three omics methods (metabolomics, proteomics, transcriptomics) use principal component analysis (PCA), partial least-squares discriminant analysis (PLS-DA), *k*-nearest-neighbor clustering, hierarchical clustering, and a variety of machine learning approaches (neural networks and support vector machines) to help interpret or process their data. Metabolomics does, however, differ from other "omics" techniques because unlike proteomics or transcriptomics, the technology to routinely and rapidly identify every metabolite is not yet available. Consequently, there is still considerable effort going into the development of hardware and software (algorithms and databases) to make this possible. The last two sections of this chapter described some of the resources (databases and modeling software) that can be used to interpret, visualize, and analyze metabolomic data at a biological or clinical level. While most of the resources described in these sections were of the open source variety, there are also a growing number of high-quality commercial tools (such as Ingenuity's Pathway Analysis and Ariadne's Pathway Studio) that can greatly assist with biological interpretation and modeling. One of the most obvious trends in computational metabolomics is the growing alignment or integration of metabolomics with systems biology. The large body of knowledge that is available about human metabolism, coupled with our growing capacity to quantitatively measure perturbations to metabolic functions – both spatially and temporally, has made metabolomics the "golden child" for many systems biology applications. As a result, there is an impressive abundance of high-quality software tools to simulate and predict the metabolic consequences of enzyme or genetic perturbations. The fact that these metabolic modeling systems are starting to play an increasingly important role in interpreting metabolomic data suggests that these tools and techniques may eventually be adapted to interpreting proteomic and transcriptomic data in the not-too-distant future.

Acknowledgments

I would like to acknowledge Genome Canada, Genome Alberta, and the Alberta Ingenuity Centre for Machine Learning (AICML) for their financial support.

References

1. German JB, Hammock BD, Watkins SM. (2005) Metabolomics: building on a century of biochemistry to guide human health. *Metabolomics* 1:3–9.
2. Wishart DS. (2007) Human Metabolome Database: completing the "human parts list." *Pharmacogenomics* 8:683–686.
3. Yang J, Xu G, Hong Q, Liebich HM, Lutz K, Schmülling RM, Wahl HG. (2004) Discrimination of Type 2 diabetic patients

from healthy controls by using metabonomics method based on their serum fatty acid profiles. *J Chromatogr B* 813:53–58.
4. Williamson MP, Humm G, Crisp AJ. (1989) ^{1}H nuclear magnetic resonance investigation of synovial fluid components in osteoarthritis, rheumatoid arthritis and traumatic effusions. *Br J Rheumatol* 28:23–27.
5. Wishart DS, Querengesser LMM, Lefebvre BA, Epstein NA, Greiner R, Newton JB. (2001) Magnetic resonance diagnostics: a new technology for high-throughput clinical diagnostics. *Clin Chemistry* 47: 1918–1921.
6. Moolenaar SH, Engelke UF, Wevers RA. (2003) Proton nuclear magnetic resonance spectroscopy of body fluids in the field of inborn errors of metabolism. *Ann Clin Biochem* 40:16–24.
7. Coen M, O'Sullivan M, Bubb WA, Kuchel PW, Sorrell T. (2005) Proton nuclear magnetic resonance-based metabonomics for rapid diagnosis of meningitis and ventriculitis. *Clin Infect Dis* 41:1582–1590.
8. Griffin JL, Bollard ME. (2004) Metabonomics: its potential as a tool in toxicology for safety assessment and data integration. *Curr Drug Metab* 5:389–398.
9. Wishart DS. (2005) Metabolomics: the principles and potential applications to transplantation. *Am J Transplant* 5:2814–2820.
10. Wishart DS. (2007) Current progress in computational metabolomics. *Brief Bioinform* 8:279–293.
11. Coley NG. (2004) Medical chemists and the origins of clinical chemistry in Britain (circa 1750–1850). *Clin Chem* 50:961–972.
12. Rosenfeld L. (2001) Clinical chemistry since 1800: growth and development. *Clin Chem* 48:186–197.
13. Tietz NW. (1995) *Clinical Guide to Laboratory Tests*, 3rd ed., WB Saunders Press, Philadelphia, PA.
14. Dunn WB, Bailey NJ, Johnson HE. (2005) Measuring the metabolome: current analytical technologies. *Analyst* 130:606–625.
15. Cotter D, Maer A, Guda C, Saunders B, Subramaniam S. (2006) LMPD: LIPID MAPS proteome database. *Nucleic Acids Res* 34(Database issue):D507–510.
16. Wishart DS, Tzur D, Knox C, Eisner R, Guo AC, Young N, Cheng D, Jewell K, Arndt D, Sawhney S, Fung C, Nikolai L, Lewis M, Coutouly MA, Forsythe I, Tang P, Shrivastava S, Jeroncic K, Stothard P, Amegbey G, Block D, Hau DD, Wagner J, Miniaci J, Clements M, Gebremedhin M, Guo N, Zhang Y, Duggan GE, Macinnis GD, Weljie AM, Dowlatabadi R, Bamforth F, Clive D, Greiner R, Li L, Marrie T, Sykes BD, Vogel HJ, Querengesser L. (2007) HMDB: the Human Metabolome Database. *Nucleic Acids Res* 35(Database issue): D521–526.
17. Smith CA, O'Maille G, Want EJ, Qin C, Trauger SA, Brandon TR, Custodio DE, Abagyan R, Siuzdak G. (2005) METLIN: a metabolite mass spectral database. *Ther Drug Monit* 27:747–751.
18. Saude EJ, Sykes BD. (2007) Urine stability for metabolomic studies: effects of preparation and storage. *Metabolomics* 3:19–24.
19. Beckonert O, Keun HC, Ebbels TM, Bundy J, Holmes E, Lindon JC, Nicholson JK. (2007) Metabolic profiling, metabolomic and metabonomic procedures for NMR spectroscopy of urine, plasma, serum and tissue extracts. *Nat Protoc* 2:2692–2703.
20. Jiye A, Trygg J, Gullberg J, Johansson AI, Jonsson P, Antti H, Marklund SL, Moritz T. (2005) Extraction and GC/MS analysis of the human blood plasma metabolome. *Anal Chem* 77:8086–8094.
21. Schnackenberg LK, Beger RD. (2006) Monitoring the health to disease continuum with global metabolic profiling and systems biology. *Pharmacogenomics* 7:1077–1086.
22. German JB, Gillies LA, Smilowitz JT, Zivkovic AM, Watkins SM. (2007) Lipidomics and lipid profiling in metabolomics. *Curr Opin Lipidol* 18:66–71.
23. Guo K, Ji C, Li L. (2007) Stable-isotope dimethylation labeling combined with LC-ESI MS for quantification of amine-containing metabolites in biological samples. *Anal Chem* 79:8631–8638.
24. Weljie AM, Dowlatabadi R, Miller BJ, Vogel HJ, Jirik FR. (2007) An inflammatory arthritis-associated metabolite biomarker pattern revealed by ^{1}H NMR spectroscopy. *J Proteome Res* 6:3456–3464.
25. van der Werf MJ, Overkamp KM, Muilwijk B, Coulier L, Hankemeier T. (2007) Microbial metabolomics: toward a platform with full metabolome coverage. *Anal Biochem* 370:17–25.
26. Trygg J, Holmes E, Lundstedt T. (2007) Chemometrics in metabonomics. *J Proteome Res* 6:469–479.
27. Weljie AM, Newton J, Mercier P, Carlson E, Slupsky CM. (2006) Targeted profiling: quantitative analysis of ^{1}H NMR metabolomics data. *Anal Chem* 78: 4430–4442.
28. Lavine B, Workman JJ, Jr. (2004) Chemometrics. *Anal Chem* 76:3365–3371.
29. Wu W, Daszykowski M, Walczak B, Sweatman BC, Connor SC, Haselden JN,

Crowther DJ, Gill RW, Lutz MW. (2006) Peak alignment of urine NMR spectra using fuzzy warping. *J Chem Inf Model* 46:863–875.
30. Kind T, Tolstikov V, Fiehn O, Weiss RH. (2007) A comprehensive urinary metabolomic approach for identifying kidney cancer. *Anal Biochem* 363:185–195.
31. Ding C, He X. (2004) K-means clustering via principal component analysis. *Proc of the International Conference on Machine Learning (ICML 2004)*, pp. 225–232.
32. Holmes E, Nicholls AW, Lindon JC, Connor SC, Connelly JC, Haselden JN, Damment SJ, Spraul M, Neidig P, Nicholson JK. (2000) Chemometric models for toxicity classification based on NMR spectra of biofluids. *Chem Res Toxicol* 13:471–478.
33. Smith IC, Baert R. (2003) Medical diagnosis by high resolution NMR of human specimens. *IUBMB Life* 55:273–277.
34. Wilson ID, Plumb R, Granger J, Major H, Williams R, Lenz EM. (2005) HPLC-MS-based methods for the study of metabonomics. *J Chromatogr B* 817:67–76.
35. Molinaro AM, Simon R, Pfeiffer RM. (2005) Prediction error estimation: a comparison of resampling methods. *Bioinformatics* 21:3301–3307.
36. Serkova NJ, Rose JC, Epperson LE, Carey HV, Martin SL. (2007) Quantitative analysis of liver metabolites in three stages of the circannual hibernation cycle in 13-lined ground squirrels by NMR. *Physiol Genomics* 31:15–24.
37. Niwa T. (1986) Metabolic profiling with gas chromatography-mass spectrometry and its application to clinical medicine. *J Chromatogr* 379:313–345.
38. La Marca G, Casetta B, Malvagia S, Pasquini E, Innocenti M, Donati MA, Zammarchi E. (2006) Implementing tandem mass spectrometry as a routine tool for characterizing the complete purine and pyrimidine metabolic profile in urine samples. *J Mass Spectrom* 41:1442–1452.
39. Kopka J, Schauer N, Krueger S, Birkemeyer C, Usadel B, Bergmüller E, Dörmann P, Weckwerth W, Gibon Y, Stitt M, Willmitzer L, Fernie AR, Steinhauser D. (2005) GMD@CSB.DB: the Golm Metabolome Database. *Bioinformatics* 21: 1635–1638.
40. Ulrich EL, Akutsu H, Doreleijers JF, Harano Y, Ioannidis YE, Lin J, Livny M, Mading S, Maziuk D, Miller Z, Nakatani E, Schulte CF, Tolmie DE, Kent Wenger R, Yao H, Markley JL. (2008) BioMagResBank. *Nucleic Acids Res* 36(Database issue):D402–408.
41. Rossé G, Neidig P, Schröder H. (2002) Automated structure verification of small molecules libraries using 1D and 2D NMR techniques. *Methods Mol Biol* 201:123–139.
42. Kind T, Fiehn O. (2007) Seven Golden Rules for heuristic filtering of molecular formulas obtained by accurate mass spectrometry. *BMC Bioinformatics* 8:105.
43. Jiang H, Somogyi A, Timmermann BN, Gang DR. (2006) Instrument dependence of electrospray ionization and tandem mass spectrometric fragmentation of the gingerols. *Rapid Commun Mass Spectrom* 20:3089–3100.
44. Fardet A, Canlet C, Gottardi G, Lyan B, Llorach R, Rémésy C, Mazur A, Paris A, Scalbert A. (2007) Whole-grain and refined wheat flours show distinct metabolic profiles in rats as assessed by a 1H NMR-based metabonomic approach. *J Nutr* 137:923–929.
45. Margalit A, Duffin KL, Isakson PC. (1996) Rapid quantitation of a large scope of eicosanoids in two models of inflammation: development of an electrospray and tandem mass spectrometry method and application to biological studies. *Anal Biochem* 235:73–81.
46. Castro IA, Barroso LP, Sinnecker P. (2005) Functional foods for coronary heart disease risk reduction: a meta-analysis using a multivariate approach. *Am J Clin Nutr* 82:32–40.
47. Kanehisa M, Araki M, Goto S, Hattori M, Hirakawa M, Itoh M, Katayama T, Kawashima S, Okuda S, Tokimatsu T, Yamanishi Y. (2008) KEGG for linking genomes to life and the environment. *Nucleic Acids Res* 36(Database issue): D480–484.
48. Romero P, Wagg J, Green ML, Kaiser D, Krummenacker M, Karp PD. (2005) Computational prediction of human metabolic pathways from the complete human genome. *Genome Biol* 6:R2.
49. Vastrik I, D'Eustachio P, Schmidt E, Joshi-Tope G, Gopinath G, Croft D, de Bono B, Gillespie M, Jassal B, Lewis S, Matthews L, Wu G, Birney E, Stein L. (2007) Reactome: a knowledge base of biologic pathways and processes. *Genome Biol* 8:R39.
50. Alves R, Antunes F, Salvador A. (2006) Tools for kinetic modeling of biochemical networks. *Nat Biotechnol* 24:667–672.
51. Materi W, Wishart DS. (2007) Computational systems biology in drug discovery and development: methods and applications. *Drug Discov Today* 12:295–303.
52. Mendes P. (1993) GEPASI: a software package for modelling the dynamics, steady states and control of biochemical and other systems. *Comput Appl Biosci* 9:563–571.

53. Kitano H, Funahashi A, Matsuoka Y, Oda K. (2005) Using process diagrams for the graphical representation of biological networks. *Nat Biotechnol* 23:961–966.
54. Sauro HM. (1993) SCAMP: a general-purpose simulator and metabolic control analysis program. *Comput Appl Biosci* 9:441–450.
55. Shapiro BE, Levchenko A, Meyerowitz EM, Wold BJ, Mjolsness ED. (2003) Cellerator: extending a computer algebra system to include biochemical arrows for signal transduction simulations. *Bioinformatics* 19:677–678.
55. Demir O, Aksan Kurnaz I. (2006) An integrated model of glucose and galactose metabolism regulated by the GAL genetic switch. *Comput Biol Chem* 30:179–192.
56. Gagneur J, Casari G. (2005) From molecular networks to qualitative cell behavior. *FEBS Lett* 579:1867–1871.
57. Joyce AR, Palsson BO. (2007) Toward whole cell modeling and simulation: comprehensive functional genomics through the constraint-based approach. *Prog Drug Res* 64:267–309.
58. Kauffman KJ, Prakash P, Edwards JS. (2003) Advances in flux balance analysis. *Curr Opin Biotechnol* 14:491–496.
59. Lee JM, Gianchandani EP, Papin JA. (2006) Flux balance analysis in the era of metabolomics. *Brief Bioinform* 7: 140–150.
60. Oliveira AP, Nielsen J, Forster J. (2005) Modeling *Lactococcus lactis* using a genome-scale flux model. *BMC Microbiol* 5:39.
61. Jin YS, Jeffries TW. (2004) Stoichiometric network constraints on xylose metabolism by recombinant *Saccharomyces cerevisiae*. *Metab Eng* 6:229–238.
62. Durmus Tekir S, Cakir T, Ulgen KO. (2006) Analysis of enzymopathies in the human red blood cells by constraint-based stoichiometric modeling approaches. *Comput Biol Chem* 30:327–338.
63. Luo RY, Liao S, Tao GY, Li YY, Zeng S, Li YX, Luo Q. (2006) Dynamic analysis of optimality in myocardial energy metabolism under normal and ischemic conditions. *Mol Syst Biol* 2:2006.0031.
64. Duarte NC, Becker SA, Jamshidi N, Thiele I, Mo ML, Vo TD, Srivas R, Palsson BØ. (2007) Global reconstruction of the human metabolic network based on genomic and bibliomic data. *Proc Natl Acad Sci USA* 104: 1777–1782.

Chapter 15

Algorithms and Methods for Correlating Experimental Results with Annotation Databases

Michael Hackenberg and Rune Matthiesen

Abstract

An important procedure in biomedical research is the detection of genes that are differentially expressed under pathologic conditions. These genes, or at least a subset of them, are key biomarkers and are thought to be important to describe and understand the analyzed biological system (the pathology) at a molecular level. To obtain this understanding, it is indispensable to link those genes to biological knowledge stored in databases. Ontological analysis is nowadays a standard procedure to analyze large gene lists. By detecting enriched and depleted gene properties and functions, important insights on the biological system can be obtained. In this chapter, we will give a brief survey of the general layout of the methods used in an ontological analysis and of the most important tools that have been developed.

Key words: Annotation databases, ontology, enrichment analysis, biomarkers, systems biology.

1. Introduction

The introduction of DNA microarrays in the mid-1990s revolutionized the field of molecular biology (1). These first high-throughput techniques allowed the expression of thousand of genes to be monitored simultaneously, which implied important means not only for the theoretical investigation of cellular function but also for many applied sciences. This technology opened new prospects, particularly in cancer research and therapy, as it lets the changes of expression levels in pathological conditions compared to normal tissues to be traced (2). In this way, it is possible to detect the genes that are significantly over- or underexpressed in, for example, cancer cells compared to healthy control cells,

R. Matthiesen (ed.), *Bioinformatics Methods in Clinical Research*, Methods in Molecular Biology 593,
DOI 10.1007/978-1-60327-194-3_15, © Humana Press, a part of Springer Science+Business Media, LLC 2010

and it can be hypothesized that many of these genes are actively involved in the formation of the pathology. Once the gene list representing the biological system is obtained, the next step consists of translating this gene list into biological knowledge under a systems biology point of view (3, 4). This means that the different properties of the genes in the list (e.g., their molecular functions, biological pathways, etc.) have to be analyzed and the most outstanding features need to be detected. In the case of cancer investigation, this analysis is an important step toward a more reliable understanding of the underlining biological mechanisms, which is an important initial step in the design of therapies and drugs.

A number of algorithms and methods have been developed to deal with the automated functional analysis (5). In the first section of this chapter, we will review the general methodology shared by almost all algorithms for functional analysis and the design of the underlying annotation databases. This includes the process of selecting an appropriate set of reference genes, assigning annotations to the genes, calculating the statistical significance of enrichment and/or depletion of all annotations assigned to the input gene list, and applying a correction for multiple testing. Furthermore, we will discuss some additional technical aspects like the mapping of different input gene identifiers and the range of generally applied annotations.

In the second section, we will give an overview of the available algorithms and web tools, briefly discussing their general functionality, particularities, and, if applicable, the improvement or innovation they introduced.

Finally, in the last section, we will present a new tool (Annotation-Modules) that notably expands the number of annotations analyzed and additionally consider the combinations between them. This is an important step toward the adaptation of this kind of ontological analysis tool to many of the newly emerging high-throughput techniques in molecular biology.

2. A Basic Outline of the Methods

As mentioned, the main goal of this type of analysis is to respond to questions such as "Which gene functions or properties (annotations) are statistically enriched or depleted among the genes in a given list compared to a statistical background (set of reference genes)?" The genes in this list are normally obtained from an experiment (sometimes in silico) and are generally important biomarkers to describe and understand the biological system under investigation (e.g., differentially expressed genes under pathological conditions). Therefore, significantly depleted or enriched gene functions or properties might give valuable hints to interpret the analyzed biological system. Crucial steps in such an analysis are the selection and assignment of the gene properties (annotations), the correct selection of the reference genes, the choice of the statistical model to calculate the *p*-values, the correction for multiple testing, and an appropriate, user-friendly presentation of the results.

2.1. Commonly Used Annotations

The fundamental of all functional annotation algorithms that have been developed over the last years is the underlying annotation database, which holds, generally speaking, all of the available information about the genes. Several functional annotation databases exist that are commonly used in this kind of analysis.

Probably the most important is the Gene Ontology (GO) project (6), which describes gene and gene product attributes in any organism (7). It includes three structured vocabularies (*ontologies*) that describe the gene products in terms of their associated biological processes, cellular components, and molecular functions in a species-independent manner. On the other hand, the GO project also facilitates the annotation of gene products associating the ontologies to the genes and gene products. Each entry in GO has been assigned a unique numerical identifier with the general nomenclature GO:*xxxxxxx*. Furthermore, each identifier is associated with a term name such as "cell," "fibroblast growth factor receptor binding," or "signal transduction. Each term belongs exclusively to one of three ontologies: molecular function, cellular component, or biological process. The ontologies are ordered in directed acyclic graphs (**Fig. 15.1**), which are hierarchical structures having the particularity that a child term (more specialized term) can have many parent terms (more general or less specialized terms).

Figure 15.1 shows a subgraph of the GO term "metabolic process." It can be seen that the term "cellular biosynthetic process" has two parents: "cellular metabolic process" and "biosynthetic process," which arises because "cellular biosynthetic process" is a subtype of "cellular metabolic process" and "biosynthetic process." "Cellular biosynthetic process" is more concrete or specialized than the more general term "biosynthetic process." This hierarchical structure has a direct and important consequence. If any of the genes is annotated to the term "cellular biosynthetic process," it is automatically also annotated to both parent terms: "cellular metabolic process" and "biosynthetic process." This occurs because the GO terms obey the true path rule.

Another commonly used vocabulary is available in the KEGG pathway database: the Kyoto Encyclopedia of Genes and Genomes (8). KEGG consists of a manually drawn collection of

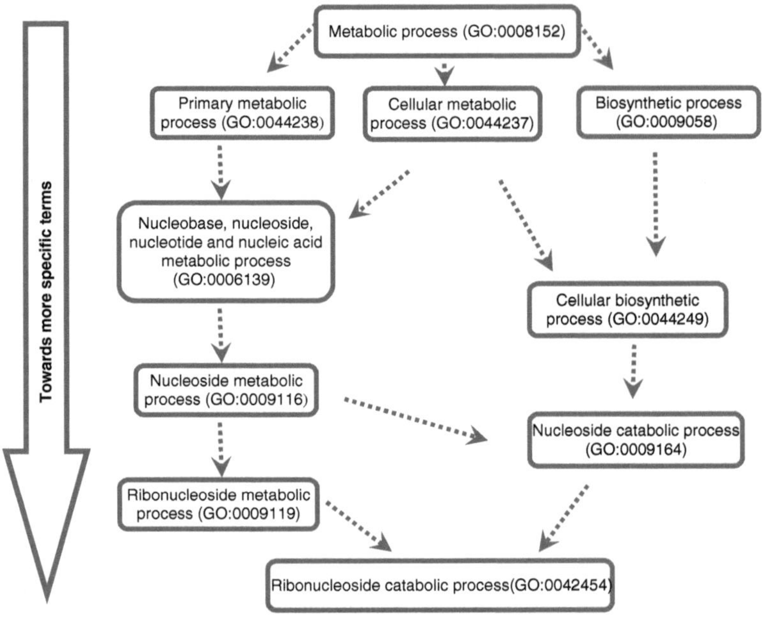

Fig. 15.1. The figure illustrates the structure in which the functional terms are organized in the Gene Ontology by means of a subgraph of the GO term "metabolic process." The terms are ordered in a hierarchical structure called a *direct acyclic graph* (DAG). The categories are ordered from more general (*top* of the graphic) to more specific terms (*bottom* of the graphic).

pathway maps and focuses on molecular interactions, chemical reactions networks, and relationships between the gene products (9). The knowledge is divided into several main categories: (1) metabolism (e.g., carbohydrate metabolism, energy metabolism, lipid metabolism, etc.), (2) genetic information processing (e.g., transcription, translation, folding, etc.), (3) environmental information processing (e.g., membrane transport, signal transduction, and signaling molecules and interaction), (4) cellular processing (e.g., cell growth and death, immune system, nervous system, etc.), and (5) human diseases, with special emphasis on cancer and neurodegenerative diseases like Alzheimer's and Parkinson. Note that all main categories are successively divided into subcategories, which leads to a structured, tree-like hierarchy of annotations.

The keywords from the Swiss-Prot/UniProt knowledge database (10) constitute a third commonly used vocabulary, which associates functional categories with gene products (11). The keywords are divided into 10 principal categories, including biological process, cellular component, coding sequence diversity, developmental stage, disease, domain, ligand, molecular function, PTM (posttranslational modifications), and technical term. The keywords themselves are also organized in a hierarchical structure similar to the GO terms. For example, the keyword "amino acid transport" (protein involved in the transport of amino acids) is also a member of the more general categories transport and biological process.

The annotations described above are by far the most commonly applied over the last years. Note, however, that the depletion/enrichment analysis that we describe in the following sections can generally be applied to any annotations that can be assigned in the form of a label or item (like the GO terms). Therefore, no limit exists on the biological annotations that can be used although the discretization of continuous values is needed in some cases. Some recently developed tools (or newest versions of older tools) took advantage of this possibility and incorporated new features such as the analysis of transcription factor-binding sites or the posttranscriptional regulation of gene expression by microRNAs (*see* **Section 3.1**). Moreover, even continuous values (such as the G+C content of the mRNA or the number of tissues where the gene is expressed) can be used as labels if they are previously classified (*see* **Section 4.1**).

2.2. Basic Workflow

Although many different algorithms have been developed in recent years, the basic procedural method is the same. **Figure 15.2** shows a schematic workflow of the most important steps that are shared by all algorithms. In general, the input data consist of a gene list that usually was obtained by a previous experiment (for example, differentially expressed genes). First, the annotations are assigned to each of the genes in the input gene list by means of an underlying annotation database. The random variable, which will be tested later, is the number of genes in the input list that belong to a given annotation, and therefore the second step consists of finding for all annotations the assigned genes (**Fig. 15.2**, step c). If the number of genes for a given item is known for the genes in both the reference set and the input list, the enrichment and depletion of this item can be tested for a given null hypothesis (see the next section). The calculated *p*-values must be corrected for multiple testing; otherwise, the wrong biological conclusions may be drawn (*see* **Section** 2.5). Finally, the last step normally consists of representing the results in a compact and user-friendly way. Given the vast amount of data that is normally produced in the output of this kind of analysis, this is not a trivial point.

2.3. Statistical Methods and Models

The first step consists of assigning all (user-chosen) annotations to the genes in the reference set and input list (**Fig. e 15.2**). After this step, we can calculate N_p and n_p, which are the number of genes assigned to a given annotation $\boldsymbol{A}$, in the input gene list and reference list, respectively. Using these numbers, a coefficient for

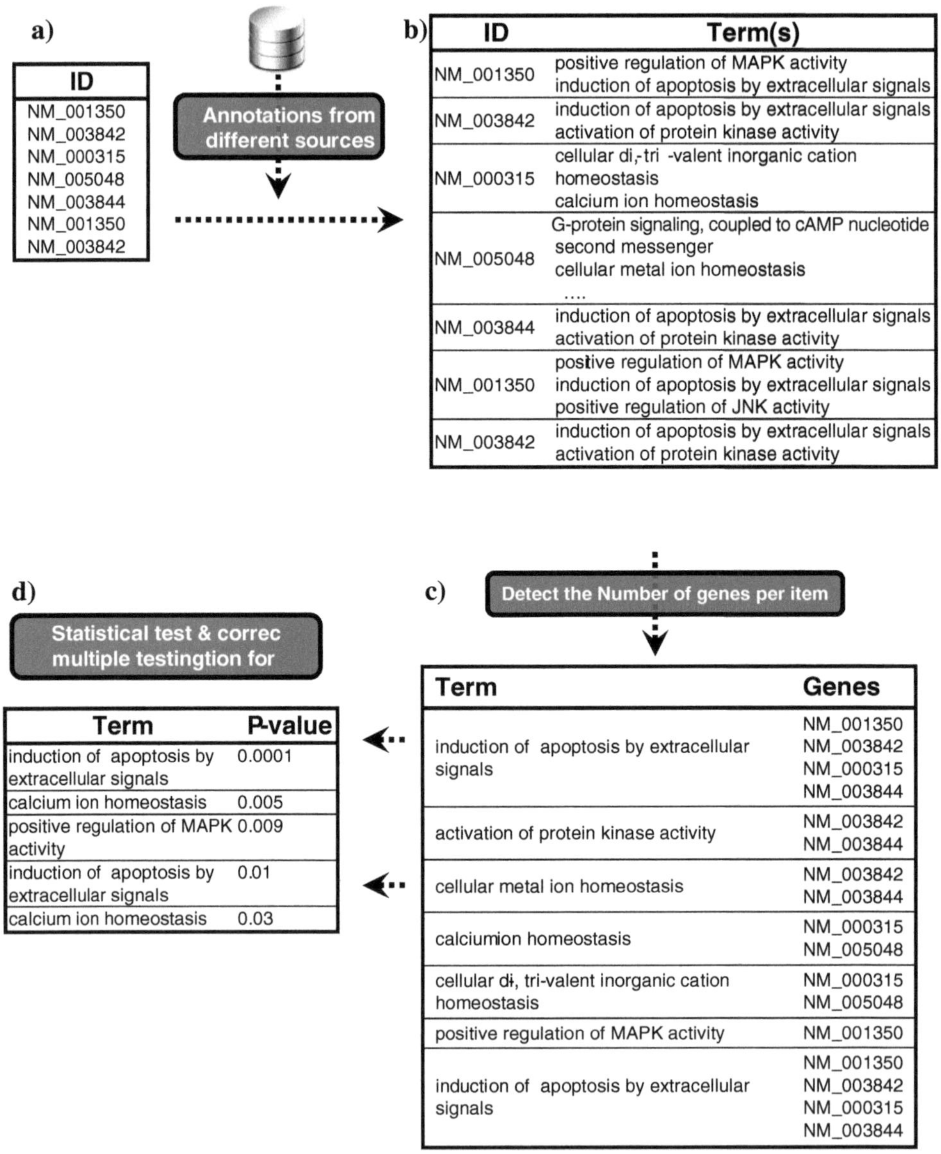

Fig. 15.2. The schema gives an overview of the most important steps in calculating the statistical significance of the enrichment or depletion of an annotation (item) for a gene list. First, the annotations are assigned to the genes in the input list (**a**, **b**). These labels can be a functional category from the GO ontology, a predicted microRNA, or any other annotation that can be assigned by a label or item. Note that even continuous values, like the expression breadth or codon usage of a gene, can be assigned by binning the values. The next step (**c**) consists of finding the number of genes assigned to each annotation. With the number of genes in the set of reference genes and the supplied gene list, the *p*-values can be calculated (**d**) and corrected for multiple testing.

the relative enrichment or depletion can be calculated for each annotation item $\boldsymbol{A}$:

$$R_e(A_i) = \frac{N_p}{N} \cdot \frac{n}{n_p}, \qquad [1]$$

where n and N are the total number of genes in the input gene list and reference list, respectively. The relative enrichment coefficient can be calculated and interpreted easily: If the coefficient is smaller than 1, then the analyzed item A_i is depleted in the input list relative to the set of reference genes; coefficients greater than 1 indicate the relative enrichment. Moreover, the random expectation is given by 1, and therefore we can say that the farther away from 1 the coefficient is, the more pronounced the relative depletion or enrichment is.

However, any particular relative enrichment can occur with a nonzero probability just by chance, and even coefficients "far away" from 1 may turn out to be not statistically significant. Therefore, the aim of the statistical test is to estimate the probability that an observed relative enrichment coefficient is statistically significant or obtained just by chance alone. To this end, many different statistical models have been implemented, including the hypergeometric (12), binomial, chi-square (13), and Fisher's exact test (5, 14). However, besides the fact that different statistical tests have been applied in the past, it can be shown that there is just one single exact null distribution, the hypergeometric distribution (15). Equation [1] shows the hypergeometric distribution, where n_n is the number of genes in the reference set lacking the annotation (number of negatives) and x is the number of genes in the input gene list assigned to the annotation A:

$$P(x = i) = \frac{\binom{n_p}{x}\binom{n_n}{N - i}}{\binom{n_p + n_n}{N}}. \qquad [2]$$

Note that by assuming this null distribution, we implicitly assume that the genes in the input list and the rest of the genes (e.g., the genes in the reference set minus those in the input list) have the same probability of belonging to a given annotation. Furthermore, for a large number of genes, the hypergeometric distribution can be approximated by the binomial distribution, which is computationally less demanding.

2.4. The p-Values

In general, null distributions are probability density functions that directly give us the probability of occurrence of a given value of the random variable (for example, the probability of observing a given number of genes for a given annotation category A). The null distribution therefore gives us a realization of the random variable, which we need to test against some alternative hypothesis H_a. In general, one chooses a priori a probability alpha, called the *significance level*, to get a Type I error (rejecting the H_0 when it is actually true) that must not be exceeded. The significance level is the maximal p-value for which H_0 would be rejected. The determination of the p-value depends largely on the choice of the alternative hypothesis, which can be (1) enrichment (one-sided test to determine if the enrichment of the annotation is statistically significant), (2) depletion (one-sided test to determine if the depletion of the annotation is statistically significant), and (3) enrichment/depletion (two-sided test to determine if the category is either significantly enriched or depleted without distinguishing between enrichment and depletion).

2.4.1. One-Sided Tests

The most common definition of the p-value for a one-sided test is given by the cumulative density function. Equation [3] shows the cumulative density function at point x (the number of genes belonging to a given category in the input gene list). The CDF_x can be interpreted as the probability of finding at most x genes by chance assigned to the category under analysis. If the alternative hypothesis is "depletion," then the CDF_x at position x corresponds directly to the p-value. Otherwise, if the alternative hypothesis is "enrichment," the p-value can be calculated as $1 - CDF_x$ (16):

$$\mathrm{CDF}_x = \sum_{i=0}^{x} \frac{\binom{n_p}{i}\binom{n_n}{N-i}}{\binom{n_p+n_n}{N}}. \qquad [3]$$

2.4.2. Two-Sided Test

If the alternative hypothesis is either enrichment or depletion, several definitions to calculate the p-value exist: (i) A first approach is the *doubling approach* (17), which defines the two-sided p-value as twice the minimum p-value from the one-sided tests for enrichment and depletion; (ii) a second approach is called the *minimum-likelihood approach* (18), which defines the p-value as the sum of all probabilities that are smaller than or equal to the probability at point x (the observed number of genes for a given category).

2.5. Correction for Multiple Testing

A crucial step that should follow the statistical analysis, preceding the interpretation of the outcomes, is the correction for multiple testing. Note that this type of correction is not specific for ontological analyses, but for all statistical tests where many different hypotheses are tested simultaneously (19). When many different hypotheses are tested at the same time, a control of the increased Type I error (rejecting a hypothesis when it is actually true) is needed. Note that an increased Type I error in this kind of ontological analysis would lead one to infer statistical significance and therefore often biological meaning to many functional annotations when actually this conclusion cannot be drawn.

Although this issue is of outstanding importance, it is still controversially discussed, and many different correction methods exist whose applicability might depend largely on the analyzed data structure (5, 20–22). In the following, we will review the most important methods, also briefly discussing their strengths and weaknesses.

2.5.1. Bonferroni, Sidak, and Holms' Step-Down Adjustment

The traditional concern in multiple-hypothesis-testing problems has been about controlling the probability of erroneously rejecting even one of the true null hypotheses, that is, controlling the family-wise error rate (FWER). If C independent null hypotheses are tested, the probability of making at least one Type I error is given by

$$\alpha = 1 - (1 - \alpha_{\text{per-comparison}})^{C}.$$

In case of a dependent null hypothesis, at least the following inequality holds:

$$\alpha \leq \alpha_{\text{per-comparison}} \cdot C.$$

The experiment-wide error increases with the number of comparisons. Therefore, in order to retain the same overall rate of false positives (the number of erroneously rejected null hypotheses), the standards for each individual comparison must be more stringent. Intuitively, reducing the size of the allowable error (alpha) for each individual comparison by the number of comparisons will result in an overall alpha that does not exceed the desired limit. This way of readjusting the significance level (multiplying the significance level for individual comparisons by $1/C$) is called the *Bonferroni correction* (23). The use of Bonferroni very often is a good choice if few hypotheses are tested (less than 50). However, it is known to be overly conservative if the number of hypotheses is large. That means that many null hypotheses fail to be rejected, and therefore interesting biology might be missed in such cases (24, 25).

Sidak correction is slightly less conservative than Bonferroni correction and is often used in microarray analysis. The p-values are corrected by the following formula:

$$p_{i,new} = 1 - (1 - p_i)^{R-(i+1)}$$

where p_i is sorted in ascending order and $p_{i,\,new}$ is the corrected p-value. The i index starts at 1.

A related method, as it also controls the FWER, is Holm's step-down group of methods, which are, in general, less conservative than the Bonferroni correction (26, 27). This method can be decomposed into two steps. First, one has to order the resulting p-values of all statistical tests from the smallest to biggest values. Second, each p-value is tested at the significance level of alpha$/(C–i)$, where i is the ith-smallest$p$-value.

2.5.2. Benjamini and Hochberg's False-Discovery Rate

A different method on how to consider errors in multiple testing was proposed by Benjamini and Hochberg (28), who proposed the false-discovery rate (FDR). The FDR is the expected proportion of erroneous rejections among all rejections. If all tested hypotheses are true, controlling the FDR controls the traditional FWER. However, when many of the tested hypotheses are rejected, indicating that many hypotheses are not true, the error from a single erroneous rejection is not always as crucial for drawing conclusions from the family tested, and the proportion of errors should be controlled instead. This implicates bearing with more errors when many hypotheses are rejected, but with less when fewer are rejected. The method works in the following way: Let $p(1) <= p(2) <= \ldots <= p(m)$ be the uncorrected and ordered *p*-values for the *m* hypotheses tested. The Benjamini and Hochberg procedure rejects all null hypotheses $H_0(i)$ for which

$$p(i) \leq \frac{i}{m} \cdot \alpha. \quad [4]$$

The practical difference between FDR and FWER is neither trivial nor small. In general, it is believed that for data sets with a high correlation between variables, the FDR method works better than methods that control the family-wise error rate (5).

2.5.3. Randomization: The p-Value of p-Values

Finally, with the increases in computational power, resampling methods like bootstrapping or Monte Carlos simulations have become more accessible in recent years. For example, Berriz et al. (29) proposed a Monte Carlo simulation that calculates a kind of "*p*-value for the *p*-values." Briefly, a gene list of the same size as the original list is drawn randomly from the set of reference genes. This random draw is performed *X* times and for each member of the random list, a *p*-value is calculated for the different annotation items. The corrected *p*-value for a given annotation item is then defined as the fraction of random *p*-values that are as good as or better than the observed *p*-value.

2.6. The Set of Reference Genes

A crucial issue in the assessment of statistical significance is the correct selection of the set of reference genes. Using the number of genes assigned to an annotation, and the corresponding number of genes that are not assigned, the *p*-values are calculated as probabilities that in the submitted gene list, more genes (enrichment) or fewer genes (depletion) are assigned to the item than expected by chance alone. The reference set determines the random expectation assuming a hypergeometric null distribution. Therefore, a wrong selection of the reference genes will lead to biased *p*-values, which, in turn, might lead to wrong biological interpretations of the statistical analysis. As a rule of thumb, the pool of reference genes should contain all genes that might appear in the input list. For example, if the analysis is on differentially expressed genes, the input gene list theoretically can be composed of all genes that are on the DNA microarray chip, and consequently the reference set must be made up of all genes on the chip.

3. Brief Discussion of Available Tools

The first automatic ontological analysis approach using Gene Ontology was published in 2002 by Khatri, Draghici, et al. (13). Since then, several dozens of ontological tools have been made available [for a review, see Khatri and Draghici (5)]. As mentioned before, the general approach is basically the same in all of the tools, but the methods differ notably in some aspects, which might not just condition the choice of the user but also influence the results of the analysis. Among the aspects that might make one tool more appropriate for a given type of analysis than other tools are the accepted input IDs (e.g., Swiss-Prot, RefSeq IDs, etc.), the scope of the analysis (e.g., the "size" of the underlying annotation database), the visualization capabilities (e.g., graphical output), and installation and performance issues. These points will most likely just influence the researcher's preferential choice; however, there are also two crucial aspects that might distort the results, leading to biological misinterpretations. These two important issues are the correct selection (and the availability of this possibility in the tools) of the reference set of genes and the correction for multiple testing.

In this section, we will discuss five different tools: Onto-Express (13, 30), FatiGO+ (20, 31), DAVID (32, 33), g:Profiler (34), and GENECODIS (35). We have chosen these tools based on two criteria: (1) tools published years ago but being still maintained and developed further, or (2) recent tools implementing new advances. We will briefly discuss the differences among these tools regarding the points mentioned above (*see* **Table 15.1** for a summary).

3.1. The Underlying Annotation Database

In theory, each biological property that can be assigned to a gene in the form of a label or item can be used to drive a depletion/enrichment analysis. Functional annotations like Gene Ontology motivated the development of tools for the ontological analysis of large gene lists, and therefore at least one of the three mayor vocabularies of functional annotations (Gene Ontology, KEGG pathways, and Swiss-Prot keywords) is available in all of the tools presented here.

However, some of the tools stand out due to the incorporation of new biological properties that might be important for a complete interpretation of some gene lists. One example is the incorporation of predicted microRNA binding sites. The posttranscriptional regulation by microRNAs is now recognized to be a key player in many important biological functions and pathways, also realizing important roles in animal evolution (36). It is estimated that at least one third of all genes are subjected to posttranscriptional regulation by microRNAs. Furthermore, many cases are known in which microRNAs are involved in the formation of cancer (37). Therefore, especially for gene lists derived from cancer tissue, information on the regulation by microRNAs may be important to understanding the key processes that lead to the observed pathology. Currently, these annotations are implemented in FatiGO+, g:profiler, and Annotation-Modules (*see* **Section 4**).

Another source of knowledge on the regulation of gene expression is given by the presence of binding sites for the transcription factors. There are several methods to detect transcription factor-binding sites (TFBS); however, the most commonly used one is probably detection via scoring position frequency matrices (PFMs). Briefly, PFMs reflect the probabilities that a nucleotide will appear at a given position within a motif and are generally derived from experimentally verified target sites. Note, however, that all methods have a serious problem with overprediction; that is, they predict too many false positives (TFBS that are actually not functional). The reason is simply that the signal on a sequence level is not strong enough and the functionality of TFBS often depends on other factors like the chromatin state or the availability of other proteins. Therefore, an analysis driven by annotations of TFBS will be quite "noisy," and the results should be treated with caution. Nevertheless, in cases where a strong enrichment of a given TFBS is found, this may uncover the participation of the corresponding transcription factor in the observed pathology and point out targets for treatment. Currently, TFBS can be analyzed in FatiGO+, g:profiler, and Annotation-Modules.

Finally, some of the presented tools incorporate information on protein–protein interactions (PPI). A large number of interactions in the gene list with respect to the reference group may indicate the existence of one or more complexes (or at least the existence of a complex interaction scheme) within the selected genes. PPIs can be analyzed with FatiGO+ and David.

3.2. Reference Set and Correction

As mentioned above, the correct choice of the reference genes is a crucial aspect in each ontological analysis, as otherwise the *p*-values will be biased. In the past, many tools used either all genes on a microarray chip or all genes in the genome as the statistical background. However, in many cases these might not be sufficient. For example, numerous current research projects are focused just on certain genomic regions like the ENCODE project (38) or the human epigenome project (39). For a gene list obtained from this kind of analysis, neither all genes on a microarray chip nor all genes in the genome would constitute an appropriate statistical background. In such cases, it is indispensable to allow the user to provide a customized set of reference genes in order to carry out a correct statistical analysis. With the exception of g:profiler, all the tools presented here have now incorporated the possibility to upload a customized gene background in their newest versions.

The second crucial step in this kind of analysis is the correction for multiple testing. In the past, many tools did not implement any corrections and presented just the uncorrected *p*-values in the output, which can lead to fatal misinterpretations of the biological background. Currently, however, all of the tools discussed implement at least one method for the correction of multiple testing.

3.3. Input IDs

One of the most important and critical issues in an ontological analysis is the mapping between different biological entities (e.g., between identifiers from different databases like Swiss-Prot, GeneBank, RefSeq, Gene Ontology, KEGG, etc.) (40).

In general, two situations exist in which we need to map between different biological entities.

1) Our current knowledge is spread out over a huge number of databases. Many of these databases host information for many different species but are specialized on a subset of biological entities: UniProt focuses on proteins, Entrez Gene on genes, or RefSeq on transcripts, etc. Furthermore, our knowledge is annotated at different levels. For example, the GO categories are annotated on a protein level (e.g., Swiss-Prot accession or IPI IDs), while microRNA target sites are normally annotated on a transcript level. Therefore, in order to be able to unify the largest possible amount of information, we need to map between different IDs.
2) The range of accepted input IDs will greatly increase the applicability of the tools; thus, having a large number of accepted input IDs is desirable. However, this again requires mapping between different IDs.

The pioneering versions of the programs just supported Affymetrix IDs plus some of the most prevalent IDs, such as GeneBank, Swiss-Prot, RefSeq, and Ensembl. Only relatively recently has the whole degree of complexity, which entails the mapping between different biological entities, been addressed. Particularly mentionable are two sophisticated mapping concepts that have recently been developed. First is Onto-Translate, which currently can perform 462 types of mappings among 29 different types of IDs concerning 53 organisms (40). The tool is also integrated into Onto-Tools (30, 47). A second tool that recently addressed this complex problem is incorporated into the DAVID knowledgebase (33).

Table 15.1
A summary of the most important features of the five tools discussed in this section

Tool	Supported input IDs	Annotations	Short description
Onto-Express/ Onto-Tools	Onto-Translate is a powerful tool incorporated int this suite. It can deal with 29 different IDs for 53 organisms.	Gene Ontology, pathways	First published in 2002, this tool is now embedded in a web-accessible software suite called Onto-Tools (41). The underlying database is just composed of the Gene Ontology categories; however, the suite contains some more programs that allow, for example, the analysis of the gene pathways. The algorithm offers six different graphical output options. It implements a "tree view," which reflects the hierarchical structure of the GO terms in the results.
Fati-GO+	The most important gene identifiers are accepted and cross-linked to Ensemble genes (if an input gene is not annotated in Ensemble, it is lost): HGNC symbol, EMBL acc, UniProt/Swiss-Prot, UniProtKB/TrEMBL, Ensembl IDs, RefSeq, EntrezGene, Affymetrix, Agilent, PDB, Protein Id, IPI.	Gene Ontology, InterPro motifs, SwissProt keywords, KEGG pathways, Biocarta, cisRed, miRBase targets, gene expression, PPI, annotations from the biomedical literature	This web tool is an evolution of the previously published FatiGO algorithm (42). It implements a method called *nested inclusive analysis* (NIA) for GOSlim terms, reporting just the deepest (most detailed) level of GO categories where significance is found. The *p*-values are calculated using a Fisher exact test and corrected with the FDR method. The tool also counts with some annotations that cannot be (or rarely are) found in any other tool like regulatory regions and sites (CisRed and TransFac TFBS), microRNA binding sites, or protein–protein interactions.
DAVID	RefSeq, UniProt, UniGene, Affymetrix, Entrez Gene ID, Gene Bank Accession, Gene Symbol, Flybase ID.	Gene Ontolgoy, KEGG, Biocarta, protein–protein interactions, InterPro, disease and sequence properties, bio-pathways, homologies, gene functional summaries, gene tissue expressions, literature	This tool suite (43), introduced in the first version of DAVID, mainly provides typical batch annotation and gene-GO term enrichment analysis to highlight the most relevant GO terms associated with a given gene list. The new version of the tool keeps the same enrichment analytic algorithm but with extended annotation content coverage, increasing from only GO in the original version of DAVID to currently over 40 annotations. It implements a modified Fisher exact test, called *EASE score*, to estimate the statistical significance of the analyses; however, it does not correct for multiple testing.

(continued)

Table 15.1 (continued)

Tool	Supported input IDs	Annotations	Short description
g:Profiler	The program supports most of the generally used IDs in the different species.	Gene Ontology (GO) terms, KEGG, and REACTOME pathways, and TRANSFAC motifs	The application is web-based [available under (44)]. It supports a great number of species (31) and allows even the input of gene lists with mixed identifiers. It corrects for multiple testing, applying a new resampling method designed for complex and structured functional profiling data. It furthermore allows the analysis of ordered gene lists, and many different output options are available, both graphical and text. The big disadvantage is that the user can neither choose between different sets of reference genes nor upload reference sets.
GENE-CODIS	GeneID,Gene Name, UniGene, Protein Accession, RefSeq, CGD ID, FlyBase ID, Locus tag, MGI ID, SGD ID.	Species-dependent, for human: GO categories, GOSlim, KEGG pathways, InterPro motifs, Swiss-Prot keywords	The application is web-based [available under (45)]. It calculates not just single annotations but also all combinations up to a user-defined support frequency and allows the use of a user-defined set of reference genes. It implements the hypergeometric distribution and chi-square test and supports correction by FDR and randomization. It supports 13 different species (46).

3.4. Visualization Capabilities

Often a graphical output may give an important summary of the results, which may help the user to better understand the biological implications. In particular, if a hierarchical structure exists among the annotations (as with Gene Ontology), the representation of the results in a hierarchical context might help to better understand the analyzed biological system. The most complete tool concerning the graphical capabilities is Onto-Express/Onto-Tools, which currently implements six different graphical output options.

4. Annotation-Modules: A New Tool for Ontological Analysis

As shown above, the ontological analysis of the gene lists obtained from DNA microarray experiments constitutes an important step in understanding the underlying biology of the analyzed system. Of outstanding importance have been functional annotations like Gene Ontology, KEGG pathways, and Swiss-Prot keywords. However, in recent years, many other high-throughput techniques have emerged, now covering basically all the "omics" fields. For some of these techniques, the generally used functional ontologies might not be sufficient to describe the biological system represented by the derived gene lists. For a more complete and correct interpretation of these experiments, it is important to substantially extend the number of annotations, adapting the ontological analysis to the newly emerging techniques. Recently, a new tool was published (Annotation-Modules) whose most outstanding feature notably extends the underlying annotation database, implementing about 60 different gene annotation features (48). The annotations are derived from many different fields (see **Table 15.2** for an overview), including gene regulation and expression, sequence properties, evolution and conservation, genomic localization, and functional categories. As a second improvement, it examines not only single annotations but also all the combinations, which is important to gain insight into the interplay of different mechanisms in the analyzed biological system.

In this section, we will briefly discuss the underlying annotation database and the method of analyzing concurrent annotations. Furthermore, we will point out the types of analysis under which this tool may have certain advantages over other methods for ontological analyses.

4.1. The Annotation-Modules Database

The Annotation-Modules database implements several features for the first time for enrichment analysis but counts also with the most important vocabularies of functional annotations like

Table 15.2
Overview of the annotations implemented in the Annotations-Modules database

Annotation "field"	Annotated entities
Regulation and expression	Transcription factor-binding sites (TFBS), CpG islands, microRNA target sites, the expression breadth (housekeeping vs. tissue-specific)
Evolution and conservation	Taxonomic depth (last common ancestor taxonomic level in the gene cluster to which the gene belongs), co-localization with phylogenetically conserved elements (PhastCons)
Functional annotations and network properties	GO terms, Swiss-Prot keywords, posttranslational modifications, disease association
Population genetics	Association with SNPs
Sequence properties (mRNA and protein)	GC-content, GC3, GC3s, mRNA length, codon usage (e.g., *Nc*: effective number of codons), protein properties
Miscellaneous	Co-localization with transposons, compositional features of the of promoter region (GC-AT classification)

Gene Ontology and Swiss-Prot keywords. **Table 15.3** shows the most important gene properties that have been implemented for the first time in Annotation-Modules. Furthermore, the tool also considers annotations on the regulation of gene expression like microRNA target sites or transcription factor binding sites, which are implemented in just one or two other tools (*see* **Section 3**).

Moreover, it was reported that the position of the TFBS respective to the TSS is important (49). Annotation-Modules takes this fact into account by binning the promoter region in different ways, assigning the TFBS to different bins as a function of distance to the TSS. In this way, it generates four different annotation sets, dividing the promoter region (from TSS-1500 bp to TSS+500 bp) into 1, 2, 4, and 10 bins.

4.2. The Methods Implemented in Annotation-Modules

As mentioned, an important feature in Annotation-Modules is the detection and analysis of concurrent features, a method that has been proposed and implemented in the GENECODIS algorithm (35). The statistically significant co-occurrence of annotations from different fields may give valuable hints on the interplay of different mechanisms in the analyzed biological system. For example, in cancer investigation, the co-occurrence of certain functional annotations and pathways together with information on the regulation of gene expression (TFBS and microRNA) and protein properties like posttranslational modification may increase the comprehension of the analyzed pathology on a cellular and molecular level and facilitate the development of new drugs.

Table 15.3
An overview on the annotations that are uniquely implemented in the Annotation-Modules database

Annotation	Biological relevance	Method
CpG islands	CpG islands associate with around threequarters of all known TSS. At least in humans, they are very important regulatory regions that are involved in both the normal and disease-related regulation of gene expression (50, 51).	We incorporated the CpG islands predicted by the *CpGcluster* algorithm (52), as those can be calculated easily for each species applying the same thresholds and might have some advantages over other prediction algorithms by not being so sensitive to spurious transposable elements.
Epigenetic state of CpG islands	The epigenetic state of a CpG island influences directly the chromatin state and in general can therefore be viewed as an indicator of activity of gene expression.	Annotation-Modules incorporates a recent prediction of "CpG island strength" based on the epigenetic states, histone modifications, and chromatin accessibility (53).
Overlap with genomic elements	The presence of certain genomic elements near or within genes has a clear biological meaning. Important examples are TFBS in the promoter region or CpG islands that overlap the TSS of most housekeeping genes. However, the presence of highly conserved elements – PhastCons (54), SNPs, or transponsable elements – may uncover interesting facts and is a source of biological knowledge.	Beside the "intrinsic", unambiguous regions like exons, introns, 5'UTR (untranslated regions), and 3'UTR, Annotation-Modules defines eight gene regions where it measures the presence or absence of the genomic element (see http://web.bioinformatics.cicbiogune.es/AM/doc.php).
Protein properties	Posttranslational modifications, like phosphorylation, are part of common mechanisms for controlling the behavior of a protein, for instance, activating or inactivating an enzyme by extending the range of functions of the protein by attaching to it other biochemical functional groups.	The information on modifications was extracted from the Swiss-Prot/UniProt KnowledgeBase (55). Furthermore, Annotation-Modules also implements features concerning transmembrane proteins (TRANS_MEM) and the disease relatedness.

(continued)

Table 15.3 (continued)

Annotation	Biological relevance	Method
Expression breadth	The expression breadth is the number of tissues in which the gene is transcribed. With this measure, we can distinguish between housekeeping and tissue-specific genes.	The expression values were derived from the human, mouse, and rat gene atlas (56), which we downloaded from the UCSC table browser. Expression values of different probes of one gene are averaged and considered as expressed if the expression value is higher than 200 units.
Genomic localization	The genomic localization of a gene is believed to be related to some very interesting properties. GC-rich genomic regions are likely to be endowed with several specific features, like high transcription levels, an open chromatin structure, and a very high density of genes, short introns, and associated CpG islands (57).	The IsoFinder algorithm (58) has been used to predict the isochores in the different genomes. Each gene is then assigned to a physical isochore. Finally, the name of the host isochore is used as the name of the annotation label.
Sequence properties	Several sequence properties are known to be related to function. For example, *Nc*, which is based on the codon homozygosities, might reveal constraints on the evolution of codon usage (59). The synonymous codon usage may be caused by various forms of natural selection, to optimize the efficiency and accuracy of translation or maintain structural features of the mRNA or DNA.	Annotation-Modules implements the synonymous codon usage, the length of the coding region, the length of the messenger RNA, the G+C content of the messenger RNA, the G+C content at third position, and the G+C content of synonymous codons at third position.

The main problem in this kind of analysis is coping with the extremely high number of possible combinations. The number of theoretic combinations is given by

$$N_{n,k} = \sum_{i=1}^{i=k} \binom{n}{i}, \quad [5]$$

where n is the number of different features and k is the number of items per combination (the size of the combination set). This equation shows that if the number of different items or annotations is substantially increased, the algorithm will become computationally unfeasible. For example, if we assume 1,000 different items (which can easily be reached by utilizing the extensive number of annotations in our database) and a maximal combination size of $k = 3$, this would lead to approximately 167 million different combinations. Therefore, it is mandatory to introduce some approximations in order to limit the number of combinations to an analyzable size. The approach applied in Annotation-Modules is based on two concepts or assumptions: (1) A combination between an enriched and a depleted set of annotations is less likely to be statistically significant, and (2) a maximum number of combinations are processed on each level k.

Briefly, the modified algorithm performs the following steps:

(1) Calculate the p-values for all single annotations, generate one set of depleted and one of enriched single annotations, initialize the sets of enriched and depleted combinations of annotations, and store the significant annotations.

(2) Combine in the following order as long as the number of combinations does not exceed the maximum number of combinations: (a) enriched single annotations vs. enriched combinations of annotations; (b) depleted single annotations vs. depleted combinations of annotations; (c) depleted single annotations vs. enriched combinations of annotations; and (d) enriched single annotations vs. depleted combinations of annotations.

(3) Calculate the p-values of all resulting combinations and save the significant ones.

(4) Generate the new sets for enriched and depleted combinations of annotations corresponding to the current level k.

(5) Repeat steps 2–4 until the threshold for k is reached.

(6) Apply the multiple testing separately for each k.

4.3. A Short Guide and Working Example

The data submission process in Annotation-Modules includes three steps on three different, dynamically generated input pages. The dynamic generation of the input pages is necessary, as the number of available annotations varies widely between the different species and depends also on the chosen gene table.

1) The user has to indicate the species for which he or she wants to carry out the enrichment/depletion analysis. The current version of Annotation-Modules is implemented for human (hg18), mouse (mm8), and rat (rn4).

2) The tool's web interface will ask for the input data and some method parameters (**Fig. 15.3**).

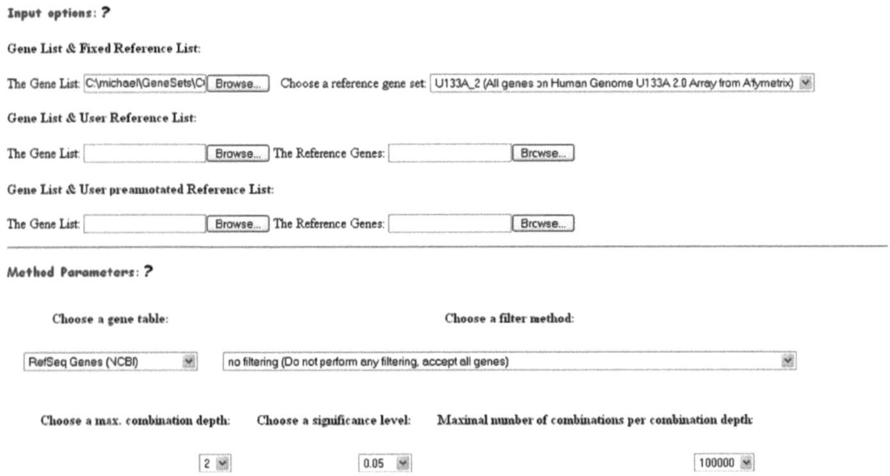

Fig. 15.3. A screenshot of the second "input page" of Annotation-Modules. Three different data input options exist (*upper part*). Worth mentioning is the possibility to provide a user-specific set of reference genes (statistical background) and the option to upload a preannotated gene list. This allows the user to analyze customized annotations and to combine them further with all the features in the annotation database. There are five method parameters for the client's use. See the documentation of the program (60) or the tutorial (61) for further details.

3) With the supplied species and gene table, the web interface dynamically generates the third page, where the different annotations can be chosen. **Figure 15.4** shows a cutout of this page for the human RefSeq genes. Depending on the species and gene table, up to 60 different annotations can be chosen.

After selecting the annotations, these are sent back to the server, and the Java program that performs the actual analysis is launched. The results will appear in the browser window; alternatively, a link is given where the results are deposited when finished. The program writes a total of four output files. Two are in HTML format (for a fast overview on the results), one is a more extensive text file, and the fourth is an overview of the process (number of mapped IDs, chosen parameters, etc.).

To show the potential of this tool, we analyzed all human RefSeq CpG island genes [those genes having a *CpGcluster* (52) CpG island overlapping their transcription start site]. **Figure 15.5** shows part of the output in HTML format. This output page contains several links that permit access to information on the

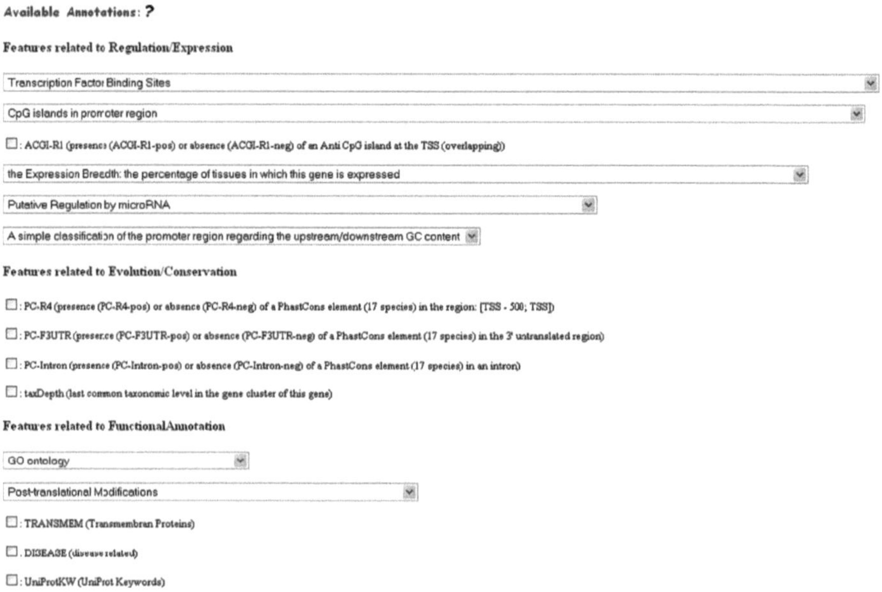

Fig. 15.4. A cutout of the available features for human RefSeq genes. The annotations available vary depending on the chosen species and the gene table. The annotations are roughly divided into six classes: features related to Regulation/Expression, Evolution/Conservation, FunctionalAnnotation, PopulationGenetics, Miscellaneous, and Sequence Properties. Within these classes, some related features are grouped and placed in lists, from which just one type of annotation can be chosen. Examples are the different predictions of microRNA binding sites or the different predictions of CpG islands.

The Annotation Module	The Module is ...	P-value	FDR-limit	# of genes	Relative Enrichment	Genes
Transducer Sensory transduction	depleted	5.192e-141	2.671e-06	5	0.02	Genes
Sensory transduction G-protein coupled receptor	depleted	4.068e-140	5.342e-06	5	0.02	Genes
Sensory transduction G-protein coupled receptor Transmembrane	depleted	4.068e-140	5.000e-07	5	0.02	Genes
Transducer Sensory transduction Transmembrane	depleted	4.068e-140	1.000e-06	5	0.02	Genes
Transit peptide	enriched	2.905e-13	2.670e-03	279	1.29	Genes
Direct protein sequencing Transcription	enriched	7.252e-13	2.965e-04	133	1.41	Genes
Direct protein sequencing Transcription regulation	enriched	8.660e-13	3.018e-04	129	1.42	Genes
Transcription Direct protein sequencing Transcription regulation	enriched	8.660e-13	9.200e-05	129	1.42	Genes

Fig. 15.5. HTML output file of Annotation-Modules for all CpG island human RefSeq genes (all in all, four different output files are written).

annotations and the genes. Furthermore, for each combination, the *p*-value, the FDR limit, the number of genes, and the relative enrichment are listed. The output shows some very significant combinations of annotations. It can be seen that CpG island genes are strongly depleted in modules related to signaling pathways and enriched in modules related to transcription.

5. Concluding Remarks

The ontological analysis of gene lists has been of outstanding importance in recent years in biomedical research. This is evidenced by the fact that just the three most important tools, FatiGO (20, 31), Onto-tools (13, 47), and David (32), are cited more than 2,000 times (by January 2009). However, the fields of molecular biology and medical biology advance rapidly, which will lead to new challenges for the bioinformatics analysis of those experiments. Moreover, current tools still have some shortcomings that will need to be addressed in the future. One critical point is the gene mapping and the redundancies that may be introduced in the analysis due to the use of different identifiers, namely, genes, mRNA, or proteins. For example, the gene ontology terms are assigned at a protein level. However, when the input consists of transcript identifiers, the terms might be assigned to all annotated splice forms, which might introduce a redundancy in the analysis. Along the same line, many splice forms will share the same promoter region, and therefore an analysis of transcription factor-binding sites may be biased, as the same regulatory regions are counted several times. The elimination of redundancies so far is just addressed in the Annotation-Modules tool (48). In the future, the concepts introduced there should be improved, and the elimination of redundancies might depend on both the gene level used (e.g., transcript, gene, or protein) and the particular annotation.

Over the last couple of years, our understanding of the regulation of gene expression has been revolutionized by recognizing the impact of epigenetic modifications and noncoding RNAs. To achieve a more complete understanding of the underlying molecular mechanisms of the pathology, these new findings will have to be incorporated in an adequate way.

Finally, many interesting gene and promoter properties cannot be assigned a label (discrete values like GO terms or TFBS) but are quantitative values (e.g., the number of interaction partners of proteins, base composition and physical properties of promoter regions). These features should also be incorporated in such kinds of analysis in the future [see, for example, the ContDist citation (62)].

Acknowledgments

This work has been partially supported by the Department of Industry, Tourism and Trade of the Government of the Autonomous Community of the Basque Country (Etortek Research Programs 2005/2006) and from the Innovation Technology Department of the Bizkaia County. Support for RM was provided from Ramon y Cajal (RYC-2006-001446). MH furthermore acknowledges support from the Junta de Andalucia Grant No. P07FQM3163

References

1. Schena M, Shalon D, Davis RW, et al. (1995) Quantitative monitoring of gene expression patterns with a complementary DNA microarray. *Science* 270:467–470.
2. Golub TR, Slonim DK, Tamayo P, et al. (1999) Molecular classification of cancer: class discovery and class prediction by gene expression monitoring. *Science* 286: 531–537.
3. Dopazo J. (2006) Functional interpretation of microarray experiments. *Omics* 10: 398–410.
4. Westerhoff HV, Palsson BO. (2004) The evolution of molecular biology into systems biology. *Nat Biotechnol* 22:1249–1252.
5. Khatri P, Draghici S. (2005) Ontological analysis of gene expression data: current tools, limitations, and open problems. *Bioinformatics* 21:3587–3595.
6. Ashburner M, Ball CA, Blake JA, et al. (2000) Gene ontology: tool for the unification of biology. The Gene Ontology Consortium. *Nat Genet* 25:25–29.
7. http://www.geneontology.org/.
8. Ogata H, Goto S, Sato K, Fujibuchi W, Bono H, Kanehisa M. (1999) KEGG: Kyoto Encyclopedia of Genes and Genomes. *Nucleic Acids Res* 27:29–34.
9. http://www.genome.jp/kegg/.
10. http://us.expasy.org/sprot/.
11. Apweiler R, Bairoch A, Wu CH, et al. (2004) UniProt: the Universal Protein knowledgebase. *Nucleic Acids Res* 32:D115–D119.
12. Cho RJ, Huang M, Campbell MJ, et al. (2001) Transcriptional regulation and function during the human cell cycle. *Nat Genet* 27:48–54.
13. Khatri P, Draghici S, Ostermeier GC, Krawetz SA. (2002) Profiling gene expression using onto-express. *Genomics* 79: 266–270.
14. Man MZ, Wang X, Wang Y. (2000) POWER_SAGE: comparing statistical tests for SAGE experiments. *Bioinformatics* 16:953–959.
15. Rivals I, Personnaz L, Taing L, Potier MC. (2007) Enrichment or depletion of a GO category within a class of genes: which test? *Bioinformatics* 23:401–407.
16. Draghici S, Khatri P, Martins RP, et al. (2003) Global functional profiling of gene expression. *Genomics* 81: 98–104.
17. Yates F. (1984) Test of significance for 2×2 contingency tables. *J. Roy Stat Soc Ser A* 147:426–463.
18. Gibbons JD, Pratt JW. (1975) P-values: interpretation and methodology. *Am Stat* 29:20–25.
19. Miller RG. (1991) Simultaneous Statistical Inference. Springer-Verlag, New York.
20. Al-Shahrour F, Diaz-Uriarte R, Dopazo J. (2004) FatiGO: a web tool for finding significant associations of Gene Ontology terms with groups of genes. *Bioinformatics* 20: 578–580.
21. Beissbarth T, Speed TP. (2004) GOstat: find statistically overrepresented Gene Ontologies within a group of genes. *Bioinformatics* 20:1464–1465.
22. Zeeberg BR, Feng W, Wang G, et al. (2003) GoMiner: a resource for biological interpretation of genomic and proteomic data. *Genome Biol* 4:R28.
23. Bonferroni CE. (1935) *Il calcolo delle assicurazioni su gruppi di teste.*, pp. 13–60.
24. Perneger TV. (1998) What's wrong with Bonferroni adjustments. *BMJ* 316:1236–1238.
25. Draghici S. (2003) Data Analysis Tools for DNA Microarrays. Chapman and Hall/CRC Press, Boca Raton, FL.

26. Hochberg Y, Benjamini Y. (1990) More powerful procedures for multiple significance testing. *Stat Med* 9:811–818.
27. Holm S. (1979) A simple sequentially rejective multiple test procedure. *Scand J Stat* 6:65–70.
28. Benjamini Y, Hochberg Y. (1995) Controlling the false discovery rate: a practical and powerful approach to multiple testing. *J Roy Stat Soc Ser B Stat Methodol* 57(1): 289–300.
29. Berriz GF, King OD, Bryant B, et al. (2003) Characterizing gene sets with FuncAssociate. *Bioinformatics* 19:2502–2504.
30. Khatri P, Voichita C, Kattan K, et al. (2007) Onto-Tools: new additions and improvements in 2006. *Nucleic Acids Res* 35:W206–W211.
31. Al-Shahrour F, Minguez P, Tarraga J, et al. (2007) FatiGO +: a functional profiling tool for genomic data. Integration of functional annotation, regulatory motifs and interaction data with microarray experiments. *Nucleic Acids Res* 35:W91–W96.
32. Dennis G, Jr., Sherman BT, Hosack DA, et al. (2003) DAVID: Database for Annotation, Visualization, and Integrated Discovery. *Genome Biol* 4:P3.
33. Sherman BT, Huang da W, Tan Q, et al. (2007) DAVID Knowledgebase: a gene-centered database integrating heterogeneous gene annotation resources to facilitate high-throughput gene functional analysis. *BMC Bioinformatics* 8:426.
34. Reimand J, Kull M, Peterson H, et al. (2007) g:Profiler–a web-based toolset for functional profiling of gene lists from large-scale experiments. *Nucleic Acids Res* 35:W193–W200.
35. Carmona-Saez P, Chagoyen M, Tirado F, et al. (2007) GENECODIS: a web-based tool for finding significant concurrent annotations in gene lists. *Genome Biol* 8:R3.
36. Niwa R, Slack FJ. (2007) The evolution of animal microRNA function. *Curr Opin Genet Dev* 17:145–150.
37. Saito Y, Liang G, Egger G, et al. (2006) Specific activation of microRNA-127 with downregulation of the proto-oncogene BCL6 by chromatin-modifying drugs in human cancer cells. *Cancer Cell* 9:435–443.
38. Birney E, Stamatoyannopoulos JA, et al. (2007) Identification and analysis of functional elements in 1% of the human genome by the ENCODE pilot project. *Nature* 447:799–816.
39. Eckhardt F, Lewin J, Cortese R, et al. (2006) DNA methylation profiling of human chromosomes 6, 20 and 22. *Nat Genet* 38: 1378–1385.
40. Draghici S, Sellamuthu S, Khatri P. (2006) Babel's tower revisited: a universal resource for cross-referencing across annotation databases. *Bioinformatics* 22:2934–2939.
41. http://vortex.cs.wayne.edu/projects.htm.
42. http://babelomics.bioinfo.cipf.es.
43. http://david.abcc.ncifcrf.gov/home.jsp.
44. http://biit.cs.ut.ee/gprofiler/.
45. http://genecodis.dacya.ucm.es/.
46. http://genecodis.dacya.ucm.es/help.html.
47. Draghici S, Khatri P, Bhavsar P, et al. (2003) Onto-Tools, the toolkit of the modern biologist: Onto-Express, Onto-Compare, Onto-Design and Onto-Translate. *Nucleic Acids Res* 31:3775–3781.
48. Hackenberg M, Matthiesen R. (2008) Annotation-Modules: a tool for finding significant combinations of multisource annotations for gene lists. *Bioinformatics* 24: 1386–1393.
49. Vardhanabhuti S, Wang J, Hannenhalli S. (2007) Position and distance specificity are important determinants of cis-regulatory motifs in addition to evolutionary conservation. *Nucleic Acids Res* 35:3203–3213.
50. Neumeister P, Albanese C, Balent B, et al. (2002) Senescence and epigenetic dysregulation in cancer. *Int J Biochem Cell Biol* 34:1475–1490.
51. Shen L, Kondo Y, Guo Y, et al. (2007) Genome-wide profiling of DNA methylation reveals a class of normally methylated CpG island promoters. *PLoS Genet* 3:2023–2036.
52. Hackenberg M, Previti C, Luque-Escamilla PL, et al. (2006) CpGcluster: a distance-based algorithm for CpG-island detection. *BMC Bioinformatics* 7:446.
53. Bock C, Walter J, Paulsen M, et al. (2007) CpG island mapping by epigenome prediction. *PLoS Comput Biol* 3:e110.
54. Siepel A, Bejerano G, Pedersen JS, Hinrichs AS, Hou M, et al. (2005) Evolutionarily conserved elements in vertebrate, insect, worm, and yeast genomes. *Genome Res* 15: 1034–1050.
55. Bairoch A, Apweiler R, Wu CH, et al. (2005) The Universal Protein Resource (UniProt). *Nucleic Acids Res* 33:D154–D159.
56. Su AI, Wiltshire T, Batalov S, et al. (2004) A gene atlas of the mouse and human protein-encoding transcriptomes. *Proc Natl Acad Sci USA* 101:6062–6067.
57. Bernardi G. (2001) Misunderstandings about isochores. Part 1. *Gene* 276:3–13.
58. Oliver JL, Carpena P, Hackenberg M, Bernaola-Galvan P. (2004) IsoFinder: computational prediction of isochores in genome sequences. *Nucleic Acids Res* 32: W287–W292.

59. Wright F. (1990) The 'effective number of codons' used in a gene. *Gene* 87:23–29.
60. http://web.bioinformatics.cicbiogune.es/AM/doc.php.
61. http://web.bioinformatics.cicbiogune.es/AM/tutorial.html.
62. Hackenberg M, Lasso G, Matthiesen R. (2009 Jan 7) ContDist: a tool for the analysis of quantitative gene and promoter properties. *BMC Bioinformatics* 10:7.

Chapter 16

Analysis of Biological Processes and Diseases Using Text Mining Approaches

Martin Krallinger, Florian Leitner, and Alfonso Valencia

Abstract

A number of biomedical text mining systems have been developed to extract biologically relevant information directly from the literature, complementing bioinformatics methods in the analysis of experimentally generated data. We provide a short overview of the general characteristics of natural language data, existing biomedical literature databases, and lexical resources relevant in the context of biomedical text mining. A selected number of practically useful systems are introduced together with the type of user queries supported and the results they generate. The extraction of biological relationships, such as protein–protein interactions as well as metabolic and signaling pathways using information extraction systems, will be discussed through example cases of cancer-relevant proteins. Basic strategies for detecting associations of genes to diseases together with literature mining of mutations, SNPs, and epigenetic information (methylation) are described. We provide an overview of disease-centric and gene-centric literature mining methods for linking genes to phenotypic and genotypic aspects. Moreover, we discuss recent efforts for finding biomarkers through text mining and for gene list analysis and prioritization. Some relevant issues for implementing a customized biomedical text mining system will be pointed out. To demonstrate the usefulness of literature mining for the molecular oncology domain, we implemented two cancer-related applications. The first tool consists of a literature mining system for retrieving human mutations together with supporting articles. Specific gene mutations are linked to a set of predefined cancer types. The second application consists of a text categorization system supporting breast cancer-specific literature search and document-based breast cancer gene ranking. Future trends in text mining emphasize the importance of community efforts such as the BioCreative challenge for the development and integration of multiple systems into a common platform provided by the BioCreative Metaserver.

Key words: Text mining, information extraction, natural language processing, pathways, cancer, diseases, gene raking, document classification, biomarkers, epigenetics.

R. Matthiesen (ed.), *Bioinformatics Methods in Clinical Research*, Methods in Molecular Biology 593,
DOI 10.1007/978-1-60327-194-3_16, © Humana Press, a part of Springer Science+Business Media, LLC 2010

1. Introduction

Current research in biological sciences is generating humongous amounts of heterogeneous experimental data at an increasing pace, especially thanks to advances in high-throughput techniques as used in the proteomics and genomics domains. To effectively manage and utilize experimentally generated results, data standards and formalized experiment descriptions, combined with the design of suitable databases collecting and storing results in the form of structured records, are crucial.

A substantial part of the current knowledge in biology and biomedical sciences is encoded as natural language data, comprising primarily scientific articles but also books, theses, reports, or even patents. Also, most of the biological annotations contain a substantial amount of natural language data such as functional keywords or specialized terminology for characterizing biologically relevant aspects of gene products.

The scientific literature provides descriptions of biological discoveries at different levels of granularity (1). At the molecular level, articles contain detailed biochemical characterizations of genes and proteins in terms of sequence and structural and functional features. To detect associations between particular genes and diseases or phenotypic properties, considerable amounts of population, epidemiologic, or even gene knockout studies using animal models have been published. The literature plays a central role in the current scientific discovery process, from the initial stage of experiment planning to the final step of result interpretation and the subsequent generation of new hypotheses. To allow more efficient access to electronic articles published by a growing number of dispersed biomedical journals and publishers, centralized repositories such as the bibliographic database PubMed, hosted by the U.S. National Library of Medicine (NLM), have been developed.

Based on the availability of literature repositories, a variety of different biological annotation databases have been constructed. The common characteristic underlying manual annotation efforts is the transformation of relevant biological descriptions from scientific papers into structured database records, not only improving information exchange, but also facilitating the analysis of larger collections of biological entities and their relationships by means of bioinformatics techniques (2). Most of the existing annotations lack the original evidence passages from articles used to derive the annotations, making both the interpretation and the reproducibility of annotations especially challenging. This drawback has been partially addressed through the use of experimental evidence qualifiers for annotations, like the Gene Ontology evidence codes for functional annotations or the interaction detection method terms of the PSI-MI ontology for protein interactions.

The rapid accumulation of literature data made purely manual curation efforts obsolete (3), resulting in a significant time delay between the actual publication date of a given article and the manual extraction of relevant information during the database literature curation process. To minimize the workload of manual curation and to improve both the efficiency and the consistency of the overall annotation process, the use of text mining tools and suitable information extraction software that assists the work of curators has been proposed as a potential solution (4). The aim of text mining systems is to find relevant pieces of information hidden within large collections of textual data (e.g., scientific articles) through computational approaches, often based on algorithms from data mining, artificial intelligence (AI), and statistical analysis.

Even with manual literature curation performed by expert database curators, a common challenge is the correct interpretation of the experimental characterizations described by the original article authors, not only due to the underlying requirement of in-depth knowledge of the associated biological subdiscipline and its specific vocabulary, but also due to the intrinsic ambiguity of certain annotation-relevant aspects, such as the unambiguous identification of the correct gene and its corresponding database identifier. Recent trends promoted by collaborations among annotation databases, journal publishers, and text mining researchers are working toward the need to integrate author-based annotations in the form of structured digital abstracts, where literature mining tools will play an important role to assist authors in annotating their own papers (5). The development of annotation databases for specific biological topics has also experienced recent changes in terms of increasing the integration of text mining approaches as part of the information extraction process (6, 7).

The significance of text mining, information retrieval, and extraction tools goes beyond the use by specialized curators or improving the quality of biological databases. Literature mining is becoming increasingly useful for enabling more efficient information access for experimental biologists: it can assist in the analysis and interpretation of large-scale experimental results, providing evidence for qualified relationships between biological entities or associations of proteins to certain biological processes or diseases (1).

The extraction of direct pointers of biological entities such as proteins or genes to their mentions and interaction descriptions in the literature permits access to biologically relevant contextual information not currently covered by structured database records, such as certain experimental conditions or parameters (e.g., used cell lines or cell types) and experimental protocols. Such literature pointers are already provided by popular systems such as iHOP (8), which allows users to navigate the biomedical literature through the proteins co-mentioned in sentences.

This chapter describes the basic features of natural language data generally exploited by computational text processing tools. A brief introduction to the most relevant databases currently storing biomedical articles and abstracts, together with short descriptions of how their content can be accessed, will be discussed in this chapter. The main natural language processing concepts and most relevant tasks currently applied to the biomedical literature will be briefly introduced.

A selected collection of existing text mining and information extraction tools specifically developed to address biological questions and to analyze associations of proteins to diseases as well as complex biological relationships like pathways or processes will be described. For certain scenarios, existing applications do not satisfy very specialized user demands. Thus, some of the basic steps required for the implementation of biomedical literature mining tools will be provided in this chapter as a kind of case study.

Both to determine the performance of a given method and to compare it to other alternative strategies, an evaluation of the generated results using a test set data collection is needed. Some of the recent community evaluation initiatives for text mining strategies applied to biologically relevant tasks as well as future trends will be discussed at the end of this chapter.

2. Electronic Texts

The growing interest in the use of text mining strategies applied to the biomedical domain is directly related to the availability of electronic texts and digitalized articles through the web. Also, methodological aspects derived from the web mining community, which try to process the growing amount of data currently accessible through the web in the form of electronic documents, but also as multimedia or image files, have influenced current efforts in biomedical text mining.

Documents can be considered as the basic data unit generally processed by literature mining systems. In principle, it is possible to distinguish between unstructured documents written in natural language (mainly English), as is the case with scientific articles, and structured documents like annotation records, or even lexical resources and controlled vocabularies. Depending on the underlying format and encoding of a given article, additional preprocessing steps might be required. Currently, most of the available articles are provided as PDF, HTML, SGML, XML, and/or plain text files. Despite the availability of both open source and commercial software for handling these document formats, journal- and publisher-specific article formatting still represents a considerable hurdle for implementing a general-purpose, efficient full-text article preprocessing pipeline.

3. Relevant Features of Free Text Data

Natural language is used as the common vehicle to satisfy communicative needs in biomedical sciences, and despite its inherit variability, flexibility, and dynamic change over time (not only in terms of vocabulary), there are features of written language that can be exploited by computational approaches to generate statistical models for certain aspects of language. Similar to bioinformatics strategies, which try to find functionally relevant sequence patterns or profiles within experimentally generated results, natural language processing (NLP) and text mining systems try to detect existing regularities encountered within the linguistic structures of natural language texts. A range of statistical methods are currently being applied to discover common patterns that occur in the literature as well as rules underlying existing constraints of syntactic and semantic structures of well-formed utterances of language (9). Many of the used algorithms and statistical machine learning methods are showing similarities to the general techniques currently applied by traditional bioinformatics approaches.

Biomedical text processing systems operate at different levels of granularity, using processing features relevant to detect special characteristics of natural language from the basic level of characters and strings to aspects associated with complex relationships derived from multidocument collections. The overall text processing levels encountered are shown in **Fig. 16.1**.

Processing at the level of characters and text strings is being used for breaking a text up into its constituent tokens (*tokenization*), implying the detection of the start and end characters for each token. For example, the most straightforward word tokenization is breaking up a sentence at its whitespace separators. To account for specific variations encountered in biomedical literature when compared to general English and newswire texts, several specially adapted word-tokenizers have been implemented (10, 11). Sentences represent grammatical units that constitute the main input for systems trying to extract syntactic or semantic relationships between words. For example, they can be used

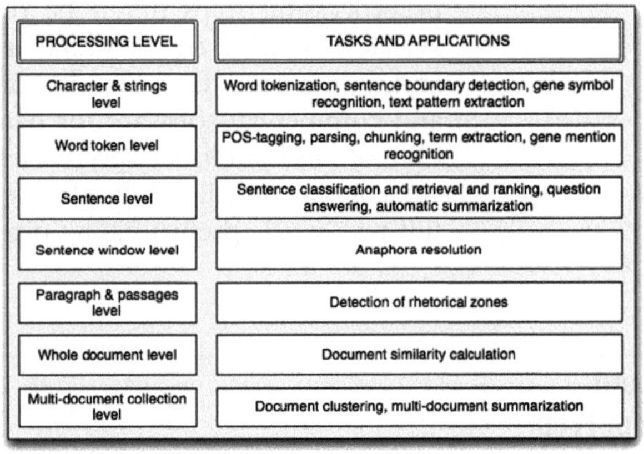

Fig. 16.1. Levels of granularity of natural language processing (NLP) approaches.

as logical boundaries to derive associations between biological objects like protein–protein interactions (12). The underlying assumption of many biological relationship extraction systems is that if two biological objects are co-mentioned in a sentence, they likely share some type of semantic association.

Sentence boundary detection is often accomplished through algorithms based on regular expressions and heuristics that consider certain character combinations like periods followed by capital letters to mark the end of a sentence. More sophisticated machine learning techniques have recently been applied to increase tokenization efficiency in the case of ambiguous sentences boundaries (11).

Morphological processing and standardization of words to their corresponding root or stem (known as *stemming*, e.g., transforming “interacts” to “interact”) are generally accomplished by analysis of the terminal words’ characters, taking into account the rules governing the correct formation of verb inflections and plurals (13). Stemming is a common initial step used by both information retrieval tools (i.e., tools for finding relevant documents from large document repositories) as well as automatic text categorization systems. It can be applied to group protein or gene names that have minor morphological differences (14).

For tagging protein mentions in text, orthographic features and analysis of character types like numbers, special characters, capital letters, and lowercase letters are frequently used (15). Not only single characters, but also combinations of consecutive characters (character *n*-grams), can lead to discriminative patterns for identifying gene symbols that usually show special character patterns (uppercase letters and numbers at the end of symbols). Similar characteristics can be used to tag protein or peptide sequence mentions in articles (7). The use of features at the character string level shows promising results for biomedical text classification and ranking tasks (16), as well as for protein mention normalization, which consists of finding correct associations between proteins mentioned in the literature and their corresponding database records (17). **Figure 16.2** illustrates the basic differences in terms of character composition and word morphology of gene symbols compared to overall words of the same length mentioned in the literature.

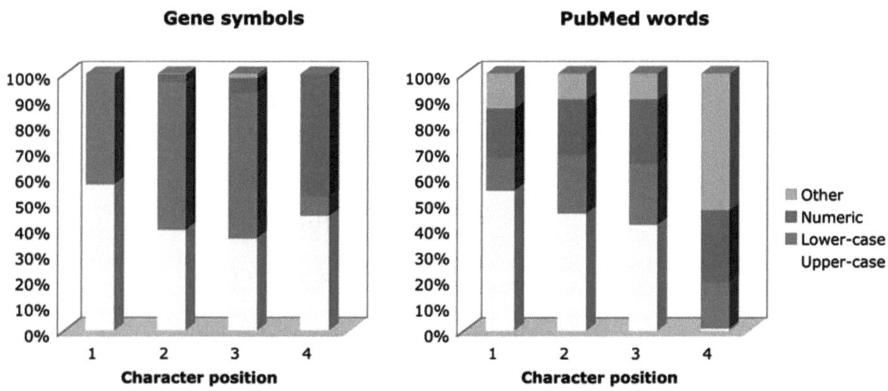

Fig. 16.2. Character types and positions of four character gene symbols from SwissProt records compared to four character words from the whole PubMed database. It becomes apparent that gene symbols are characterized by a high fraction of numeric and uppercase characters in the terminal position when compared to words of the same length derived from PubMed.

An important part of research in computational language processing is devoted to the study of words, labeling them given their context with the corresponding part-of-speech tag (18) or extracting relationships between words in order to build syntactic analysis of relationships between words (*parses*) (19). Such linguistic relationships between words can be useful to facilitate the extraction of semantic associations between certain biological entities and have been used to detect protein–protein interactions (20) protein transport information (21), functional annotation of gene products with gene ontology terms (22), or gene regulation events (23). There are some general aspects often considered for the computational analysis of biomedical texts at the level of word tokens, both to address more linguistic and syntactic aspects as well as for semantic relationships. In a subject–verb–object language like English, these aspects basically relate to the study of words, terms, or phrases considering (i) their relative position within sentences, (ii) their order or relative order of appearance within sentences, (iii) the actual directionality of the relationship between them, and (iv) the analysis of distances between words.

To build statistical distributions of the words from documents linked to a gene of interest, to determine the terms relevant for a group of genes based on their associated documents, to calculate document similarity, or even to determine whether the co-mention of two genes in a sentence is significant, it is important to assign weights and quantitative descriptors for each term or word. Most of these are based on particular types of word counts (9). The most important term counts include the term frequency *tf*, consisting of the number of times a term *t* occurs in a given document. This value is often normalized based on the document length. The term frequency *tf* for a term t_i within the particular document d_J can be calculated by Equation [**1**]:

$$tf_{i,j} = \frac{n_{i,j}}{\Sigma_k n_{k,j}}, \quad [1]$$

$$idf_i = \log \frac{|D|}{|\{d_j : t_i \in d_j\}|}, \quad [2]$$

$$tfidf_{i,j} = tf_{i,j} \cdot idf_i, \quad [3]$$

where $n_{i,j}$ is the number of times the term occurs in document d_J, and $n_{k,j}$ is the number of occurrences of all terms in that document. Other important numerical descriptors include the document frequency, that is, the number of documents from the document collection of size $|D|$, where the term t_i occurs, $|\{d_J : t_i \in d_J\}|$, which is used to calculate common term weights like the inverse document frequency idf_i shown in Equation [**2**] and the $tf^{*} idf_{i,j}$ shown in Equation [**3**]. Term weighting is commonly used in information retrieval systems to weight the words that are used to query a document collection according to how informative they are, or to score words used as features by automatic document classification systems.

Terms that describe biologically relevant aspects of genes are often composed of several words corresponding to *collocations*. Collocations are contiguous words that co-occur more often than expected by chance and that have generally limited compositionality; that is, the meaning of the expression can be poorly guessed from the actual meaning of its components. Collocations are usually detected using probability of co-occurrence models, and, statistical methods like the *t*-test or Pearson's chi-square test, together with certain POS-based filters. Example collocations include compound terms like "spindle body" or phrasal verbs (e.g., "build up") and are often included in important lexical resources such as Gene Ontology terms. Systems like McSyBi make use of collocations to improve the recognition of protein names (24).

Taking into account the growing number of manually curated terminological resources and ontologies relevant to describe a particular biological or medical domain, and their use to annotate, analyze, and interpret experimental results, recent efforts were devoted to identify technical terms automatically from the literature using dictionary-based, rule-based, and machine learning techniques (25). Technical terms cover not only expressions of clinical relevance like "myocardial infarction" or "breast cancer," but also standard experimental methodologies like "mass spectrometry". To avoid repetition when referring to multiword expressions in text, abbreviations are used. Acronyms are specialized forms of abbreviations, commonly constructed using the initial letters of a multiword expression. Acronyms can be extracted from the biomedical literature using specialized systems like Acromine that provide relationships between acronyms and their corresponding expanded forms extracted from the PubMed database (26).

An important research area in biomedical text mining is currently devoted to labeling text with gene (and protein) mentions, namely, correctly identifying the start and end positions of gene names in articles. Most of the currently available software tools for gene mention recognition, such as ABNER (A Biomedical Named Entity Recognizer), use machine learning algorithms like Conditional Random Fields (CRFs) trained on manually labeled text collections (corpora) to tag biological entities mentioned in the literature (15). Recent initiatives such as the BioCreative Metaserver (BCMS) are trying to integrate several gene mention recognition tools to exploit the advantages offered by combining predictions from multiple systems (27).

Concordances are aligned occurrences of a given term together with surrounding text, using a fixed window of characters or words. Knowledge of word concordances can be useful to collect information about patterns of occurrence of verbs or for constructing a dictionary of terms. Searching the biomedical literature for multiword concordances and word co-occurrence-based relationships connected through Boolean operators can be achieved through the MedEvi online tool (28).

When analyzing the frequency of words in natural language text collections, most of them occur extremely infrequently. Many even occur only once in the text collection; these words are called *hapax legomena*. This characteristic represents a general challenge for text mining applications, as it implies that when a new document has to be processed, a considerable number of its words did not appear before in the text corpus used for building the system. This explains why machine learning systems applied to text mining tasks have difficulties generalizing efficiently when trained on small data collections. As manual data labeling is very time-consuming, recent strategies such as active learning propose more efficient selection criteria for finding informative training examples for statistical machine learning systems (29).

4. Literature Databases, Lexical Resources, Ontologies, and Applications

The most important digital library storing biomedical articles is PubMed, available primarily via the Entrez retrieval system maintained by the National Center for Biotechnology Information (NCBI) (30). PubMed contains bibliographic information including titles, abstracts, publication dates, and author names for over 4,800 scientific journals, most of them from the biomedical domain, but also related to engineering, chemistry, environmental sciences, or psychology. A number of records are linked to gene symbols and molecular sequence databank identifiers or are indexed with Medical Subject Headings (MeSH) terms. MeSH is a hierarchically structured thesaurus of controlled vocabulary terms relevant for the biomedical domain used for indexing each PubMed record to improve literature search. MeSH terms have also been applied for clustering genes using a gene-MeSH term matrix extracted from the literature based on co-citation analysis (31). Currently, the PubMed database holds over 18 million citations and the number is rapidly growing, accumulating over 600,000 new entries every year. Most of the articles (over 14 million) are in English, but there is also a significant number in other languages, as seen in **Fig. 16.3**.

A

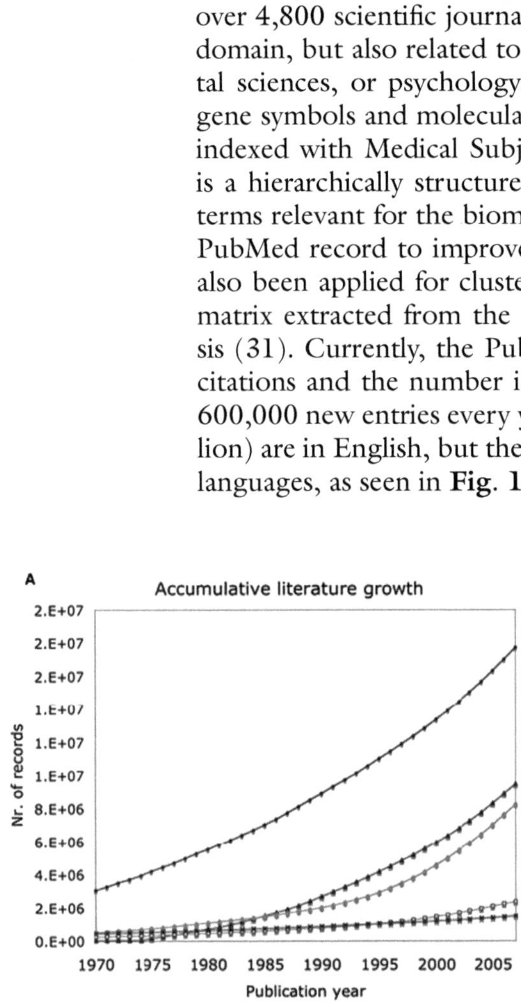

B

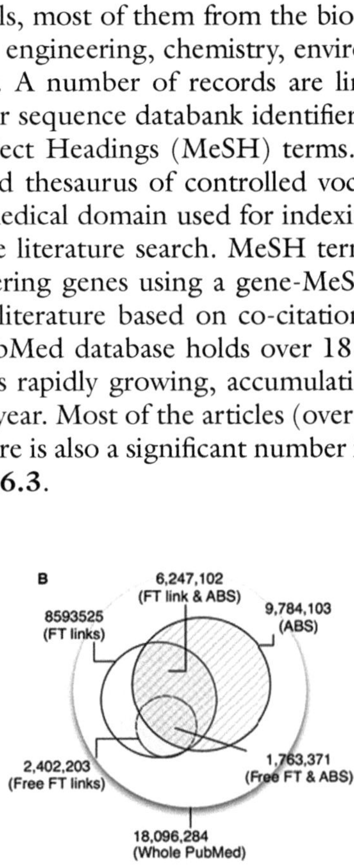

C

Language	Nr. Records
English	14,376,762
French	631,486
German	766,738
Italian	285,432
Japanese	369,915
Russian	632,650
spanish	255,086

Fig. 16.3. PubMed database content. (**a**) Accumulative growth of the PubMed database and the PubMed Central (PMC) database of full-text articles. (**b**) Diagram showing the number of PubMed records with links to abstracts (ABS) and full-text articles (FT). (**c**) Number of records in different languages (*alphabetically ordered*).

Each entry in the PubMed database is characterized by a unique identifier, the PubMed identifier (PMID), which is also widely used as a literature evidence identifier by most of the current annotation databases like SwissProt or GOA (32). **Figure 16.4** shows the distribution of PubMed records in terms of their length. More than half of them (over nine million) have abstracts, the main text resources for literature mining systems, and also often links to the full-text articles, are displayed through the LinkOut system. The biomedical community heavily uses this database, with over 82 million queries in March 2007, and periodic e-mail alert systems serve to improve access to relevant information.

For text mining systems, a more systematic access to PubMed records is required. This can be achieved using the additional programming utilities provided by the NCBI, called Entrez Programming Utilities (eUtils), which are provided together with Perl scripts facilitating the systematic retrieval of records. Through the Batch Citation Matcher, it is also possible to retrieve a list of records uploading a file as input that specifies citations in a predefined format. To retrieve PubMed or PMC UIs for fewer than 100 citations, each citation string is entered on a separate line using that input format. Community programming initiatives like BioPython and BioPerl provide modules for the automatic retrieval of PubMed records. Certain user scenarios require a local copy of the PubMed database. The National Library of Medicine (NLM) also leases the content of the PubMed/Medline database on a yearly basis; the whole set of records can be directly downloaded from the NCBI as XML-formatted files after signing a license agreement specifying the intended use. Alternative search engines to the Entrez PubMed system have recently been implemented; these include Relemed (33), which enables sentence-level searches, PubMed Reader (34), which allows export of the retrieved citations into several formats, including EndNote and Bibtex, or PubReMiner (35), which structures the hits into a summary table and provides information on co-occurring terms and counts.

Enhanced navigation, visualization, and grouping of search results can be obtained by using the HubMed search interface (36). It allows visualizing clusters of related articles as well as of links between articles using TouchGraph.

To find records similar to a given article not indexed in PubMed or any other document, the eTBLAST server (37) can be used, which calculates the document similarity of the query texts to each record in PubMed using a term-weighting approach similar to the one introduced earlier in this chapter.

Although abstracts provide a summary of the most relevant aspects of a paper, most of the experimental details are described in the body or figure legends of the corresponding full-text article. The NCBI therefore developed the open access PubMed Central database, containing a collection of digital full-text articles of a set of life sciences journals (38). Another source of full-text articles

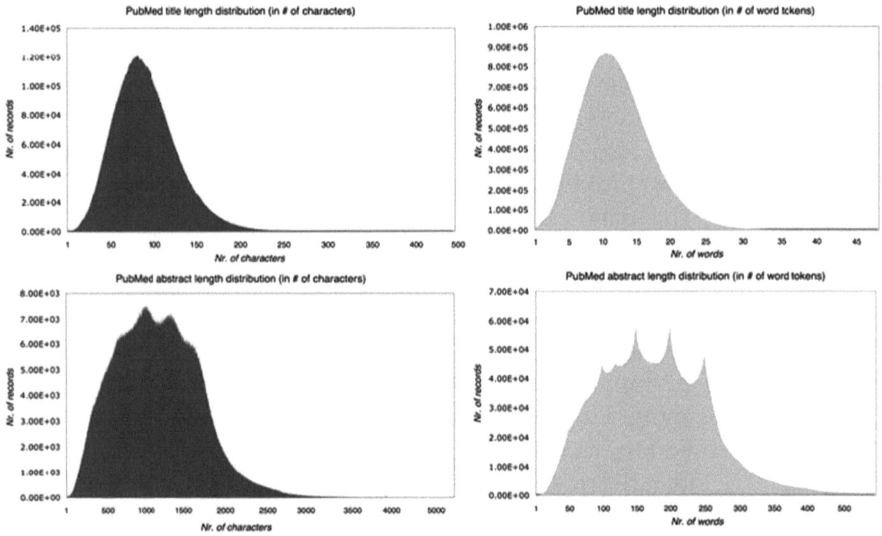

Fig. 16.4. PubMed database content. Distribution of title and abstract length measured in characters and words. The set of peaks in the abstract length distribution can be explained by the initial cutoff in terms of record length posed for older records, as well as the limits of abstract lengths posed by certain journals (e.g., 150 or 200 words).

from peer-reviewed life sciences journals is the Highwire Press literature database (39), hosting over 4.6 million full-text articles and allowing the creation of graphical visualizations of the article's citation map.

Often, figure and table captions contain experimentally relevant information (12), describing, for instance, the content of figures that correspond to images generated by diverse experiments (e.g., 2D-gel spots or mass spectrometry results), described in detail in other chapters of this book. A web application called BioText Search Engine facilitates searching article figures and their captions (40).

Lexical resources for text mining systems basically consist of machine-readable texts, thesauri, dictionary, term collections, and ontologies. They serve for linking gene or protein database records to corresponding mentions in the literature (17) or to structure citations according to biological terminologies of interest (41). The Open Biomedical Ontologies initiative is promoting the development of specialized biological ontologies using a standardized format (42). In the case of lexical resources for gene names and symbols, existing gene and protein databases like SwissProt, EntrezGene, or model organism databases are commonly adapted, removing highly ambiguous or obsolete names either through manual inspection or using statistical analysis of gene mention frequencies. To integrate various lexical resources for gene names, efforts like the BioThesaurus can be useful (43). For clinical and medically relevant terms, the Unified Medical Language System (UMLS) constitutes a valuable lexical resource integrating a thesaurus and multilingual vocabulary database of health-related concepts as well as the semantic relationships between them (44). Lists of uninformative words – referred to as *function words* or *stop words* like determiners or prepositions – in certain scenarios get filtered out when calculating document-based similarity scores or to speed up search engines.

A recurrent problem in biology is characterizing functional aspects of genes and gene products. Attempts have been made to develop consistent functional descriptions using ontologies composed of controlled vocabulary terms. Gene Ontology terms have been used not only to allow a more accurate annotation and efficient exchange of information across multiple annotation efforts, but also to directly interpret and describe large-scale experimental data, in particular clusters of differentially expressed genes obtained through gene expression microarray experiments. Microarray experiments are widely used to detect genes that are differentially expressed in tumors when compared to normal tissues, allowing the simultaneous analysis of thousands of genes. The interpretation of these experimental setups is obviously associated with experimental noise intrinsic to this technology as well as difficulties in providing a biologically coherent explanation for the obtained results. Annotations of the analyzed genes with pathway or GO information are used for human interpretation of gene expression data. As manually curated resources are generally incomplete, text mining techniques have been implemented to directly extract GO annotations from the literature. The online application CoPub allows the retrieval of co-occurrences of biomedical concepts in Medline abstracts, such as genes, GO terms, and disease names (45).

When searching with the term "breast cancer," CoPub returns a summary table of all the selected concepts together with links to the actual papers where the concepts that are co-mentioned are also visually highlighted in the papers. For the individual genes, links to database identifiers are provided (Entrez Gene ID). The top five returned genes for this query include GREB1, TSP50, BCAS3, SCGB2A2, and H41. The top-ranked co-occurring pathway mentions include cell cycle and g1/s check point, estrogen metabolism, and akt signaling pathway, all known to play a role in breast cancer. To rank the co-occurring concepts, CoPub uses an *R*-scaled score that is based on the individual frequencies of each term and the mutual information measure. This system also allows the analysis of microarray data by uploading a file containing the Affymetrix gene identifiers of the genes used in the experiment and then calculates overrepresented keywords (e.g., GO biological process terms, pathways, or liver pathology terms) extracted from the literature co-occurrence analysis. A graphical output visualizes the generated literature network from the co-occurrence analysis.

Beyond using GO terms for biological interpretations of microarray data, they have also been used directly to index PubMed abstracts and extract protein annotations using text mining methods. GoPubMed facilitates navigating PubMed abstracts based on the relationships between terms defined in the ontology structure (41). Other literature retrieval systems that integrate GO terms include GenNav for finding GO terms and gene products in PubMed (46), GOCat (Gene Ontology Categorizer), a text classifier that automatically annotates any PMID or query text with associated GO terms (47), or BioLit, a web server that provides a web-based article viewer highlighting gene ontology terms contained in full-text articles from PubMed Central retrieved previously based on user-specified queries (48). Whatizit is a web service that returns an XML-tagged document for user-defined input text, labeling GO terms but also pathways, diseases, and protein mentions (49). Another recently published system called PhenoGO provides for a given query gene a summary table of associated GO terms and phenotypic context information like cell type, tissue, and phenotype terms derived from biological ontologies (UMLS, Cell Ontology, and Mammalian Phenotype Ontology) (50).

Although controlled vocabulary terms show clear advantages for annotation purposes, they are also associated with limitations when used by literature mining tools. One of them is the difficulty of detecting these terms or their corresponding typographical variants in the literature, as functional terms like those contained in GO have been primarily designed for annotation purposes and not text indexing (51). The other limitation is related to the heterogeneous levels of functional description specificity, as some of the upper-level terms are too general to provide biologically relevant information for detecting biological differences between gene groups. Therefore, alternative strategies to analyze gene groups using information from the literature tried to directly extract relevant terms from the articles associated to the collection of genes (52). Raychaudhuri and co-workers introduced a computational method named ***neighbor divergence per gene*** (NDPG) that allows the quantitative assessment of the functional coherence of gene groups, an aspect that can be applied to determine the quality of differential expression-based gene clusters (53) (**Fig. 16.5**).

To analyze the functional similarity of two or more proteins through their associated GO terms, it can be useful to quantify the semantic similarity between each term, that is, how similar they are in meaning. Metrics to provide a quantitative relationship of how close words are in meaning based on their information content were originally developed in the domain of computational linguistics (54) and were only later adapted for bioinformatics analysis to correlate the functional and sequence similarity of proteins (55).

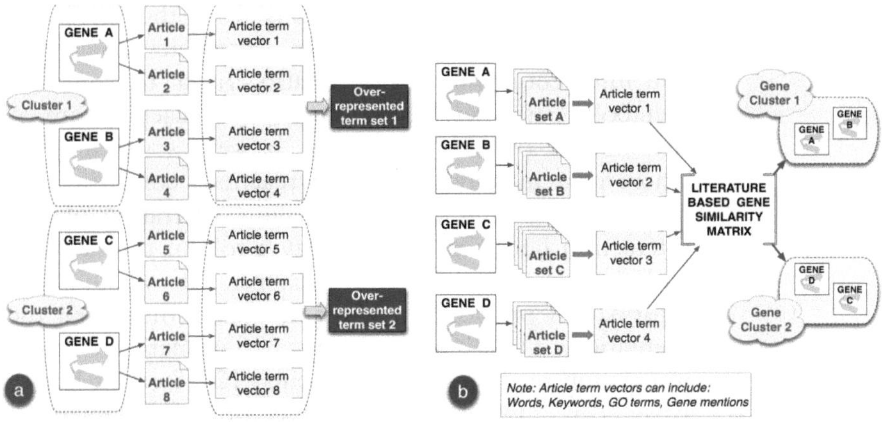

Fig. 16.5. Literature-based gene group analysis. (a) Extraction of descriptive words, keywords, or GO terms for gene clusters based on statistical detection of terms significantly overrepresented in documents associated to each group of genes. The set of documents linked to each gene cluster can also serve to score the functional coherence of each gene group. (b) Automatic clustering of genes based on associated literature calculating the similarities between genes through an analysis of how similar their associated documents (words, keywords, GO terms) are.

Most of the original work proposing alternative strategies for semantic similarity weighting was developed using WordNet, a hierarchically organized lexicon of general English words structured in nodes of words that have similar meaning (called *synsets*). Some attempts have been made to adapt WordNet for the medical domain (MedicalWordNet), integrating additional medical terminology, medical facts, and medical beliefs (56). Recent developments also exploited the use of external information (e.g., proteins instantiated with GO term annotations) to quantify functional similarities (57).

Recent studies showed that the lexical analysis of ontologies can be used not only to retrieve relationships between concepts within a given ontology but also among multiple different ontologies (58), therefore being able to integrate and relate information provided by heterogeneous annotation types.

5. Extraction of Biological Relationships: Protein Interactions and Pathways

The study of interactions between proteins (*protein–protein interactions*, PPI) has captured considerable interest not only in the life sciences domain but also in the fields of computational biology and literature mining. Bioinformatics methods generally attempt to predict interactions between proteins using sequence, structural, or evolutionary information about proteins, or even explore functional annotations to determine potential interaction partners. They rely on the availability of interaction databases, like MINT or IntAct, which host a large collection of manually extracted interactions from the literature as well as data derived from large-scale proteomics experiments. Protein interaction information is also crucial to understand both metabolic pathways, where biochemical reactions are often carried out by protein complexes, as well as signaling pathways, where protein phosphorylation reactions constitute a common mechanism for intracellular signaling. Alteration of the protein interaction behavior can also play a role in the development of pathological conditions, such as cancer. The wealth of interaction information provided in scientific papers motivated the implementation of text mining systems to automatically extract binary interaction relationships of proteins (12). Systems like PreBIND have been implemented to detect protein interaction-relevant abstracts using text classification methods (59), and the BioCreative MetaServer provides interaction information from several applications on the level of PubMed abstracts. These systems rely on manually labeled interaction-relevant abstracts to derive features (words, terms, or text patterns) that can be used by machine learning techniques like support vector machine (SVM) algorithms to automatically categorize interaction-relevant abstracts from unlabeled document collections.

Most of the protein interaction extraction systems use co-mention of proteins in text units as the underlying approach followed by the extraction process, assuming that if two proteins or bio-entities are mentioned together in the literature, they should have some biological interaction relationship. These text units range from single sentences or sentence passages to whole documents. Computational techniques that can be useful for deriving interaction information include (1) statistical approaches trying to exploit co-mention frequencies (e.g., statistically significant co-occurrences), (2) techniques analyzing the context of co-mentions in terms of finding interaction-relevant textual patterns or verb frames (e.g., using patterns like "protein A *interacts with* protein B") (60), (3) machine learning–based sentence classifiers to determine if the context of a co-mention is interaction-relevant (61), or (4) systems integrating syntactic information and sentence parse trees to derive semantic relationships between co-mentioned proteins (21). **Figure 16.6** provides a schematic view of the protein interaction network extraction process.

For a given query protein, the PubGene system automatically constructs a literature-derived protein co-mention network together with numbers corresponding to the documents where the two entities co-occur. Using this tool for extracting the literature network of BCAR3 (breast cancer anti-estrogen resistance protein 3) results in a collection of co-occurring proteins that include, for instance, SPECC1, SH2DC3, RTN2, SH2D3A, and DDX19A. PubGene also allows searches where protein interaction keywords are mentioned in the sentences (62).

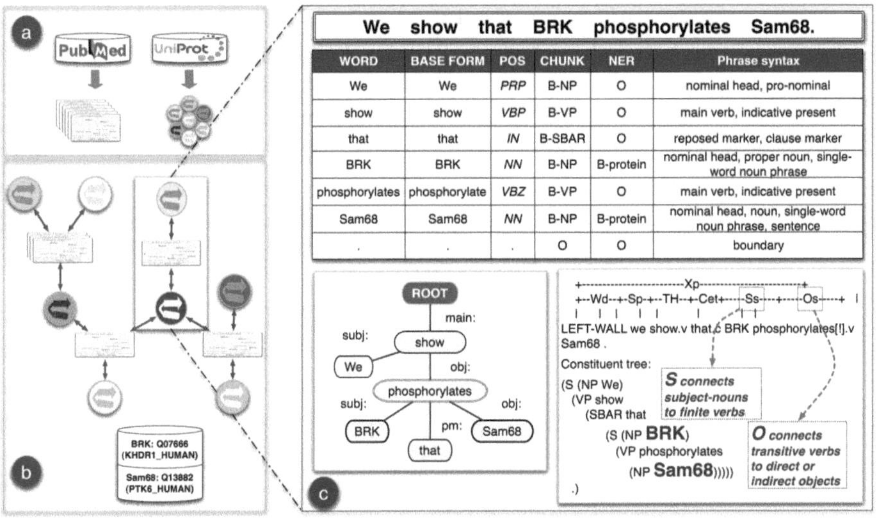

Fig. 16.6. Simplified view of the protein interaction and pathway extraction process followed by text mining systems. (a) Initially, a collection of interaction-relevant articles is assembled and preprocessed. Existing biological databases like UniProt are used to build a protein name dictionary. (b) Based on co-occurrences in articles or sentences, an initial protein association network is constructed. (c) Either statistical analysis is used to determine whether a given co-occurrence between two proteins is significant, or the extraction of textual or linguistic associations is used to further detect interactions. Linguistic, syntactic, and semantic information provided by tools like the GENIA or Connexor parser, as shown in the table in part 3 of this figure, could be useful for this purpose. To determine the directionality of relations, syntactic or link grammar parse trees (shown on the bottom left and right) can be useful.

Another popular system that extracts protein interactions from PubMed is iHOP, offering the possibility to retrieve the collection of co-mentioned proteins together with the corresponding evidence sentences and linking each protein to its corresponding database identifier (8). Rather than automatically generating the interaction network, iHOP allows, based on the resulting sentence collection and their links to experimentally confirmed interactions, the construction of a manually curated protein interaction network. Searching iHOP with the same human protein (BCAR3) results in a set of interaction sentences, where the synonyms of this symbol, like NSP2, are also used to find interaction information.

The EBIMed tool developed at the European Institute for Bioinformatics (EBI) (63) returns a total of 17 Medline abstracts when searching for BCAR3, together with a summary table of co-occurring proteins, GO terms, and drugs. Partially different protein associations are retrieved, with BCAR1, NSP1, and p130, among others. Links to the evidence sentences for these associations are provided by this web application. EBIMed also supports querying with a list of PMIDs, for which proteins and their relationships should be extracted. A widget-like system that allows dragging retrieved query proteins into a content view for finding interaction relationships and supporting sentences has also recently been developed; called Info-PubMed, it was developed at the University of Tokyo (64). For human BCAR3, it retrieves several interaction partners (CCND1, PAK1, ITPA, PRKCL1, PKN1). There are cases where rather than retrieving a long list of potential interaction partners, the end user is more interested in retrieving supporting evidence sentences for a user-defined interaction pair from the literature. Chilibot facilitates the retrieval of a set of qualified interaction relationships for a given pair of proteins through an NLP-based text mining application (65).

Each experimental interaction detection method has a certain degree of reliability, some techniques being more accurate than others (e.g., crystallographic characterization of protein complexes vs. yeast two-hybrid screens). Thus, it is also important to know the experimental qualifier/technique described in the paper for detecting a given interaction. Some initial studies used a text pattern-matching approach to extract interaction detection methods from full-text articles and link them to their corresponding controlled vocabulary from the Molecular Interaction (PSI-MI) ontology (66). Promoted by the BioCreative community assessment of biomedical text mining systems, the extraction of normalized protein interaction pairs (i.e., interactors characterized by their unique database identifiers) from full-text papers has attracted increasing attention both from interaction databases as well as from the developers of text mining systems. One additional aspect when extracting interactions from full-text articles is the retrieval and ranking of those text passages that best summarize interaction descriptions (67).

Certain transient protein interactions like phosphorylation reactions are critical in signal transduction pathways and represent a regulatory mechanism in central biological processes, such as cell division, often significantly altered in cancer cells. Due to the biological importance of phosphorylation reactions, specialized literature mining systems devoted to the extraction of this interaction type have been constructed. The approach proposed by Narayanaswamy et al. combined pattern matching of manually defined template patterns with a collection of predefined rules to retrieve phosphorylation relationships (68).

When extracting metabolic or signaling pathways, one general hurdle is that the whole pathway is in general either not mentioned within a single paper or had been only characterized so far in part, some of the steps still being poorly understood (69). In the first case, pathways need to be reconstructed from multiple papers, posing the obvious challenge of correctly linking each bio-entity across multiple papers. This is usually addressed by first extracting binary relationships between proteins and/or compounds from the document collection and then constructing a network from the resulting set of interaction pairs. Another challenge for text-derived pathway detection is to determine the directionality and causal relationships between elements underlying pathways reactions. In the case of the NF-kB pathway, Oda and colleagues used a manually annotated corpus with event information relevant for this model pathway to determine the main classes of bio-inferences that need to be addressed (70).

In the GENIES system, which was constructed to retrieve cellular pathways from full-text papers (71), the pathway is generated by iteratively retrieving known genes in the regulatory hierarchy immediately above or below the original gene. The resulting network can then be edited and visualized. It also defines some basic semantic classes for the extracted relationships, such as "createbond" (methylate, phorsphorylate) or "attach" (bind, form complex), and makes use of 125 different verbs relevant for pathway relationships. Sentence parsing systems (similar to the example shown in **Fig. 16.7**) have also been used to derive relationships between biological objects by assuming that the syntactic structure of a sentence can be used to determine pathway-relevant semantic associations. Similarly, the Medscan technology has been used to reconstruct biological association networks for mammalian tissues' signaling pathways from the literature. This system distinguished between two relationship types: direct physical interactions (binding, protein modification, and promoter binding) and indirect interactions (regulation, expression regulation, protein transport regulation, and molecular synthesis regulation) (72).

Among proteins participating in signaling pathways, protein kinases have been especially well characterized, as some of them are known to be involved in human cancer development and regulation. For instance, the human tyrosine kinases FAC and Src important for cell migration and adhesion are known to play a role in breast cancer. The cell cycle checkpoint kinase CHEK2 is associated with an increased risk of developing both female and male breast cancer. The Kinase pathway database uses natural language processing methods like the extraction of phrase patterns (i.e., regular expressions of noun and preposition phrases) and the construction of light syntactic representations of sentences to automatically extract protein interactions and pathways (73). The resulting template phrase patterns together with a lexical resource for finding protein names in texts are used to generate the resulting database, which also integrates additional information such as functional conservation information and ortholog tables derived from sequence information. When querying for pathway information, the user can either enter the start point protein of a pathway of interest or specify the start and end point proteins. Additional pathway construction options include the possibility to specify the maximum number of connection steps or whether the interactions are direct. The web tool PathBinderH also allows specification of two query terms for which associations should be extracted from PubMed sentences (74). First, the initial query term is entered, and then, in a second step, from a list of associated enzymes or compound names PathBinderH returns the passages providing associations between the two terms. This system also integrated terms derived from MeSH and the Enzyme Nomenclature to facilitate the retrieval of disease-relevant pathway information.

Regulatory events modulating central biological processes are related to control not only at the level of gene expression or phosphorylation, but also at the level of protein degradation and turnover.

One of the most important protein degradation mechanisms is ubiquitination, a posttranslational modification, where ubiquitin is transferred to a target protein, and where ubiquitin-protein ligase (E3) is one of the key enzymes involved. An enhanced online access to E3-relevant information provided by a text mining tool called E3Miner is available, also providing links to target substrates, other proteins participating in the ubiquitination pathway, as well as E3-related human diseases (75).

As text mining approaches only cover part of the rich biological information that can be used to analyze pathways and associations of genes to pathological conditions, systems like Babelomics that integrate heterogeneous knowledge sources into a bioinformatics suite of web applications and combine data derived from both curated databases like KEGG and Biocarta pathways with text mining-based functional terms are also of practical relevance (76).

The analysis of interactions between proteins as well as that of proteins with small molecule chemical compounds are needed to understand metabolic pathways or inhibitory mechanisms of drugs used in cancer therapy. The detection of these relationships requires the correct labeling of chemical compounds in the literature. STITCH offers the possibility to extract the protein–protein and protein–compound relationship network, currently covering more than 68,000 different chemicals (77). The chemical compounds are linked to their corresponding PubChem record, and relationships between different compounds are also extracted. STITCH not only takes into account information extracted using text mining techniques but also provides relationships derived from other knowledge sources such as experiments or biological databases. When searching for compound–protein relationships, SMILES strings of the chemical molecules can also be used as a query. When searching for interaction partners for HER2 (ERBB2), an important oncogene in invasive breast cancer, STITCH, in addition to protein interactions with ERBB3, SHC1, GRB2, EGFR, or NRG1, also extracts interactions with a set of compounds including tamoxifen and lapatinib, corresponding to drugs used in the treatment of breast cancer.

The BRENDA (BRaunschweig ENzyme DAtabase) database of manually curated enzymatic pathway information integrates data collections generated by text processing techniques: FRENDA (Full Reference ENzyme DAta) and AMENDA (Automatic Mining of ENzyme DAta). FRENDA offers automatically generated links of enzyme-relevant information (i.e., enzyme names, their synonyms, and EC numbers) to PubMed records. Ambiguous enzyme names have been previously filtered to increase the precision of this system. The other component, AMENDA, generates associations of enzymes to subcellular locations, tissues, and organisms extracted using text mining (78).

In order to integrate dispersed information sources relevant to apply a systems biology type of analysis of metabolomics data, the availability of controlled vocabularies and ontologies is needed. For text mining and relationship extraction, high-quality lexical resources that can be used to structure and index the literature relevant for the metabolomics field are required. Motivated by the interest of the Metabolomics Standard Initiative (MSI) in the development of standard ontologies and controlled vocabulary for describing nuclear magnetic resonance spectroscopy and gas chromatography experiments, Spasić and colleagues designed a pipeline for automatically recognizing and filtering relevant terms for these two topics (79),

6. Association of Diseases, Mutations, and Epigenetic Information with Genes Through Literature Mining

To understand complex diseases such as cancer, diverse biological relationships at multiple contextual levels are currently being studied. These include tissue and anatomical information, knowledge of cellular components and protein localization, analysis of interactions and signaling pathways, as well as cancer-associated polymorphisms, mutations, SNPs, and epigenetic modifications. All of these aspects are described in the literature, and text mining systems have recently been implemented addressing these and other cancer-relevant aspects. A number of literature mining systems use domain-focused term dictionaries as lexical resources to uncover genes and proteins involved in cancer. Widely used biomedical term sets include UMLS, GO, OMIM, and HUGO. Various types of string-matching algorithms are usually applied to identify individual lexicon terms. To complement these strategies, the MTag named-entity recognizer based on machine learning methods (Conditional Random Fields, CRFs) has been applied to automatically tag malignancy mentions like "neoplasm" and "neuroblastoma" in biomedical texts from the cancer genomics domain (80). Other machine learning techniques (maximum-entropy classifiers) have been used to filter out false recognitions of ambiguous disease names from UMLS mapped through dictionary-based, longest-matching algorithms to PubMed sentences for the extraction of gene–disease relationships (81).

Several studies showed the usefulness of text mining approaches in analyzing and finding candidate genes for diseases. Pospisil et al. combined different data sources, including pathway information derived from the Ingenuity system (Ingenuity Pathway Analysis) and text mining systems like iHOP, to identify candidate lists of cancer-related human hydrolases present in the extracellular space of cancer cells (82).

Natarajan and colleagues used text mining techniques to analyze gene expression data from gioblastoma in response to sphingosine-1-phosphate (SIP), inferring gene–gene interaction networks for differentially expressed genes. The extracted gene relationships triggered by SIP induction have been generated based on text mining applied to full-text articles, obtaining results that underline the importance of the matrix metalloproteinase MMP-9 in glioma invasion and angiogenesis (83).

TP53 is one of the most relevant regulatory proteins of the human cell cycle and has been extensively described in the literature. A recent article in the *J ournal of Biomedical Informatics* described the extraction of the gene–protein interaction network for this protein (84). The TP53 interaction network had been automatically extracted from the PubMed database using the Arizona Relation Parser (a syntactic-semantic relationship extraction system) to detect the interaction pairs (85). The authors additionally carried out a topological analysis of the resulting network. Other domains where text mining techniques have been applied to analyze relevant genes and/or disease information include cardiovascular diseases like atherosclerosis (86), neurosciences, synapse biology and brain disease-associated genes (87), infectious disease outbreaks (88), retinal diseases (89), or asthma candidate genes (90).

Co-occurrence analysis of particular items has been well studied for a range of different fields, such as web mining and social networks, text mining, as well as in the domains of comparative genomics and promoter analysis. In comparative genomics, co-occurrence studies of genes in completely sequenced genomes have been carried out, while in the analysis of promoters, co-occurrences of transcription factor-binding motifs in collections of co-regulated genes have been characterized in detail.

When considering co-occurrence-based approaches in biomedical text mining, co-occurrences between genes and proteins and of genes with functional terms and disease mentions have captured considerable interest.

Several text mining systems rely on the analysis of co-occurrences between genes and disease terms in order to retrieve disease-associated genes. A system that allows the extraction of protein–disease relationships using co-mention frequency with disease concepts is FACTA (Finding Associated Concepts with Text Analysis) (91). When searching FACTA with the query HER2, it returns a list of passages where the used query is mentioned. In this particular case, a total of 2,362 document hits were retrieved (November 2008). For each query, this system returns links to co-occurring concepts, including other genes or proteins, but also disease mentions, symptoms, drugs, enzymes, and compounds. In the case of HER2, a total of 1,950 relevant concepts were provided, ordered by default using the frequency of co-mentions. Also, alternative ranking criteria can be selected, including the pointwise mutual information (pmi), a common statistic used in information theory for measuring association between items, or using the frequency times pmi. Concepts integrated in FACTA were previously extracted from several biomedical databases and lexical resources, including UniProt, BioThesaurus, UMLS, KEGG, and DrugBank. This system assigns to each concept an identifier and groups names and synonyms. FACTA recovers relationships for HER2 to several disease concepts; among the top-ranking ones are breast cancer, metastatic breast cancer, adenocarcinoma, and ovarian cancer. It provides associations to the compound tamoxifen, also detected by STITCH, a tool described in the previous section, and links to several drugs such as Herceptin or Gefitinib.

A detailed statistical analysis to determine models suitable for evaluating the significance of associations between entities co-occurring in the literature was carried out by Müller and Mancuso (92) and resulted in a software tool called NetCutter. The study shows that under certain circumstances, using a bipartite graph model of co-occurrences and *z*-scores of bi-binomial distribution (BBD) is more suitable than other measures like Jaccard and uncertainty coefficients (92) (**Table 16.1**).

Sometimes it can be useful to characterize a document not by a single term or binary relation of co-occurring concepts but rather by constructing a concept profile of related terms, providing a relevant weight to each concept to signify its importance. This idea has been implemented in the software tool Anni 2.0 (93), which integrates an automatic concept recognition software, extracts concept profiles from papers, and also exploits the ontological relationships and semantic types underlying each of the used concepts (UMLS terms and genes). A user can query for direct associations (based on co-occurrences), to match concept profiles, or to apply hierarchical clustering to structure the obtained results. Anni 2.0 has been used to examine a data set resulting from a DNA microarray experiment for characterizing

Table 16.1
Overview of existing literature missing Systems useful for extracting human disorder relevant gene-associations

System	Publication	1st Author	Description	PMID
Anni 2.0	Genome Biol., 9(6):R96.	Jelier et al. (2008)	Ontology-based interface to MEDLINE, extracts associations for genes, drugs and diseases	18549479
Arrowsmith	Comput Met. Prog. Biomed., 57:149–53.	Smalheiser et al. (1998)	Knowledge discovery & hypothesis generation tool linking queries through overlapping terms shared between their documents.	9822851
CARGO	Nucleic Acids Res., 35:W16–20.	Cases et al. (2007)	Web portal for cancer genes, integrates and visualizes results from different resources incuding IHOP	17483515
CGMIN	BMC Bioinformatics, 6:78	Bajdik et al. (2005)	Identifies genetically-related cancers and cancer-related genes from OHIm records.	1579677
CoPub Mapper	BMC Bioinformatics, 6:51	Alako et al. (2005)	Extracts co-occuring biomedical concepts (genes, GO terms, liver pathologies, diseases, drugs and pathways).	15760478
E3Miner	Nucleic Acids Res., 36:W416–W22.	Lee et al. (2008)	Extracts ubiquitin-protein ligases (E3s) information	18483079
ENDEAVOUR	Nucleic Acids Res., 36:W377–84.	Tranchevent et al. (2008)	System for prioritization of candidate genes also using textming information	16680138
FACTA	Bioinfomatics, 24(21):2559–60.	Tsuruoka et al. (2008)	Retrieves co-appearing concepts for a given query (genes/proteins, disease, enzymes and chemical compounds).	18772154
G2D	BMC Genet., 6:45.	Perez-Iratxeta et al. (2005)	Find candidate genes related to a genetic disease of their interest.	16115313
HuGE Navigator	Net Genet., 40(2):124–5.	Yu et al. (2008)	Human genome epidemiology literature mining system.	18227866

Table 16.1 (continued)

System	Publication	1st Author	Description	PMID
LitMiner	Biomed Digit Libr., 3:11.	Demaine et al. (2006)	Extracts relations of terms based on statistical co-citation analysis.	18227866
MarkerInfo Finder	Bioinformatics, 23(18):247–84.	Xuan et al. (2007)	Extracts information about genetic markers or genetic variations in Medline records.	17823133
MeInfoText	BMC Bioinformatics, 9:22.	Fang et al.	Retrived gene methylation and cancer associations from teh literature.	18194557
MTag	BMC Bioinformatics,	Jin et al. (2006)	Entity tagger for recognizing clinical descriptions of malignancy using CRFs.	17090325
NetCutter	PLoS ONE, 3(9):e3178.	Müller et al. (2008).	Co-occurrence analysis system based on biparties graph.	18781200
OSIRIS	BMC Bioinformatics, 9:84.	Furlong et al. (2008)	Retrieves literature about sequence variation (SNPs) of a gene.	18251998
PhenoGO	Pac Symp Biocomput., 64–75.	Lussier et al. (2006)	Provides cell type, disease, tissue and organ information for genes and proteins.	17094228
PolySearch	Nucleic Acids Res., 36:W399–W405.	Cheng et al. (2008)	Retrieves relations between genes/protein, diseases, tissues, cell compartments, SNPs, mutations, drugs & metabolities.	18487273
PubGene	Nat Genet., 28(1):21–8.	Jenssen et al. (2001)	PubGene extracts relations between genes, proteins, protein sequences, compounds, GO terms and diseases (MeSH).	11326270
PubMeth	Nucleic Acids Res., 36:D842–846.	Ongenaert et al. (2007)	Cancer gene methylation database including using text mining and manual curation.	17932060

differentially expressed genes in metastatic and localized prostate cancer, which was useful in this context to understand the pathways involved in progression to the metastatic state.

The open source programming language R offers a large collection of modules used for the statistical analysis of microarray and gene expression data, providing algorithms that can be used for extensive statistical inference. To be able to invoke this statistical analysis programming language and connect its utilities to information provided by literature data mining results, the R library MedlineR can be used (94), offering the possibilities to carry out PubMed queries with Boolean operators, to construct a concept co-occurrence association matrix, and to carry out further network analysis by exporting the retrieved data in the format used by Pajek, a visualization software to examine network topology.

The integration of text mining systems to other software applications can enhance not only the analysis of biological data, but also the integration of literature-derived information with sequence and structural knowledge. To achieve this, CARGO, a visualization tool based on small software agents (widgets), combines a widget for literature mining that retrieves information from iHOP with other biologically relevant aspects such as information related to SNPs (95), genetically inherited diseases (from OMIM), or structural information (from PDB).

There is increasing interest in detecting SNP alleles linked to complex disorders, in particular to certain types of human cancers. This interest is promoted by recent technological advances resulting in the development of high-density SNP scanning platforms, important for carrying out genome-wide association studies. The identification of new genetic biomarkers may lead to promising results for both diagnostic and prognostic purposes in clinical research. The biomedical literature contains a large amount of studies where genetic variations have been associated with diseases, relevant to facilitate the interpretation of large-scale SNP characterizations. The web tool MarkerInfoFinder provides an interface to information extracted from the literature for polymorphism markers (SNP, STS) integrating also genomic marker locations from different databases and connections to diseases (96). Another system that focuses on the extraction of SNPs from the literature is OSIRISv1.2, which extracts and normalizes sequence variants for human genes using a pattern-matching algorithm together with a sequence variant nomenclature dictionary. According to the authors of this system, it achieved a relatively high performance of 99% precision rate with a recall rate of 82% (97). In general, the automatic recognition of mutation mentions in text can be useful as a component of a more specialized text mining system, for instance, for building a knowledge base for nonsynonymous coding SNPs of breast cancer–relevant genes.

The MutationFinder software is a rule-based system implemented in various programming languages (e.g., Python and Java) that can be used as a standalone script to easily extract mutations contained in large text collections (98). Besides rule- and pattern-based approaches, popular machine learning algorithms like CRFs have been used to automatically identify mentions of acquired genomic variation in texts, both at the level of amino acid variations and at the nucleic acid level. VTag is a CRF-based software for tagging genomic variation mentions in texts. It uses a model generated through training this machine learning approach on a manually labeled document collection that contains variation mentions related to cancer, labeled using the WordFreak package (99).

Using text mining–assisted results for producing manually annotated mutation information can considerably reduce the human workload, not only speeding up the annotation process and producing more systematically extracted information, but also allowing easy maintenance of a mutation database. Mutations in proteins involved in the coagulation process are known often to result in bleeding disorders like hemophilia B or FX deficiency. The authors of the coagulation protein database CoagMDB used the automatic extraction of mutations from text (abstracts and full-text articles) with regular expressions to identify amino acid and numeric elements followed by a manual inspection and validation process. The extraction system had a recall rate of 99.6% and a precision rate of 87.4%, also competitive with other published strategies (Rebholz-Schuman et al., 2004, reported a performance of 74.7% recall and 98.6% precision rate and Lee et al., 2007, a recall rate of 77.3% and a precision rate of 97.7%). The resulting database contained a total of 832 mutation records (100).

Although most newly constructed biomedical databases are characterized by using well-structured records to allow more systematic access to the hosted information, there are also cases where records consist of unstructured text descriptions. This is the case with OMIM (Online Mendelian Inheritance in Man), a collection of referenced overviews of Mendelian disorders providing descriptions of the phenotypic and genotypic aspects of human genes. This database, initiated in the early 1960 s, contains valuable information on genes and their associations to diseases described in the form of free text. To fully exploit information on cancer-relevant aspects contained in OMIM records, the text mining system CGMIM was implemented, with the aim of identifying genetically related cancers and cancer-related genes. This system analyzed mentions of 21 major cancer types in OMIM records. To account for the various ways a given cancer might be referred to in the text (e.g., "breast cancer" can also be mentioned as "breast tumor," "breast carcinoma," "mammary gland tumor," or even using "cancer of the breast"), the developers of CGMIM created a list of synonyms for each cancer type using the International Classification of Disease for Oncology (ICD-O) and added familiar lay terminology. The resulting system identified 1,943 genes related to cancer, such as BRCA2, BRAF, and CDKN2A, related to 14 cancer types (101).

Information on human genome epidemiology can also be efficiently extracted directly from the literature using text mining approaches. This has been done by the HuGE Navigator system, which integrates a knowledge base for genetic associations and candidate gene selection with a search engine for finding literature relevant for genome epidemiology (102). The HuGE Navigator supports searches with several terms, including diseases, environmental factors, and genes. It can be used to prioritize genes using PubMed abstracts based on evidence from animal models. The literature search system adapted for the HuGE Navigator is based on an SVM text classifier application called GAPscreener (103). When searching for genes associated with breast neoplasms, the top-ranking hits returned by the HuGE Navigator are BRCA1, BRCA2, TP53, and GSTM1. HER2 (ERBB2) ranked at position 15 and had several disease-associated MeSH terms assigned (e.g., "breast neoplasms," "lung neoplasms," and "Carcinoma, Non-Small-Cell Lung").

A common type of user request for biological relationships can be summarized as "given *X*, find all *Y*s," where *X* and *Y* are the biological entities or terms of interest. To be able answer these kinds of questions, the integration of multiple databases with literature information can generate more comprehensive results.

The web tool PolySearch for extracting text-derived relationships (104) allows users, in an initial step, to select two basic types of biological entities of interest (e.g., genes, pathways, metabolites, or disease), and then provides the option to retrieve and rank not only informative abstracts or sentences from the literature but also information from a set of multiple databases (e.g., DrugBank, SwissProt, HGMD, and Entrez SNP).

PolySearch uses a rule-based pattern recognition system and allows user-specified synonyms to be added to the original query (query synonym expansion) returning ranked text and sequence data. For the HER2–pathway relationship, among the top-scoring terms, PolySearch returns "survival," "apoptosis," "cell cycle," "Akt pathway," and "migration."

Recent research trends underlined the importance of epigenetic modifications like DNA methylation for the development of human cancers. Aberrant DNA methylation profiles encountered in certain cell types may constitute useful biomarkers of both diagnostic and prognostic importance. Hypomethylation events frequently contribute to chromosomal instability and an increase in the transcriptional gene expression of oncogenes, while the hypermethylation of promoter regions can result in abnormal repression of the transcription of tumor suppressor genes. Increasing numbers of papers are publishing results on experimental characterizations of gene methylation profiles of specific cancer types. To provide systematic access to descriptions of cancer-relevant gene methylation events from the literature, the MeInfoText system has been developed (105). To detect associations between gene methylation and cancer, this system uses term co-occurrences in abstracts and sentences together with association rules, a method used in data mining to detect relationships between items in large data collections. Genes are detected in texts by MeInfoText using a dictionary of gene names and symbols derived from the SwissProt and NCBI Entrez Genes databases, while terms referring to cancer types are obtained through MeSH vocabulary after some preprocessing of the cancer-type vocabulary to increase detection efficiency. To detect methylation expressions, MeInfoText uses keywords for methylation such as "methyl-," "hypermethyl-," "hypomethyl," and "histone," and methods for DNA methylation such as "MSP" and "COBRA." Additional information covered by this system includes protein–protein interaction information (derived from the databases from HPRD and IntAct) and pathway information (obtained from HPRD and KEGG). MeInfoText calculates gene methylation-related pathway clusters automatically through literature mining results together with information from pathway databases. When searching this application using the gene symbol HER-2 (ERBB2), 38 methylation, 10 hypermethylation, and three hypomethylation papers are retrieved, also providing 27 sentences with "HER-2," "methylation," and "cancer." One example evidence sentence retrieved is, "We generated the DNA methylation profiles of 143 human breast tumors and found significant differences in HER-2/neu expression and DNA methylation of five genes [PMID:16397211] (*Cancer Res.* 2006)." When looking at gene methylation and cancer associations for this gene, the terms "breast cancer" and "mammary cancer" are detected. MeInfoText associated HER-2 to several KEGG pathways, including "Androgen Receptor," "EGFR1," "Pancreatic cancer," "Notch," and "Focal adhesion."

Another resource for methylation information is the PubMeth database (106), where text mining followed by manual curation is used to build a comprehensive collection of cancer–gene methylation associations.

The initial selection of the abstracts used by PubMeth is based on an NCBI eUtils search against PubMed using 15 methylation-relevant keywords and their variants. To index these articles, a gene dictionary derived from the GeneCards database is used that covers both gene names from Ensembl as well as those from Entrez Gene. Variants of this initial dictionary are also added to account for typographical variations such as the alternative use of hyphens. The cancer-related keywords added to PubMeth were extracted from the National Cancer Institute (NCI), providing textual variants of these terms. The resulting resource is accessible through a gene-centric and a cancer-centric search interface.

The usage of text mining approaches for the computational prioritization of candidate genes involved in human disorders showed promising results and motivated the implementation of several tools for candidate gene selection that exploit literature-derived information: ENDEAVOUR (107), G2D (108), and CAESAR (110).

ENDEAVOUR integrates multiple data sources, including annotations of pathways (KEGG), interactions (BIND), functional terms (GO), and many other relevant aspects together with text mining–derived information to build a model useful for gene ranking. G2D is a web server that facilitates the prioritization of genes of a human genome region in terms of their relevance for genetically inherited diseases (108). Through G2D it is possible to carry out searches providing chromosomal locations (e.g., bands, markers, or positions) as the input. Finally, CAESAR ranks genes using a combination of text and data mining by their relevance for complex traits (109). It requires user-defined input text and also does text processing on OMIM records, extracting ontology terms from the text. CAESAR uses information from the GO (molecular function and biological process terms), the mammalian phenotype ontology, and the anatomical ontology eVOC. **Figure 16.7** provides a gene-centric overview of the different information types currently extracted using text mining, most of them having relevance for the analysis of disease and pathway association of human genes.

6.1. Implementing a Text Mining System: From Polymorphisms to Breast Cancer Genes

In a number of biological scenarios, manually curated databases do not cover all the demands of information. In these cases, new literature mining tools can provide an alternative valid solution. We describe here some of the most relevant steps for developing biomedical text mining systems by analyzing two real application cases: (i) the extraction of human gene polymorphism information and cancer-related mutations, and (ii) the navigation and prioritization of breast cancer–relevant genes. We recently applied similar strategies to biological relationships in three other topics: mitotic spindle–related proteins, biodegradation pathways, and for building a knowledge base for cell division–related processes in the plant model organism *Arabidopsis thaliana* (110). The efforts to extract literature-derived information for mitotic spindle and cell division proteins have been developed in the framework of two European projects, ENFIN and DIAMONDS, with the aim of directly using this information as input for experimental studies.

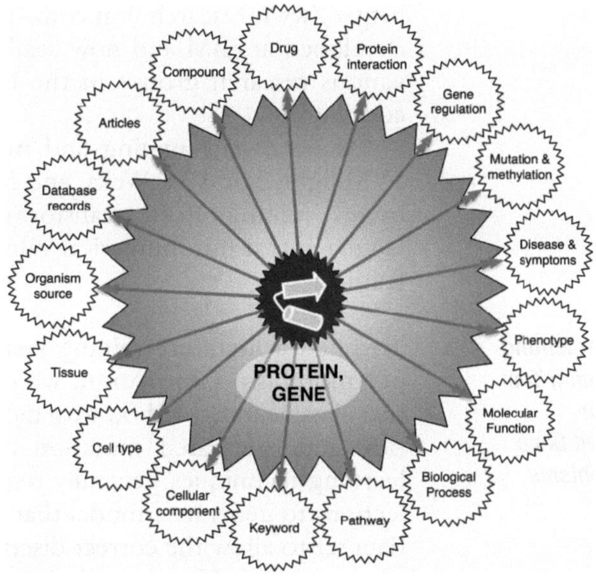

Fig. 16.7. Summary figure providing an overview of the different types of biological information that have been extracted for genes and proteins from the literature.

An initial step to develop text mining systems is to retrieve the textual data collection that should be mined. It is possible to obtain a local copy of the PubMed database or articles from other literature repositories like PMC or HighWire press. An important aspect of text mining and information retrieval systems is efficient text indexing and preprocessing, especially if the resulting system should be available through web services. Systems like Lucene (http://lucene.apache.org) or Sphinx (http://www.sphinxsearch.com) can be used for this purpose, as well as commercial systems like Oracle that integrate full-text indexing software.

Additional aspects such as text encoding and the conversion of common full-text formats like PDF to plain text (e.g., using pdftotext) can affect the performance of the resulting text mining system. For tokenization, stemming, and sentence splitting, several freely available software tools can be adapted. A range of available natural language processing software frameworks can be adapted to handle biomedical text; GATE (General Architecture for Text Engineering, Natural Language Processing system, http://gate.ac.uk), NLTK (Natural Language Toolkit, http://nltk.sourceforge.net), and MALLET (MAchine Learning for LanguagE Toolkit, http://mallet.cs.umass.edu/) are recommended as being very well-rounded and complete. To implement more efficient text mining pipelines able to integrate and connect various language processing modules, UIMA (Unstructured Information Management Architecture, http://www.research.ibm.com/UIMA), a scalable platform first developed at IBM and now available as open source, is used by various research groups in the biomedical literature processing community.

General data mining and machine learning software (e.g., SVM'light, LibSVM, Weka, and Matlab) are also commonly used by text mining efforts, transforming data in the form of natural language text into numerical values representing relevant textual features.

6.2. Implementing a Text Mining Tool for Cancer-Associated Gene Polymorphisms

To build a literature mining system useful for extracting gene-polymorphism information, we combined a supervised learning text classification method with dictionary-based gene name detection and rule-based mutation mention extraction. Supervised learning techniques generally require manually labeled data collections to generate a model that exploits characteristics from the data set to allow the correct discrimination of relevant from non-relevant items. Using simple keyword searches for preparing relevant training articles can result in a biased system in terms of collecting all the relevant information. To minimize the manual workload while retaining a robust training data selection strategy, we used regular expression to detect SNP mentions in the whole PubMed database. The resulting set of abstracts was enriched with positive instances and subjected to manual classification. An equal-sized set of negative (not polymorphism-relevant abstracts) training instances was prepared by a domain expert through the manual inspection of a random sample of PubMed abstracts. Preparing a negative training set is generally a challenging task, as it should reflect the heterogeneous types of articles contained in the PubMed database.

This resulted in a balanced training collection of 841 positive and negative training items (abstracts). Using an SVM classifier (radial basis kernel function) and word tokens as features, we generated a classifier model that implements a feature dictionary of 9,677 words (out of the initial set of 22,987 words present in the training set). Each of the PubMed records was then converted into a feature vector and classified using the previously mentioned model. The resulting polymorphism-relevant PubMed subset (polymorphism bibliome) contained a total of 223,779 abstracts with 1,986,841 sentences. In order to link each of the sentences of the polymorphism-relevant bibliome to sequence information, we constructed a manually filtered human protein symbol/name dictionary from the SwissProt database and used a maximum-length dictionary look-up method to find protein mention in the sentences. A postprocessing step to disambiguate certain gene mentions including species association was carried out. Additionally, all the mutation mentions encountered in the sentence collection were tagged using MutationFinder. A mapping/validation of the extracted mutations to their corresponding positions in the sequences for the proteins mentioned in the article context was carried out. Finally, cancer terms co-mentioned in the collection of abstracts containing both proteins and mutations were detected. **Figure 16.8** shows the user interface for browsing the results generated by this information pipeline.

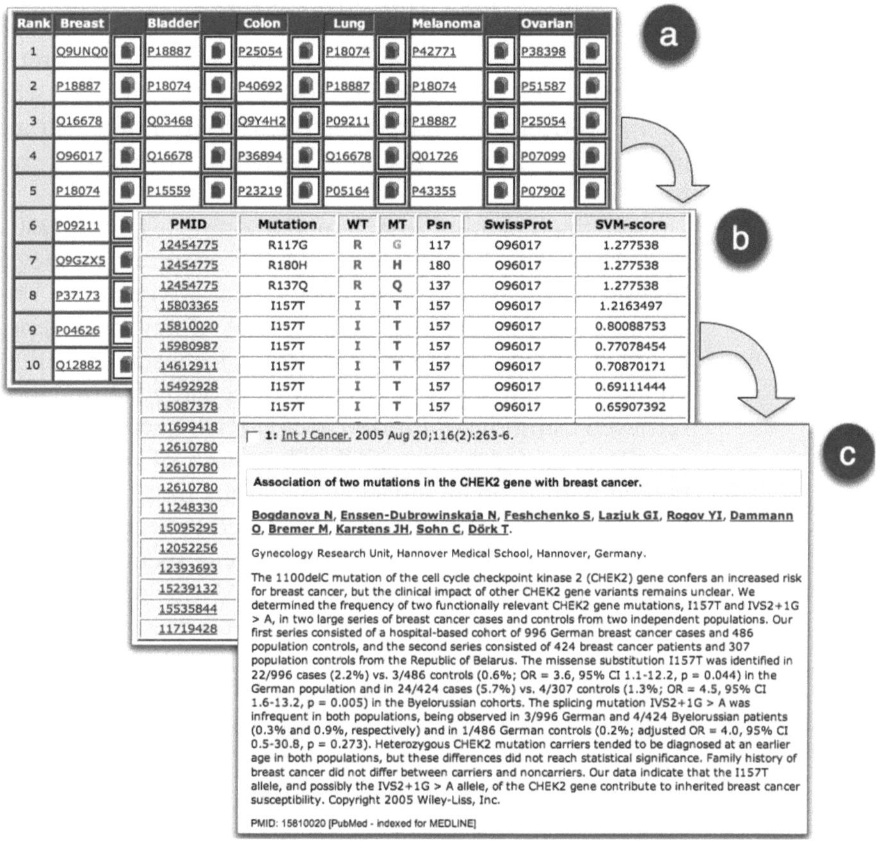

Fig. 16.8. Extraction of polymorphism information and mutations for human genes and proteins and their association to cancer types using a text mining and information extraction pipeline. (a) For each gene/protein, a co-occurrence network with cancer terms is generated (including breast, bladder, colon, lung, and ovarian cancer as well as melanoma). Each protein is linked to its corresponding UniProt database identifier. (b) From the papers linked to each protein, all the mutations are extracted using the MutationFinder system and matched to the corresponding protein sequence to remove false positives. (c) To address polymorphism-specific information, a text categorization system based on support vector machines to classify abstracts relevant to human SNPs has been implemented. This system provides a score for each abstract, indicating its association to polymorphism relevance.

For each of the human proteins extracted by the pipeline, a collection of abstracts ranked by their association to polymorphism relevant information is provided. This facilitates the retrieval of SNP and mutation information for human proteins.

6.3. Navigating the Breast Cancer Bibliome and Prioritizing Human Breast Cancer Genes

Most of the current text mining systems aiming to detect disease-associated genes rely on the use of specialized terminologies and thesauri. The use of such lexical resources for finding topic-specific associations has several limitations. These strategies heavily depend on the quality of the lexical resources, both in terms of completeness (covering most of the vocabulary relevant to a given domain) as well as in terms of usefulness to textual indexing and term mapping. It is often challenging to retrieve expressions corresponding to the meaning of a complex term within the text, such as biological pathway terms.

Another complication is associated with the difficulty in providing robust co-occurrence-based gene rankings, where different co-occurring terms and their relationships should result in a single score reflecting its importance for a biological topic.

We implemented a system not only to find breast cancer–relevant genes, but also to improve literature retrieval going beyond keyword co-occurrence-based navigation. This setup allows a flexible, topic-specific ranking of any term appearing in the breast cancer literature, including, for instance, pathways and GO terms. Initially, a collection of records from PubMed was retrieved using the optimized MeSH term query for breast cancer articles provided by the NCI. Then each of these abstracts was tagged with protein and gene mentions using ABNER, a CRF-based gene tagger previously introduced in **Section 3**. Only those abstracts that contained at least five gene mentions were used as positive training examples for an SVM document classifier. We used the gene mention filtering step to ensure that the resulting classifier would specifically detect those articles associated with the topic of breast cancer at the genetic level and not those referring to breast cancer surgery. This generated a training set of 8,150 PubMed abstracts. A semisupervised SVM classifier was trained using a balanced collection of randomly selected PubMed records, yielding a precision rate of 98.55% and a recall rate of 96.38% on a test set collection. The resulting breast cancer bibliome can be searched using any keyword or gene name, returning a breast cancer–relevant ranked collection of hits, as shown in **Fig. 16.9**. In a similar way, all human genes and GO biological process terms – associated with the corresponding genes – were ranked using this context-based document categorization strategy.

We have selected these two examples because they show the possibilities of classification tools to obtain system specific information, which can be of direct interest for users – experimental biologists – but also because they show the limitations of simple keyword extraction, the need for positive and negative data sets curated with expert advice, and the need to provide information connected directly to the corresponding sources of textual information – underlying information – accessible to the users and displayed in user-friendly systems with facilities to inspect the information.

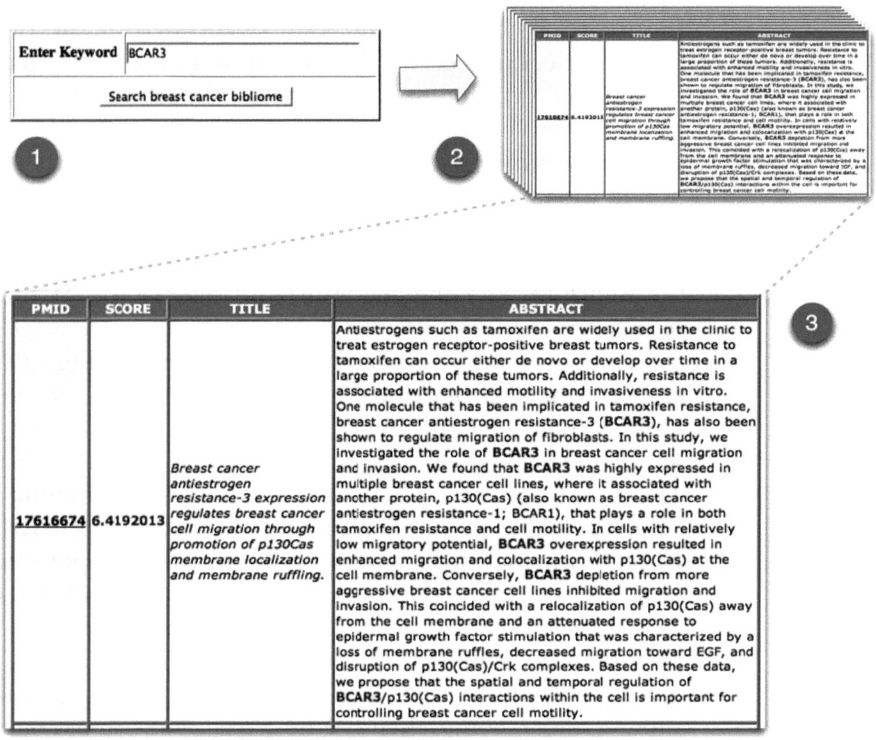

Fig. 16.9. Navigating the breast cancer bibliome. For a given query term or gene name, a set of articles is returned ranked according to the score provided by the breast cancer document classifier.

7. Future Trends in Biomedical Text Mining

From the example results generated by different systems introduced in this chapter, it becomes clear that results are often difficult to compare and interpret; reasons include that different underlying resources are integrated (mapping to alternative gene or protein database), heterogeneously formatted results are returned, or results are combined from systems that provide a rank with others that do not. Therefore, carrying out community evaluations like the BioCreative initiative (12) to define comparable evaluation criteria and common data formats is crucial to determine the performance of different techniques and algorithms applied in biomedical text mining tasks. The importance of integrating results generated by different online systems allowing visualization and direct comparison using a common infrastructure was also realized from the BioCreative assessment. This motivated the implementation of the BioCreative Metaserver (BCMS), the first text mining initiative to integrate multiple annotation servers into a common metaserver infrastructure. **Figure 16.10** illustrates the results generated by the BCMS for a given query, integrating results from a range of different systems under a common user interface. Almost 80% of all the articles related to biomedical text mining have been published within the last five years, and the current trend points toward a duplication in the number of new articles for this topic from one year to another, resulting as well in a considerable number of new online applications. Considering this growing number of new systems, the availability of a unifying framework like the BCMS that allows results returned by different online tools to be overseen is even more important. Future developments in text mining still depend on the availability of manually labeled text corpora, although only a few of these are currently accessible without restrictions, such as several data collections provided by the BioCreative effort (111). The recent diversification process in terms of text mining systems being applied to new biological topics and the importance of literature information for analyzing experimental data point toward a near future where text mining systems will be part of the standard computational analysis carried out in biomedical research.

Using text mining applications together with other bioinformatics tools in the process of solving biological problems should allow a more comprehensive understanding of biological data collections.

Fig. 16.10. The BioCreative MetaServer. The integration of multiple text mining services and a combined visualization of the generated results.

8. Notes

1. Most of the current text mining systems are based on processing PubMed records (often being restricted to data derived from titles and abstracts, but some also use associated MeSH terms). Despite its practical relevance, only a few methods are able to process full-text articles, in part because of difficulties related to the automatic retrieval of full-text articles and copyright issues. Also, preprocessing full-text articles is a challenging task. Some recent initiatives based on direct collaboration with publishers could facilitate the future use of full-text articles. It is thus important to examine what literature collection is being used by a given text mining system.
2. As a general recommendation, it is important to try out several text mining systems using the same or similar queries, to determine which tool is most suitable for a particular task. Aspects related to the supported query input and the generated output can also serve to benchmark different systems (e.g., if they allow queries using protein symbols, lists of proteins, or gene names; if they allow the direct use of database identifiers such as Uniprot database accession numbers, Unigene IDs). For an overview of different online systems, see http://zope.bioinfo.cnio.es/bionlp˙tools.
3. Text mining and information extraction techniques based on supervised machine learning techniques are heavily dependent on the quality and selection criteria of the underlying training data used. Therefore, it is important to know the selection criteria and basic properties of the used training data and validation strategies in order to understand the generated results. There are only a few manually labeled training collections, such as the GENIA corpus (based on the selection of abstracts related to human blood transcription factors).
4. One aspect that can indicate a given system's performance is an independent evaluation of the method through participation at community evaluation challenges such as BioCreative.
5. For the development of specific text mining applications, a range of existing NLP and data mining applications can be a useful starting point, including systems like SVMLight, LIBSVM, Weka, the MALLET toolkit, NLTK, or GATE.
6. A common error of literature mining tools trying to automatically link genes to abstracts is related to gene symbol

ambiguity. Highly ambiguous symbols or gene names are often filtered by text mining systems to improve precision. Also, distinguishing between gene mentions from two species with substantially overlapping gene and protein nomenclature (e.g., between human and mouse genes) is especially difficult.

References

1. Krallinger M, Valencia A, Hirschman L. (2008) Linking genes to literature: text mining, information extraction, and retrieval applications for biology. *Genome Biol* 9(Suppl 2):S8.
2. Braconi Quintaje S, Orchard S. (2008) The annotation of both human and mouse kinomes in UniProtKB/Swiss-Prot: one small step in manual annotation, one giant leap for full comprehension of genomes. *Mol Cell Proteomics* 7(8): 1409–1419.
3. Baumgartner WA, Cohen KB, Fox LM, Acquaah-Mensah G, Hunter L. (2007) Manual curation is not sufficient for annotation of genomic databases. *Bioinformatics* 23(13):i41–i48.
4. Leitner F, Valencia A. (2008) A text-mining perspective on the requirements for electronically annotated abstracts. *FEBS Lett* 582(8):1178–1181.
5. Ceol A, Chatr-Aryamontri A, Licata L, Cesareni G. (2008) Linking entries in protein interaction database to structured text: the *FEBS Letters* experiment. *FEBS Lett* 582(8):1171–1177.
6. Aerts S, Haeussler M, van Vooren S, Griffith OL, Hulpiau P, Jones SJ, Montgomery SB, Bergman CM. Open Regulatory Annotation Consortium. (2008) Text-mining assisted regulatory annotation. *Genome Biol* 9(2):R31.
7. Shtatland T, Guettler D, Kossodo M, Pivovarov M, Weissleder R. (2007) PepBank – a database of peptides based on sequence text mining and public peptide data sources. *BMC Bioinformatics* 8:280.
8. Hoffmann R, Valencia A. (2005) Implementing the iHOP concept for navigation of biomedical literature. *Bioinformatics* 21(Suppl 2):ii252–ii258.
9. Manning CD, Schütze H. (2003) *Foundations of Statistical Natural Language Processing*, MIT Press, Cambridge, MA.
10. Jiang J, Zhai CX. (2007) An empirical study of tokenization strategies for biomedical information retrieval. *Inform Retr* 10: 341–363.
11. Tomanek K, Wermter J, Hahn U. (2007) Sentence and token splitting based on conditional random fields. *Proceedings of the 10th Conference of the Pacific Association for Computational Linguistics*, pp. 49–57.
12. Krallinger M, Leitner F, Rodriguez-Penagos C, Valencia A. (2008) Overview of the protein-protein interaction annotation extraction task of BioCreative II. *Genome Biol* 9(Suppl 2):S4.
13. Porter MF. (1980) An algorithm for suffix stripping. *Program* 14(3):130–137.
14. Crim J, McDonald R, Pereira F. (2005) Automatically annotating documents with normalized gene lists. *BMC Bioinformatics* 6(Suppl 1):S13.
15. Settles B. (2005) ABNER: an open source tool for automatically tagging genes, proteins and other entity names in text. *Bioinformatics* 21(14):3191–3192.
16. Wang H, Huang M, Ding S, Zhu X. (2008) Exploiting and integrating rich features for biological literature classification. *BMC Bioinformatics* 9(Suppl 3):S4.
17. Hakenberg J, Plake C, Leaman R, Schroeder M, Gonzalez G. (2008) Inter-species normalization of gene mentions with GNAT. *Bioinformatics* 24(16):i126–i132.
18. Smith L, Rindflesch T, Wilbur WJ. (2004) MedPost: a part-of-speech tagger for bioMedical text. *Bioinformatics* 20(14):2320–2321.
19. Pyysalo S, Salakoski T, Aubin S, Nazarenko A. (2006) Lexical adaptation of link grammar to the biomedical sublanguage: a comparative evaluation of three approaches. *BMC Bioinformatics* 7(Suppl 3):S2.
20. Rinaldi F, Schneider G, Kaljurand K, Hess M, Andronis C, Konstandi O, Persidis A. (2007) Mining of relations between proteins over biomedical scientific literature using a deep-linguistic approach. *Artif Intell Med* 39(2):127–136.
21. Bethard S, Lu Z, Martin JH, Hunter L. (2008) Semantic role labeling for protein transport predicates. *BMC Bioinformatics* 9:277.

22. Koike A, Niwa Y, Takagi T. (2005) Automatic extraction of gene/protein biological functions from biomedical text. *Bioinformatics* 21(7):1227–1236.
23. Rodríguez-Penagos C, Salgado H, Martínez-Flores I, Collado-Vides J. (2007) Automatic reconstruction of a bacterial regulatory network using Natural Language Processing. *BMC Bioinformatics* 8:293.
24. Yamamoto Y, Takagi T. (2007) Biomedical knowledge navigation by literature clustering. *J Biomed Inform* 40(2):114–130.
25. Krauthammer M, Nenadic G. (2004) Term identification in the biomedical literature. *J Biomed Inform* 37(6):512–526.
26. Okazaki N, Ananiadou S. (2006) Building an abbreviation dictionary using a term recognition approach. *Bioinformatics* 22(24):3089–3095.
27. Leitner F, et al. (2008) Introducing meta-services for biomedical information extraction. *Genome Biol* 9(Suppl 2):S6.
28. Kim JJ, Pezik P, Rebholz-Schuhmann D. (2008) MedEvi: retrieving textual evidence of relations between biomedical concepts from Medline. *Bioinformatics* 24(11): 1410–1412.
29. Tomanek K, Wermter J, Hahn U. (2007) An approach to text corpus construction which cuts annotation costs and maintains reusability of annotated data. *Proceedings of EMNLP-CoNLL 2007*, pp. 486–495.
30. http://www.ncbi.nlm.nih.gov.
31. Natarajan J, Ganapathy J. (2007) Functional gene clustering via gene annotation sentences, MeSH and GO keywords from biomedical literature. *Bioinformation* 2(5):185–193.
32. Camon E, et al. (2004) The Gene Ontology Annotation (GOA) database: sharing knowledge in Uniprot with Gene Ontology. *Nucleic Acids Res* 32:262–266.
33. Siadaty MS, Shu J, Knaus WA. (2007) Relemed: sentence-level search engine with relevance score for the MEDLINE database of biomedical articles. *BMC Med Inform Decis Mak* 7:1.
34. http://www.pubmedreader.com.
35. http://bioinfo.amc.uva.nl/human-genetics/pubreminer/.
36. Eaton AD. (2006) HubMed: a web-based biomedical literature search interface. *Nucleic Acids Res* 34(Web server issue):W745–W747.
37. Lewis J, Ossowski S, Hicks J, Errami M, Garner HR. (2006) Text similarity: an alternative way to search MEDLINE. *Bioinformatics* 22(18):2298–2304.
38. http://www.pubmedcentral.nih.gov/.
39. http://highwire.org/.
40. Hearst MA, Divoli A, Guturu H, Ksikes A, Nakov P, Wooldridge MA, Ye J. (2007) BioText Search Engine: beyond abstract search. *Bioinformatics* 23(16):2196–2197.
41. Doms A, Schroeder M. (2005) GoPubMed: exploring PubMed with the Gene Ontology. *Nucleic Acids Res* 33(Web server issue):W783–W786.
42. Smith B, et al. (2007) The OBO Foundry: coordinated evolution of ontologies to support biomedical data integration. *Nat Biotechnol* 25(11):1251–1255.
43. Tsuruoka Y, McNaught J, Ananiadou S. (2008) Normalizing biomedical terms by minimizing ambiguity and variability. *BMC Bioinformatics* 9(Suppl 3):S2.
44. http://www.nlm.nih.gov/research/umls/.
45. Frijters R, Heupers B, van Beek P, Bouwhuis M, van Schaik R, de Vlieg J, Polman J, Alkema W. (2008) CoPub: a literature-based keyword enrichment tool for microarray data analysis. *Nucleic Acids Res* 36(Web server issue):W406–W410.
46. http://mor.nlm.nih.gov/perl/gennav.pl.
47. http://129.194.97.165/GOCat/.
48. Fink JL, Kushch S, Williams PR, Bourne PE. (2008) BioLit: integrating biological literature with databases. *Nucleic Acids Res* 36(Web server issue):W385–W389.
49. Rebholz-Schuhmann D, Arregui M, Gaudan S, Kirsch H, Jimeno A. (2008) Text processing through web services: calling Whatizit. *Bioinformatics* 24(2):296–298.
50. Lussier Y, Borlawsky T, Rappaport D, Liu Y, Friedman C. (2006) PhenoGO: assigning phenotypic context to gene ontology annotations with natural language processing. *Pacific Symposium on Biocomputing*, pp. 64–75.
51. Blaschke C, Leon EA, Krallinger M, Valencia A. (2005) Evaluation of BioCreAtIvE assessment of task 2. *BMC Bioinformatics* 6(Suppl 1):S16.
52. Oliveros JC, Blaschke C, Herrero J, Dopazo J, Valencia A. (2000) Expression profiles and biological function. *Genome Inform Ser Workshop Genome Inform* 11:106–117.
53. Raychaudhuri S, Chang JT, Imam F, Altman RB. (2003) The computational analysis of scientific literature to define and recognize gene expression clusters. *Nucleic Acids Res* 31(15):4553–4560.
54. Resnik P. (1995) Using information content to evaluate semantic similarity in a taxonomy. *Proceedings of the International Joint Conference on Artificial Intelligence*, pp. 448–453.
55. Lord PW, Stevens RD, Brass A, Goble CA. (2003) Semantic similarity measures as tools

for exploring the gene ontology. *Pacific Symposium on Biocomputing*, pp. 601–612.
56. Fellbaum C, Hahn U, Smith B. (2006) Towards new information resources for public health – from WordNet to MedicalWordNet. *J Biomed Inform* 39(3):321–332.
57. del Pozo A, Pazos F, Valencia A. (2008) Defining functional distances over gene ontology. *BMC Bioinformatics* 9:50.
58. Johnson HL, Cohen KB, Baumgartner WA, Lu Z, Bada M, Kester T, Kim H, Hunter L. (2006) Evaluation of lexical methods for detecting relationships between concepts from multiple ontologies. *Pacific Symposium on Biocomputing*, pp. 28–39.
59. Donaldson I, Martin J, de Bruijn B, Wolting C, Lay V, Tuekam B, Zhang S, Baskin B, Bader GD, Michalickova K, Pawson T, Hogue CW. (2003) PreBIND and Textomy – mining the biomedical literature for protein-protein interactions using a support vector machine. *BMC Bioinformatics* 4:11.
60. Blaschke C, Valencia A. (2001) The potential use of SUISEKI as a protein interaction discovery tool. *Genome Inform* 12:123–134.
61. Krallinger M, Malik R, Valencia A. (2006) Text mining and protein annotations: the construction and use of protein description sentences. *Genome Inform* 17(2): 121–130.
62. Jenssen TK, Laegreid A, Komorowski J, Hovig E. (2001) A literature network of human genes for high-throughput analysis of gene expression. *Nat Genet* 28(1):21–28.
63. Rebholz-Schuhmann D, Kirsch H, Arregui M, Gaudan S, Riethoven M, Stoehr P. (2007) EBIMed – text crunching to gather facts for proteins from Medline. *Bioinformatics* 23(2):e237–e244.
64. https://www-tsujii.is.s.u-tokyo.ac.jp/info-pubmed/.
65. Chen H, Sharp BM. (2004) Content-rich biological network constructed by mining PubMed abstracts. *BMC Bioinformatics* 5:147.
66. Rinaldi F, Kappeler T, Kaljurand K, Schneider G, Klenner M, Clematide S, Hess M, von Allmen JM, Parisot P, Romacker M, Vachon T. (2008) OntoGene in BioCreative II. *Genome Biol* 9(Suppl 2):S13.
67. Baumgartner WA, Lu Z, Johnson HL, Caporaso JG, Paquette J, Lindemann A, White EK, Medvedeva O, Cohen KB, Hunter L. (2008) Concept recognition for extracting protein interaction relations from biomedical text. *Genome Biol* 9(Suppl 2):S9.
68. Narayanaswamy M, Ravikumar KE, Vijay-Shanker K. (2005) Beyond the clause: extraction of phosphorylation information from Medline abstracts. *Bioinformatics* 21(Suppl 1):i319–i327.
69. Hoffmann R, Krallinger M, Andres E, Tamames J, Blaschke C, Valencia A. (2005) Text mining for metabolic pathways, signaling cascades, and protein networks. *Sci STKE* 283:pe21.
70. Oda K, Kim JD, Ohta T, Okanohara D, Matsuzaki T, Tateisi Y, Tsujii J. (2008) New challenges for text mining: mapping between text and manually curated pathways. *BMC Bioinformatics* 9(Suppl 3):S5.
71. Friedman C, Kra P, Yu H, Krauthammer M, Rzhetsky A. (2001) GENIES: a natural-language processing system for the extraction of molecular pathways from journal articles. *Bioinformatics* 17(Suppl 1):S74–S82.
72. Yuryev A, Mulyukov Z, Kotelnikova E, Maslov S, Egorov S, Nikitin A, Daraselia N, Mazo I. (2006) Automatic pathway building in biological association networks. *BMC Bioinformatics* 7:171.
73. Koike A, Kobayashi Y, Takagi T. (2003) Kinase pathway database: an integrated protein-kinase and NLP-based protein-interaction resource. *Genome Res* 13(6A):1231–1243.
74. Ding J, Viswanathan K, Berleant D, Hughes L, Wurtele ES, Ashlock D, Dickerson JA, Fulmer A, Schnable PS. (2005) Using the biological taxonomy to access biological literature with PathBinderH. *Bioinformatics* 21(10):2560–2562.
75. Lee H, Yi GS, Park JC. (2008) E3Miner: a text mining tool for ubiquitin-protein ligases. *Nucleic Acids Res* 36(Web server issue):W416–W422.
76. Al-Shahrour F, Carbonell J, Minguez P, Goetz S, Conesa A, Tárraga J, Medina I, Alloza E, Montaner D, Dopazo J. (2008) Babelomics: advanced functional profiling of transcriptomics, proteomics and genomics experiments. *Nucleic Acids Res* 36(Web server issue):W341–W346.
77. Kuhn M, von Mering C, Campillos M, Jensen LJ, Bork P. (2008) STITCH: interaction networks of chemicals and proteins. *Nucleic Acids Res* 36(Database issue):D684–D688.
78. Chang A, Scheer M, Grote A, Schomburg I, Schomburg D. (2008) BRENDA, AMENDA and FRENDA the enzyme information system: new content and tools in 2009. *Nucleic Acids Res* 37(Database issue):D588–D592.
79. Spasić I, Schober D, Sansone SA, Rebholz-Schuhmann D, Kell DB, Paton NW. (2008) Facilitating the development of controlled vocabularies for metabolomics technologies

with text mining. *BMC Bioinformatics* 9(Suppl 5):S5.
80. Jin Y, McDonald RT, Lerman K, Mandel MA, Carroll S, Liberman MY, Pereira FC, Winters RS, White PS. (2006) Automated recognition of malignancy mentions in biomedical literature. *BMC Bioinformatics* 7:492.
81. Chun HW, Tsuruoka Y, Kim JD, Shiba R, Nagata N, Hishiki T, Tsujii J. (2006) Extraction of gene-disease relations from Medline using domain dictionaries and machine learning. *Pacific Symposium in Biocomputing*, pp. 4–15.
82. Pospisil P, Iyer LK, Adelstein SJ, Kassis AI. (2006) A combined approach to data mining of textual and structured data to identify cancer-related targets. *BMC Bioinformatics* 7:354.
83. Natarajan J, Berrar D, Dubitzky W, Hack C, Zhang Y, DeSesa C, Van Brocklyn JR, Bremer EG. (2006) Text mining of full-text journal articles combined with gene expression analysis reveals a relationship between sphingosine-1-phosphate and invasiveness of a glioblastoma cell line. *BMC Bioinformatics* 7:373.
84. Li X, Chen H, Huang Z, Su H, Martinez JD. (2007) Global mapping of gene/protein interactions in PubMed abstracts: a framework and an experiment with P53 interactions. *J Biomed Inform* 40(5): 453–464.
85. McDonald DM, Chen H, Su H, Marshall BB. (2004) Extracting gene pathway relations using a hybrid grammar: the Arizona Relation Parser. *Bioinformatics* 20(18):3370–3378.
86. Gonzalez G, Uribe JC, Brophy C, Baral C. (2007) Mining gene-disease relationships from biomedical literature: weighting protein-protein interactions and connectivity measures. *Pacific Symposium of Biocomputing*, pp. 28–39.
87. Croning MD, Marshall MC, McLaren P, Armstrong JD, Grant SG. (2008) G2Cdb: the Genes to Cognition database. *Nucleic Acids Res* 37(Database issue):D846–D851.
88. Collier N, Doan S, Kawazoe A, Goodwin RM, Conway M, Tateno Y, Ngo QH, Dien D, Kawtrakul A, Takeuchi K, Shigematsu M, Taniguchi K. (2008) BioCaster: detecting public health rumors with a web-based text mining system. *Bioinformatics* 24(24):2940–2941.
89. Srinivasan P, Libbus B. (2004) Mining MEDLINE for implicit links between dietary substances and diseases. *Bioinformatics* 20(Suppl 1):i290–i296.
90. Tremblay K, Lemire M, Potvin C, Tremblay A, Hunninghake GM, Raby BA, Hudson TJ, Perez-Iratxeta C, Andrade-Navarro MA, Laprise C. (2008) Genes to diseases (G2D) computational method to identify asthma candidate. *PLoS ONE* 3(8):e2907.
91. Tsuruoka Y, Tsujii J, Ananiadou S. (2008) FACTA: a text search engine for finding associated biomedical concepts. *Bioinformatics* 24(21):2559–2560.
92. Müller H, Mancuso F. (2008) Identification and analysis of co-occurrence networks with NetCutter. *PLoS ONE* 3(9):e3178.
93. Jelier R, Schuemie MJ, Veldhoven A, Dorssers LC, Jenster G, Kors JA. (2008) Anni 2.0: a multipurpose text-mining tool for the life sciences. *Genome Biol* 9(6):R96.
94. Lin SM, McConnell P, Johnson KF, Shoemaker J. (2004) MedlineR: an open source library in R for Medline literature data mining. *Bioinformatics* 20(18):3659–3661.
95. Cases I, Pisano DG, Andres E, Carro A, Fernández JM, Gómez-López G, Rodriguez JM, Vera JF, Valencia A, Rojas AM. (2007) CARGO: a web portal to integrate customized biological information. *Nucleic Acids Res* 35:W16–W20.
96. Xuan W, Wang P, Watson SJ, Meng F. (2007) Medline search engine for finding genetic markers with biological significance. *Bioinformatics* 23(18):2477–2484.
97. Furlong LI, Dach H, Hofmann-Apitius M, Sanz F. (2008) OSIRISv1.2: a named entity recognition system for sequence variants of genes in biomedical literature. *BMC Bioinformatics* 9:84.
98. Caporaso JG, Baumgartner WA, Randolph DA, Cohen KB, Hunter L. (2007) MutationFinder: a high-performance system for extracting point mutation mentions from text. *Bioinformatics* 23(14):1862–1865.
99. McDonald RT, Winters RS, Mandel M, Jin Y, White PS, Pereira F. (2004) An entity tagger for recognizing acquired genomic variations in cancer literature. *Bioinformatics* 20(17):3249–3251.
100. Saunders RE, Perkins SJ. (2008) CoagMDB: a database analysis of missense mutations within four conserved domains in five vitamin K-dependent coagulation serine proteases using a text-mining tool. *Hum Mutat* 29(3):333–344.
101. Bajdik CD, Kuo B, Rusaw S, Jones S, Brooks-Wilson A. (2005) CGMIM: automated text-mining of Online Mendelian Inheritance in Man (OMIM) to identify genetically-associated cancers and candidate genes. *BMC Bioinformatics* 6:78.

102. Yu W, Gwinn M, Clyne M, Yesupriya A, Khoury MJ. (2008) A navigator for human genome epidemiology. *Nat Genet* 40(2):124–125.
103. Yu W, Clyne M, Dolan SM, Yesupriya A, Wulf A, Liu T, Khoury MJ, Gwinn M. (2008) GAPscreener: an automatic tool for screening human genetic association literature in PubMed using the support vector machine technique. *BMC Bioinformatics* 9:205.
104. Cheng D, Knox C, Young N, Stothard P, Damaraju S, Wishart DS. (2008) PolySearch: a web-based text mining system for extracting relationships between human diseases, genes, mutations, drugs and metabolites. *Nucleic Acids Res* 36:W399–W405.
105. Fang YC, Huang HC, Juan HF. (2008) MeInfoText: associated gene methylation and cancer information from text mining. *BMC Bioinformatics* 9:22.
106. Ongenaert M, Van Neste L, De Meyer T, Menschaert G, Bekaert S, Van Criekinge W. (2008) PubMeth: a cancer methylation database combining text-mining and expert annotation. *Nucleic Acids Res* 36:D842–D846.
107. Tranchevent LC, Barriot R, Yu S, Van Vooren S, Van Loo P, Coessens B, De Moor B, Aerts S, Moreau Y. (2008) ENDEAVOUR update: a web resource for gene prioritization in multiple species. *Nucleic Acids Res* 36:W377–W384.
108. Perez-Iratxeta C, Bork P, Andrade-Navarro MA. (2008) Update of the G2D tool for prioritization of gene candidates to inherited diseases. *Nucleic Acids Res* 35: W212–W216.
109. Gaulton KJ, Mohlke KL, Vision TJ. (2007) A computational system to select candidate genes for complex human traits. *Bioinformatics* 23(9):1132–1140.
110. Krallinger M, Rojas A, Valencia A. (2008) Creating reference datasets for systems biology applications using text mining. *Ann NY Acad Sci*, accepted for publication. 1158:14–28.
111. Krallinger M, Morgan A, Smith L, Leitner F, Tanabe L, Wilbur J, Hirschman L, Valencia A. (2008) Evaluation of text-mining systems for biology: overview of the Second BioCreative community challenge. *Genome Biol* 9(Suppl 2):S1.

Subject Index

R. Matthiesen (ed.), *Bioinformatics Methods in Clinical Research*, Methods in Molecular Biology 593,
DOI 10.1007/978-1-60327-194-3, © Humana Press, a part of Springer Science+Business Media, LLC 2010

R

S

T

MIX
Papier aus verantwortungsvollen Quellen
Paper from responsible sources
FSC® C105338

If you have any concerns about our products,
you can contact us on
ProductSafety@springernature.com

In case Publisher is established outside the EU,
the EU authorized representative is:
Springer Nature Customer Service Center GmbH
Europaplatz 3, 69115 Heidelberg, Germany

Printed by Libri Plureos GmbH
in Hamburg, Germany